# Artificial Intelligence-Based State-of-Health Estimation of Lithium-Ion Batteries

# Artificial Intelligence-Based State-of-Health Estimation of Lithium-Ion Batteries

Editors

**Remus Teodorescu**
**Xin Sui**

Basel • Beijing • Wuhan • Barcelona • Belgrade • Novi Sad • Cluj • Manchester

*Editors*
Remus Teodorescu
Aalborg University
Aalborg, Denmark

Xin Sui
Aalborg University
Aalborg, Denmark

*Editorial Office*
MDPI
St. Alban-Anlage 66
4052 Basel, Switzerland

This is a reprint of articles from the Special Issue published online in the open access journal *Batteries* (ISSN 2313-0105) (available at: https://www.mdpi.com/journal/batteries/special_issues/State_Health_Estimation_Lithium_Batteries).

For citation purposes, cite each article independently as indicated on the article page online and as indicated below:

Lastname, A.A.; Lastname, B.B. Article Title. *Journal Name* **Year**, *Volume Number*, Page Range.

**ISBN 978-3-0365-9875-8 (Hbk)**
**ISBN 978-3-0365-9876-5 (PDF)**
**doi.org/10.3390/books978-3-0365-9876-5**

© 2024 by the authors. Articles in this book are Open Access and distributed under the Creative Commons Attribution (CC BY) license. The book as a whole is distributed by MDPI under the terms and conditions of the Creative Commons Attribution-NonCommercial-NoDerivs (CC BY-NC-ND) license.

# Contents

**About the Editors** . . . . . . . . . . . . . . . . . . . . . . . . . . . . . . . . . . . . . . . . . . . . . . . . . . . . . . . . . **vii**

**Ao Li, Anthony Chun Yin Yuen, Wei Wang, Timothy Bo Yuan Chen, Chun Sing Lai, Wei Yang, et al.**
Integration of Computational Fluid Dynamics and Artificial Neural Network for Optimization Design of Battery Thermal Management System
Reprinted from: *Batteries* **2022**, *8*, 69, doi:10.3390/batteries8070069 . . . . . . . . . . . . . . . . . . . **1**

**Peter Kurzweil, Bernhard Frenzel and Wolfgang Scheuerpflug**
A Novel Evaluation Criterion for the Rapid Estimation of the Overcharge and Deep Discharge of Lithium-Ion Batteries Using Differential Capacity
Reprinted from: *Batteries* **2022**, *8*, 86, doi:10.3390/batteries8080086 . . . . . . . . . . . . . . . . . . . **19**

**Jiwei Wang, Zhongwei Deng, Jinwen Li, Kaile Peng, Lijun Xu, Guoqing Guan and Abuliti Abudula**
State of Health Trajectory Prediction Based on Multi-Output Gaussian Process Regression for Lithium-Ion Battery
Reprinted from: *Batteries* **2022**, *8*, 134, doi:10.3390/batteries8100134 . . . . . . . . . . . . . . . . . . **35**

**Jiazhi Miao, Zheming Tong, Shuiguang Tong, Jun Zhang and Jiale Mao**
State of Charge Estimation of Lithium-Ion Battery for Electric Vehicles under Extreme Operating Temperatures Based on an Adaptive Temporal Convolutional Network
Reprinted from: *Batteries* **2022**, *8*, 145, doi:10.3390/batteries8100145 . . . . . . . . . . . . . . . . . . **49**

**Yukai Tian, Jie Wen, Yanru Yang, Yuanhao Shi and Jianchao Zeng**
State-of-Health Prediction of Lithium-Ion Batteries Based on CNN-BiLSTM-AM
Reprinted from: *Batteries* **2022**, *8*, 155, doi:10.3390/batteries8100155 . . . . . . . . . . . . . . . . . . **67**

**Remus Teodorescu, Xin Sui, Søren B. Vilsen, Pallavi Bharadwaj, Abhijit Kulkarni and Daniel-Ioan Stroe**
Smart Battery Technology for Lifetime Improvement
Reprinted from: *Batteries* **2022**, *8*, 169, doi:10.3390/batteries8100169 . . . . . . . . . . . . . . . . . . **83**

**Sebastian Matthias Hell and Chong Dae Kim**
Development of a Data-Driven Method for Online Battery Remaining-Useful-Life Prediction
Reprinted from: *Batteries* **2022**, *8*, 192, doi:10.3390/batteries8100192 . . . . . . . . . . . . . . . . . . **101**

**Alessandro Falai, Tiziano Alberto Giuliacci, Daniela Anna Misul and Pier Giuseppe Anselma**
Reducing the Computational Cost for Artificial Intelligence-Based Battery State-of-Health Estimation in Charging Events
Reprinted from: *Batteries* **2022**, *8*, 209, doi:10.3390/batteries8110209 . . . . . . . . . . . . . . . . . . **113**

**Jichang Peng, Jinhao Meng, Dan Chen, Haitao Liu, Sipeng Hao, Xin Sui and Xinghao Du**
A Review of Lithium-Ion Battery Capacity Estimation Methods for Onboard Battery Management Systems: Recent Progress and Perspectives
Reprinted from: *Batteries* **2022**, *8*, 229, doi:10.3390/batteries8110229 . . . . . . . . . . . . . . . . . . **135**

**Qiang Sun, Shasha Wang, Shuang Gao, Haiying Lv, Jianghao Liu, Li Wang, et al.**
A State of Charge Estimation Approach for Lithium-Ion Batteries Based on the Optimized Metabolic EGM(1,1) Algorithm
Reprinted from: *Batteries* **2022**, *8*, 260, doi:10.3390/batteries8120260 . . . . . . . . . . . . . . . . . . **161**

**Zhong Ren, Changqing Du and Weiqun Ren**
State of Health Estimation of Lithium-Ion Batteries Using a Multi-Feature-Extraction Strategy and PSO-NARXNN
Reprinted from: *Batteries* **2023**, *9*, 7, doi:10.3390/batteries9010007 . . . . . . . . . . . . . . . . . . **183**

**Kang Liu, Longyun Kang and Di Xie**
Online State of Health Estimation of Lithium-Ion Batteries Based on Charging Process and Long Short-Term Memory Recurrent Neural Network
Reprinted from: *Batteries* **2023**, *9*, 94, doi:10.3390/batteries9020094 . . . . . . . . . . . . . . . . . . **205**

**Yinfeng Jiang and Wenxiang Song**
Predicting the Cycle Life of Lithium-Ion Batteries Using Data-Driven Machine Learning Based on Discharge Voltage Curves
Reprinted from: *Batteries* **2023**, *9*, 413, doi:10.3390/batteries9080413 . . . . . . . . . . . . . . . . . . **221**

# About the Editors

**Remus Teodorescu**

Remus Teodorescu (Prof. Dr.) received a Dipl.Ing. degree in electrical engineering from the Polytechnical University of Bucharest, Bucharest, Romania, in 1989, a Ph.D. in power electronics from the University of Galati, Romania, in 1994, and Dr.HC in 2016 from Transilvania University of Brasov. In 1998, he joined the Department of Energy Technology at Aalborg University where he is currently a Full Professor. Between 2013 and 2017, he was a Visiting Professor at Chalmers University. He has been an IEEE/PELS Fellow since 2012 for contributions to grid converter technology for renewable energy systems. In 2022, he became a Villum Investigator and the leader of the Center of Research for Smart Battery (CROSBAT) at Aalborg University. His main current research areas are modular multilevel converters (MMCs) for HVDC/FACTS, Li-Ion battery SOH estimation with AI and smart batteries.

**Xin Sui**

Xin Sui (Dr.) received a B.Eng. degree from Northeast Electric Power University, Jilin, China, in 2015, and an M.Sc. degree from the Institute of Electrical Engineering, Chinese Academy of Sciences, Beijing, China, in 2018, both in electrical engineering. In 2022, Xin received a Ph.D. in machine learning for battery state of health estimation from Aalborg University, Aalborg, Denmark. She is currently a postdoctoral researcher with the Center for Research on Smart Battery (CROSBAT), AAU Energy, Aalborg University. Her research interests include battery state of health estimation, lifetime extension, feature engineering, and machine learning.

*Article*

# Integration of Computational Fluid Dynamics and Artificial Neural Network for Optimization Design of Battery Thermal Management System

Ao Li [1], Anthony Chun Yin Yuen [1,*], Wei Wang [1], Timothy Bo Yuan Chen [1], Chun Sing Lai [2], Wei Yang [3], Wei Wu [4], Qing Nian Chan [1], Sanghoon Kook [1] and Guan Heng Yeoh [1]

1. School of Mechanical and Manufacturing Engineering, University of New South Wales, Sydney, NSW 2052, Australia; ao.li@unsw.edu.au (A.L.); wei.wang15@unsw.edu.au (W.W.); timothy.chen@unsw.edu.au (T.B.Y.C.); qing.chan@unsw.edu.au (Q.N.C.); s.kook@unsw.edu.au (S.K.); g.yeoh@unsw.edu.au (G.H.Y.)
2. Brunel Interdisciplinary Power Systems Research Centre, Department of Electronic and Electrical Engineering, Brunel University London, Kingston Lane, London UB8 3PH, UK; chunsing.lai@brunel.ac.uk
3. School of Energy, Materials and Chemical Engineering, Hefei University, 99 Jinxiu Avenue, Hefei 230601, China; yangwei@hfuu.edu.cn
4. Department Materials Science and Engineering, City University of Hong Kong, Tat Chee Avenue, Hong Kong 999077, China; weiwu39-c@my.cityu.edu.hk
* Correspondence: c.y.yuen@unsw.edu.au; Tel.: +61-2-9385-5697

**Citation:** Li, A.; Yuen, A.C.Y.; Wang, W.; Chen, T.B.Y.; Lai, C.S.; Yang, W.; Wu, W.; Chan, Q.N.; Kook, S.; Yeoh, G.H. Integration of Computational Fluid Dynamics and Artificial Neural Network for Optimization Design of Battery Thermal Management System. *Batteries* **2022**, *8*, 69. https://doi.org/10.3390/batteries8070069

Academic Editors: Carlos Ziebert and Pascal Venet

Received: 7 June 2022
Accepted: 7 July 2022
Published: 8 July 2022

**Publisher's Note:** MDPI stays neutral with regard to jurisdictional claims in published maps and institutional affiliations.

**Copyright:** © 2022 by the authors. Licensee MDPI, Basel, Switzerland. This article is an open access article distributed under the terms and conditions of the Creative Commons Attribution (CC BY) license (https://creativecommons.org/licenses/by/4.0/).

**Abstract:** The increasing popularity of lithium-ion battery systems, particularly in electric vehicles and energy storage systems, has gained broad research interest regarding performance optimization, thermal stability, and fire safety. To enhance the battery thermal management system, a comprehensive investigation of the thermal behaviour and heat exchange process of battery systems is paramount. In this paper, a three-dimensional electro-thermal model coupled with fluid dynamics module was developed to comprehensively analyze the temperature distribution of battery packs and the heat carried away. The computational fluid dynamics (CFD) simulation results of the lumped battery model were validated and verified by considering natural ventilation speed and ambient temperature. In the artificial neural networks (ANN) model, the multilayer perceptron was applied to train the numerical outputs and optimal design of the battery setup, achieving a 1.9% decrease in maximum temperature and a 4.5% drop in temperature difference. The simulation results provide a practical compromise in optimizing the battery configuration and cooling efficiency, balancing the layout of the battery system, and safety performance. The present modelling framework demonstrates an innovative approach to utilizing high-fidelity electro-thermal/CFD numerical inputs for ANN optimization, potentially enhancing the state-of-art thermal management and reducing the risks of thermal runaway and fire outbreaks.

**Keywords:** thermal management; lithium-ion batteries; CFD modelling; ANN; optimization design

## 1. Introduction

With recent advancements in electric technology as well as the growing global concern of energy crisis and environmental pollution, a lot of research interests are devoted to the search for alternative energy sources, including nuclear, wind, or solar energy. Battery energy storage systems have caught the public eye due to their many advantages: fast responsiveness, controllability, structural independence, and widespread application range [1]. Lithium-ion based battery energy storage systems have become the most competitive choice for various applications [2–5]. Lithium-ion batteries (LIBs) as a source of alternative energy through renewable energies have been proposed in industries for many portable consumer electronic devices, including cell phones and laptops. Moreover, LIBs have begun to enter the automotive market as power packs for hybrid and battery electric vehicles (HEVs and EVs) due to their enormous power, efficiency, and durability of a charge cycle.

Temperature plays a critical role in many aspects of the performance of LIBs, including charge acceptance [6], energy capability [7], reliability [8], and so on. In comparison with other battery technologies, LIB performs relatively poor thermal stability, and many accidents happened in recent years [9–11]. A typical LIB comprises four main components: an anode, a cathode, a separator, and an electrolyte. All the parts form a closed system separated from the air, so there is no explosion or fire danger at the normal working temperature [12]. However, the abuse of LIB will generate the threat of thermal runaway and overheating. Both positive and negative electrode decomposition are exothermic processes. Also, oxygen can be generated during the decomposition reactions. The generated heat and oxygen are the contributions to the combustion triangle. If the battery experiences harsh working conditions during electric transportation, the generated heat triggers electrodes' decomposition. As a result, the battery potentially faces thermal issues. Suppose the cell temperature is rising over a certain threshold. In that case, a thermal runaway may turn up, leading to a quick temperature rise and potentially other related undesirable consequences such as the generation of toxic gas and smoke. With the rising battery temperature over a critical point, the other chain exothermic reactions happen. The temperature and pressure in the LIB are cumulated until it exceeds the battery endurance. Eventually, the fire and rupture/explosion are inescapable. Therefore, the thermal management of LIB is essential during the battery working process or battery application. It is also crucial to investigate the LIB thermal runaway process by accurately monitoring and predicting temperature dynamics during thermal propagation and implementing functional methods to improve the cooling efficiency of the battery itself and the battery system.

There are two key topics of concern in battery thermal management: handling the charge/discharge cycle and governing the battery heat growth [13,14]. Many pieces of research focused on the battery thermal management system of EVs have been conducted [15–17]. The heat produced during the operating process has been established as the major rise in the working temperature. Computation Fluid Dynamics (CFD) is a practical tool for investigating thermal properties and simulating multiple physics fields [18]. CFD simulations could provide detailed information about the electrical and thermal areas inside the battery during the work process that is often challenging to assess and extract by experimental approaches. Model-based investigations promote a theoretical and comprehensive understanding of battery physics beyond what is possible from practical methods only. For example, Kirad and Chaudhari [19] applied numerical models for studying the selection of the battery module spacing with an improvement in cooling performance. Due to the development of computing capability, numerical simulations are gradually applied in battery models, battery components and materials studies, and battery safety engineering [20–22]. Most numerical studies rely on the thermal models, which predict the average surface temperature for a LIB cell [23–26], and lots of experimental investigations on thermal propagation have been carried out [27–30]. Nevertheless, to achieve a proper estimation of the thermal behaviour of a battery, many aspects, including the shape, layout, and physical and electrochemical properties, should be illustrated as closely as possible in the simulation. For instance, the asymmetric surface temperature of a battery cell should be considered in the model. The non-uniform temperature distribution in the LIBs leads to an electrical imbalance, lower battery performance, and shorter battery life [31,32]. Regarding the detailed temperature distribution, an electric-thermal model with the non-uniform feature should be built.

Moreover, battery thermal management systems have been classified in various ways based on different criteria [33–35]. For example, battery thermal management systems can be branched into three kinds based on various mediums: air-based, liquid-based, and phase change materials-based. Several optimization studies on the battery thermal management system have been previously conducted [36–39]. The air-cooling method is considered the most traditional approach and is a favoured option for HEVs and EVs. It is clear that the optimization of battery packs or systems depends on many parameters, such as geometry structure, coolant properties, operating conditions, and so force. Still, few researchers focus on multiple parameters simultaneously.

As a part of artificial intelligence, machine learning focuses on the study of accuracy improvement by computer algorithms and data to imitate how humans learn [40,41]. The application of machine learning techniques, particularly artificial neural networks (ANN), can be a potential method to optimize the battery system. Because LIBs are highly complex, nonlinear systems, applying a probabilistic approach allows for quantification of uncertainty, which positively impacts making decisions in design and control. ANN model is a kind of model that characterizes the interrelation between inputs and outputs by using a collection of interconnected nodes (perceptron). The application of ANN to battery research is still relatively new. Wu et al. [42] generated a design map that fulfills both specific energy and specific power requirements using a systematic approach based on ANN. Qian et al. [43] optimized the battery spacing by neural network model and demonstrated the interrelationship between layout and temperature of battery packs. Feng et al. [44] developed an electrochemical-thermal-neural-network method used for the co-estimation of LIB state of charge (SOC) and state of temperature. Shi et al. [45] applied the fully connected deep network approach to study air-based cooling $LiFePO_4$ cuboid battery packs and optimize the U-type structure. However, pioneering studies have highlighted the possibility of using ANN for battery thermal problems. These studies include modelling battery spacing, specific format, and some battery performances. The detailed temperature distribution of LIB and battery pack have not been fully investigated. Besides, the ambient temperature and natural ventilation should be considered during the battery working process. Therefore, the combination between ANN analysis and the electro-thermal battery model is proposed to investigate further the battery system's cooling efficiency and battery fire safety performance. Figure 1 shows the schematic figure of the integrated CFD-ANN model proposed in this study.

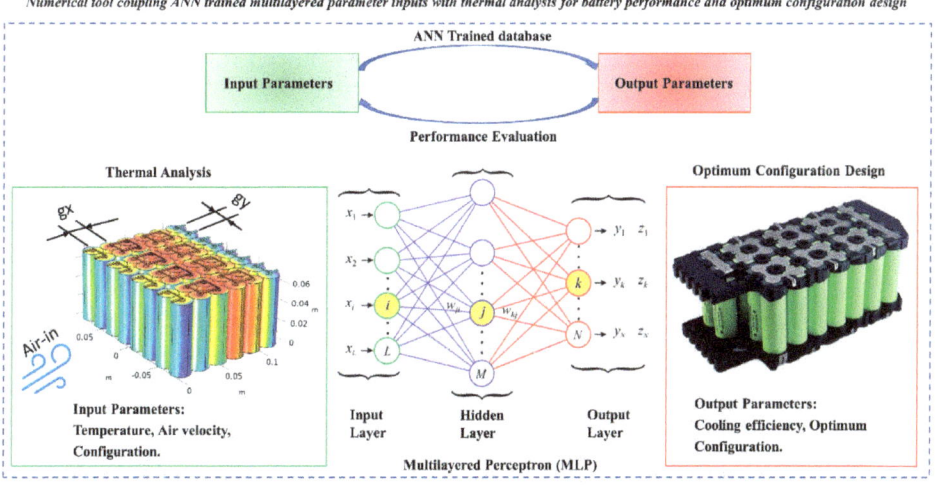

**Figure 1.** Schematic of the proposed CFD-ANN model.

To this end, the non-uniform distributions of the battery cell should be investigated for the thermal analysis, which can be treated as a measurement and prediction at the early stage of the Li-ion battery thermal runaway fires. Furthermore, with the numerical analysis, a better understanding of battery pack configuration design can be achieved. In this article, the contributions are:

(i) Establishment and development of a three-dimensional electro-thermal model capable of considering temperature distribution of battery packs and heat exchange with the ambient environment.
(ii) Utilize the numerical results to comprehensively describe and predict the battery system's thermal behaviour to improve battery safety during the designing and working stages.
(iii) Coupled the electro-thermal model with the ANN model to optimize the battery system configuration design and enhance the cooling performance of the battery system.

The outline of this paper is summarized as follows: Section 2 introduces the numerical models applied in this paper, including electrochemical model, thermal model, and ANN model. Section 3 demonstrates the numerical results of the proposed model with validation and verification. Also, the training process and optimization results are listed in this section. Finally, the author presented some conclusions and proposed the future perspectives on this field in the conclusion section.

## 2. Numerical Models Applied in the Battery Pack

### 2.1. Electrochemical Model

The electrochemical model applied in this study could be seen as a lumped version of a single particle model [46], simulating the transport of intercalated lithium in one of the electrodes. The single particle model predicts the temperature distribution and voltage changes in a single LIB cell during galvanostatic operations. The simplification of this model can be conducted when the battery is mainly controlled by the diffusion process in one of the electrodes only. The model is based on a complete model of a LIB working process cycle [47]. In this model, the lumped battery interfaces are utilized, and the battery cell voltage $E_{cell}$ is calculated by applying time-dependent cell current $I_{cell}$. Additionally, the battery open circuit voltage data, named $E_{OCV}$, is estimated from SOC.

The three-dimensional electro-thermal model built in this paper is based on a typical cylindrical LiFePO$_4$/Carbon power battery, considering the physical and electrical conservations, as well as thermal principles and electrochemical kinetics. The electrochemical reactions of common LIBs can be described as the following Equations (1)–(3), where $M$ stands for a metal, which is used as a cathode material such as cobalt or nickel, and C is recognized as the anode materials.

The reaction at the positive electrode is described as:

$$LiMO_2 \leftrightarrow Li_{1-x}MO_2 + xLi^+ + xe^- \quad (1)$$

The chemical reaction at the negative electrode is expressed as:

$$C + xLi^+ + xe^- \leftrightarrow Li_xC \quad (2)$$

The overall reaction can be presented as:

$$LiMO_2 + C \leftrightarrow Li_{1-x}MO_2 + Li_xC \quad (3)$$

Figure 2 demonstrates the working process of a typical LIB, and the fundamental cell unit is considered a sandwich structure, including the positive electrode, the separator, the negative electrode, and the current collectors located at both electrodes. The metal tab is joined at each correlated current collector and electrode. The separator is located between cathode and anode, a porous polymer membrane to prevent physical contact of electrodes. The electrolyte is the medium that enables the ion transport mechanism between electrodes. It requires specific working conditions, such as significant ion conductivity, low-set electrical conductivity, extended temperature range of operation, thermo-dynamically stable at a certain range of voltages, environmentally friendly, etc.

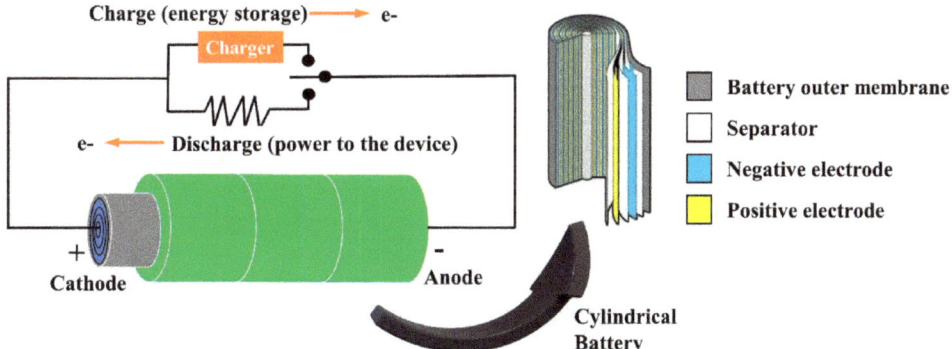

**Figure 2.** The principle of operation for a typical LIB and its structure.

Each electrode and separator is impregnated with electrolyte, achieving transportation of lithium ions. The material parameters for the electrolyte refer to a plasticized ethylene carbonate/dimethyl carbonate (EC/DMC) electrolyte remaining in a polymer matrix. Therefore, the stated electrolyte volume fraction points to this model's total liquid electrolyte and polymer matrix volume fractions. In this model, the potential losses $\eta_{IR}$ due to ohmic and charge transfer processes are given as follows:

$$\eta_{IR} = \eta_{IR,1C} \frac{I_{cell}}{I_{1C}} \tag{4}$$

where $\eta_{IR,1C}$ represents the potential losses under the 1C current. The 1C current $I_{1C}$ means that the discharge current will discharge the entire battery in one hour, and it is calculated as:

$$I_{1C} = \frac{Q_{cell,0}}{3600 \text{ s}} \tag{5}$$

The dimensionless charge exchange current $J_0$ is applied for the integrated voltage dissipation accompanied by the charge delivery reactions on the two electrodes' surfaces, shown as:

$$\eta_{act} = \frac{2RT}{F} \text{asinh}\left(\frac{I_{cell}}{2J_0 I_{1C}}\right) \tag{6}$$

The diffusion processes also lead to potential loss, which is represented by $\eta_{act}$. Derived from diffusion in an idealized particle or by applying a resistor-capacitor combination, concentration overpotential effects can be explained among the lumped battery interfaces. In this model, particle diffusion is calculated. Fickian diffusion of a dimensionless $SOC$ parameter is calculated in this case, using spherical symmetry, according to:

$$\tau \frac{\partial SOC}{\partial t} = -\nabla \cdot (-\nabla SOC) \tag{7}$$

The interval stands for a common particle of the electrode controlling the cell, where $X = 0$ and $X = 1$ mean the particle centre and particle surface accordingly.

The operating conditions at the particle boundary are as follows:

$$\nabla SOC = 0|_{X=0} \tag{8}$$

$$\nabla SOC = \frac{\tau I_{cell}}{N_{shape} Q_{cell,0}}\bigg|_{X=1} \tag{9}$$

where $N_{shape}$ equals three for spherical particles in this model. The $SOC$ of the surface, $SOC_{surface}$, is identified at the particle surface. The average $SOC$, named $SOC_{average}$, is prescribed by lumping the particle volume, appropriately considering spherical coordinates, and is defined as:

$$SOC_{average} = \frac{\int_0^1 SOC 4\prod X^2 dX}{\int_0^1 4\prod X^2 dX} = 3\int_0^1 SOCX^2 dX \qquad (10)$$

The integrated voltage dissipation accompanied by concentration overpotential is represented by $\eta_{conc}$ and defined as:

$$\eta_{conc} = E_{ocv}\left(SOC_{surface}\right) - E_{ocv}\left(SOC_{average}\right) \qquad (11)$$

Lastly, the battery cell voltage $E_{cell}$ is defined as:

$$E_{cell} = E_{ocv}(SOC_{average}) + \eta_{IR} + \eta_{act} + \eta_{conc} \qquad (12)$$

Establishing the explanation of $\eta_{conc}$ and $E_{cell}$ is also calculated as:

$$E_{cell} = E_{ocv}\left(SOC_{surface}\right) + \eta_{IR} + \eta_{act} \qquad (13)$$

The battery model involves these three steps. At first, a lumped battery model is set up and run for a time-dependent battery current. Then, parameter estimation of the parameters $\eta_{IR,1C}$, $\tau$, and $J_0$ is demonstrated using experimental data. This is achieved using the Global Least-Squares Objective node in the optimization interface, combined with the optimization study step using a Levenberg-Marquardt optimization solver. Lastly, cell voltage prediction is performed using the optimized lumped parameter values obtained in the previous parameter estimation study compared with experimental data.

### 2.2. Thermal Model

In this study, the thermal model is based on a previous two-dimensional axial symmetry approach, simulated by the Heat Transfer in Solids module. A spirally wound type of battery is chosen for this simulation, and the simplification of the heat conduction can be achieved along the spiral direction. Moreover, instead of simulating the heat conduction in each layer along the radial direction, the wound sheets are acted as a combination cell material domain. These approximations are understandable for spiral wound battery cells cooled under natural convection. The model configuration comprises three connection sections: (1) Battery outer can; (2) A combination cell material domain; (3) Center axis (mandrel where the battery cell sheets are wound).

For this model, several equations and parameters are considered. Considering the anisotropic thermal conductivities in this model and differences among various directions [48], the thermal conductivities along the radial path, $k_{T,r}$, and along the cylinder length direction, $k_{T,ang}$, are defined separately as follows:

$$k_{T,r} = \frac{\sum L_i}{\sum L_i / k_{T,i}} \qquad (14)$$

$$k_{T,ang} = \frac{\sum L_i k_{T,i}}{\sum L_i} \qquad (15)$$

The density $\rho_{batt}$ and heat capacity $C_{p,batt}$ for the combination cell material domain is defined as stated by the following equations:

$$\rho_{batt} = \frac{\sum L_i \rho_i}{\sum L_i} \qquad (16)$$

$$C_{p,batt} = \frac{\sum L_i C_{p,i}}{\sum L_i} \qquad (17)$$

The heat source produced by the combination cell material domain is identified by employing the Electrochemical Heating Multiphysics coupling module. However, the heat source term in the combination cell material domain is scaled to solve the lack of heat generation in the current collectors and the canister thickness. This mounted heat source is acquired by multiplying two factors of the volumetric heat source from the 1D Li-ion battery model. The former factor is the fraction of the total 1D model in which heat is produced. That is the total length value of electrodes and the separator, divided by the total battery length, including the measurements of both current collectors. The latter factor is the fraction of the entire 3D cylindrical battery. The volume in which heat is produced is the cell's total volume, including the homogenized wound layers of the cell material, the centre axis, and the battery case, minus the volume of the outer case and the volume of the battery centre axis. This heat source is then divided by the total volume of the battery cell domain, which is the difference between the whole battery volume and the centre axis volume. Thereby, the following equation for the 3D heat source is demonstrated:

$$Q_{h,3D} = Q_{h,1D} \frac{L_{neg} + L_{sep} + L_{pos}}{L_{batt}} \frac{((r_{batt} - d_{can})^2 - r^2_{mandrel})(h_{batt} - 2d_{can})}{(r^2_{batt} - r^2_{mandrel})h_{batt}} \quad (18)$$

For the lumped battery interface model, Arrhenius expression is applied to model each battery cylinder, with temperature-dependent ohmic, exchange current, and diffusion time-constant parameters.

The thermal conductivity in the combination cell material domain is anisotropic due to the spiral type of the battery. The orthotropic thermal conductivity in the combination cell material domain is solved by introducing a cylindrical coordinate system in the model. The zero-Mach-number limit of the compressible conservation equations was applied to depict the flow movement and heat transfer. Regarding the enthalpy equation and the temperature equation, the energy conservation for incompressible fluid can be simplified to:

$$\rho C_p \frac{\partial T}{\partial t} + \rho C_p \mathbf{v} \cdot \nabla T = \nabla \cdot (k \nabla T) + Q \quad (19)$$

*2.3. ANN Model*

ANN has rapidly developed in recent decades as a common tool to model a broad range of engineering systems due to its capability to learn and adapt to find potential correlations among different properties, mainly to map the nonlinear relationship of inputs and outputs. A hidden layer is a layer located between the input and output of the ANN model, in which artificial neurons applies a set of weights to the inputs and directs them through an activation function as the output. Hidden layers of ANN allow for a neural network's function to be taken apart for specific data transformations. For example, images and documents are treated as initial inputs from external data. The ultimate outcomes complete the task, such as recognizing objects in a snap.

A typical ANN model compromises five main components: inputs, summation functions, weights, activation functions, and outputs. The artificial neuron in the hidden layer works as a biological neuron in the brain. To form a directed, weighted graph, the network is formed by linking the output of specific neurons to the input of other neurons. A learning process can adapt the activation functions and weights. The learning rule or training approach controls the certain learning process. The summation function (denoted by $E$) is a function that calculates the net inputs, considering adjustable weights $w_{ij}$ and bias $w_{bi}$, expressed as:

$$E = \sum_{j=1}^{n} w_{ij} x_j + w_{bi} \quad (20)$$

The activation function of a node governs the output of that node, or "neuron," provided input or set of inputs. The activation function presents a functional relationship between the input and output layers. Some frequently applied activation functions are step activation, threshold function, sigmoid function, and hyperbolic tangent function [49]. The logistics sigmoid function is adopted in this study and is presented as:

$$f(E) = \frac{1}{1+e^{-E}} \tag{21}$$

The number of hidden neurons determined by the formulation according to neural network design [50] is given by:

$$N_h = \frac{N_s}{(\alpha(N_i + N_o))} \tag{22}$$

where the subscripts $h$ denotes the number of input neurons, $s$ is the sample amount in the training dataset, $o$ is the number of outputs, and $\alpha$ is a scaling factor ranging from 2–10. In this study, $\alpha$ is prescribed as 2 to achieve an optimal solution without overfitting, and ten hidden neurons were applied in the ANN model.

### 3. Results and Discussions

#### 3.1. Electro-Thermal Model Simulation Results

In this paper, a corporate software applied finite element, COMSOL Multiphysics 5.5, was employed to study the three-dimensional electro-thermal coupled model. The default mesh component was applied for generating the mesh, and mesh independence was examined as well. The MUMPS time-dependent solver was applied for battery variables, and the PARDISO solver was chosen for heat transfer variables. The mesh applied for this study consists of triangular and quadrilateral elements developed by COMSOL Multiphysics 5.5. In order to obtain the thermal behaviour and boundary layer spread, the refined mesh is achieved at the connections of the battery boundary. Mesh independence verification was performed to avoid the grid number and mesh quality impact on the simulation results, shown in Table 1. According to Table 1, 43,486 elements show reliable and efficient results, and more elements lead to larger computation time. Therefore, the total number of meshes is about 43,486 elements chosen for the validation case, considering the efficiency and accuracy. For the battery pack simulation, the most appropriate grid amount is around 211,907 elements. Subsequently, more simulation results are produced to feed the machine learning model to training.

**Table 1.** This is a table. Tables should be placed in the main text near to the first time they are cited.

| Grid Resolution | Elements Number | Calculation Time | Maximum Electrolyte Temperature |
|---|---|---|---|
| Finer | 114,273 | 75.6 min | 20.250 °C |
| Fine | 43,486 | 30.5 min | 19.829 °C |
| Normal | 23,986 | 18.7 min | 19.810 °C |
| Coarse | 9708 | 10.6 min | 18.910 °C |

The validation of the electrochemical model and thermal model is established on the Type 38,120 battery cell, in which the nominal voltage is 3.2 V and capacity is 10 Ah, as well as the thickness of the cathode, the separator, and the anode are 91 µm, 40 µm and 142 µm, respectively. The type we choose is one of the most commonly used in the current commercial market. It is developed based on physical and electrical conservations, as well as thermal principles and electrochemical kinetics. The battery pack investigation for the three-dimensional electro-thermal model is built by Type 21,700 battery cylinders with a nominal capacity of 4 Ah and a nominal voltage of 3.6 V. The battery pack is constructed by coupling two cylindrical batteries in parallel. Then the mated battery pairs are connected in series. The geometry and battery parameters are listed in Table 2. The geometry specifications are used to build the battery domain for thermal simulations, while the battery parameters are applied for the simulation of the electrochemical model.

Table 2. Geometry dimension and electrochemical parameters of the battery cell.

| Geometry Parameters | | | Battery Parameters | | |
|---|---|---|---|---|---|
| d_batt | 21 [mm] | Battery diameter | C_rate | 4 | C rate |
| r_batt | d_batt/2 | Battery radius | Q_cell | 4 [A·h] | Battery cell capacity |
| h_batt | 70 [mm] | Battery height | I_1C | Q_cell/1 [h] | 1C current |
| h_term | 1 [mm] | Terminal thickness | kT_batt_ang | 30 [W m$^{-1}$ K$^{-1}$] | Thermal conductivity, in plane |
| r_term | 3 [mm] | Terminal radius | kT_batt_r | 1 [W m$^{-1}$ K$^{-1}$] | Thermal conductivity, cross plane |
| d_sc | 2 [mm] | Serial connector depth | Ea_eta1C | 24 [kJ mol$^{-1}$] | Activation energy |
| h_sc | 1 [mm] | Serial connector height | Ea_J0 | −59 [kJ mol$^{-1}$] | Activation energy |
| h_pc | 0.5 [mm] | Parallel connector height | Ea_Tau | 24 [kJ mol$^{-1}$] | Activation energy |
| w_pc | 1 [mm] | Parallel connector width | T0 | 20 [°C] | Reference temperature |
| | | | J0_0 | 0.85 | J0 at reference temperature |
| | | | tau_0 | 1000 [s] | tau at reference temperature |
| | | | eta_1C | 4.5 [mV] | eta_1C at reference temperature |
| | | | rho_batt | 2000 [kg m$^{-3}$] | Battery density |
| | | | Cp_batt | 1400 [J kg$^{-1}$ K$^{-1}$)] | Battery heat capacity |
| | | | ht | 30 [W m$^{-2}$ K$^{-1}$] | Heat transfer coefficient |
| | | | T_init | 20 [°C] | Initial/external temperature |

The initial battery state is fully charged. The discharge process at different current densities is simulated, and the discharge curves during the process are demonstrated. The battery capacity under various discharge rates is built through the modelling. The simulation will be stopped when the cell potential decreases under 3 V, which is the state of end-of discharge. The simulation result of the nominal discharge current density representing case 1C, is shown in Figure 3a. The numerical result shows a good agreement with the experiment data. Meanwhile, there are a few deviations in the usual discharge voltage plateau related to the thermodynamic analytics and battery prototypes. The thermal model is validated, and the results are shown in Figure 3b. The experimental data is extracted from the surface of the battery along the axis to track the surface temperature development. The simulation results in the same location of the testing point have similar growth trends. The slight difference between the experimental and numerical results is because of the temperature rise of the experiment due to the local ohmic heat generation, where the electrical contact resistance among the connectors and terminals of the battery.

Considering the electrochemical performance, the current flows inside the battery cell and battery pack remain similar due to applying a single particle model. The Single Particle Battery interface answers for solid diffusion in the electrode particles and the intercalation reaction kinetics. A lumped solution resistance term is used for covering the ohmic potential drop inside the electrolyte. The cell capacity is specified through fractional volumes of both electrodes in the battery. The individual electrode operational state-of-charges are used to identify the initial charge distribution in the battery. The temperature contour and heat flux streamline of the proposed battery pack shows the temperature distribution on the whole battery pack during its working process by the proposed electro-thermal model. The maximum temperature of the battery's innermost parts is around 2 °C higher than the outermost parts. It also provides the temperature difference of the whole battery pack. With the utilization of this lumped model, the broad temperature distribution of battery cell surfaces can be represented. Besides, the heat flux generated inside the battery cell can be simulated, as well as the heat exchange between battery surfaces and the ambient cooling air can be simulated numerically.

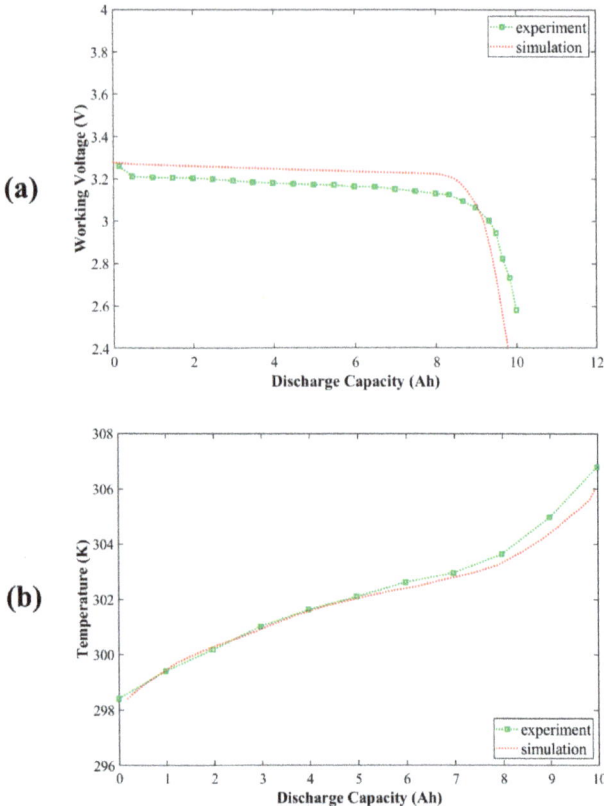

**Figure 3.** Comparison of numerical results of working voltage (**a**) and temperature (**b**) with experimental results [23] during 1C galvanostatic discharge under natural convection conditions.

Through the numerical study for the whole battery pack, the configuration setup of the battery pack is also investigated. The two-dimensional parameters are defined by different directions with various gaps, which are no gap (0 m), half of a cell (0.01 m) and one cell (0.02 m). Through permutation and combination, nine sets of collocations are formed. Figure 4 shows that all the battery cells are constructed together with various gap setups among the parallel-coupled battery pairs, which are the first nine cases. The gap enhances the convective and conductive between battery cells and ambient air from the battery safety perspective, improving battery pack cooling efficiency and fire safety.

To further understand the battery temperature distribution, air velocity and ambient environment are considered as well. Buoyancy forces cause natural convection as a consequence of density changes ascribed to temperature differences in the fluid. At heating, the fluid will rise because of the density variation in the boundary layer. Meanwhile, the cooler fluid, which will heat and increase, will replace the raised fluid. This continuous phenomenon is named free or natural convection. Thus, this study selected four sets of air velocity. Moreover, the temperature difference (The difference between the highest and the lowest temperature of the battery pack) and maximum temperature (The maximum temperature of a battery cell) are also considered. Figure 5 plots the maximum temperature and temperature difference profiles under various operating conditions, where the geometry conditions remain the same. From Figure 5a,c, air velocity positively impacts decreasing the maximum temperature and temperature difference. Under the same ambient temperature, shown in Figure 5b, increasing the air velocity can enhance the cooling efficiency, but the drop in maximum temperature is not much. Figure 5c, cases with air velocity 4 m s$^{-1}$ achieve 36%, which is the maximum percentage of temperature difference drop compared to other cases in this configuration. It is demonstrated that when the minimum values of maximum temperature and temperature difference are reached, the format set up is the best and optimization results.

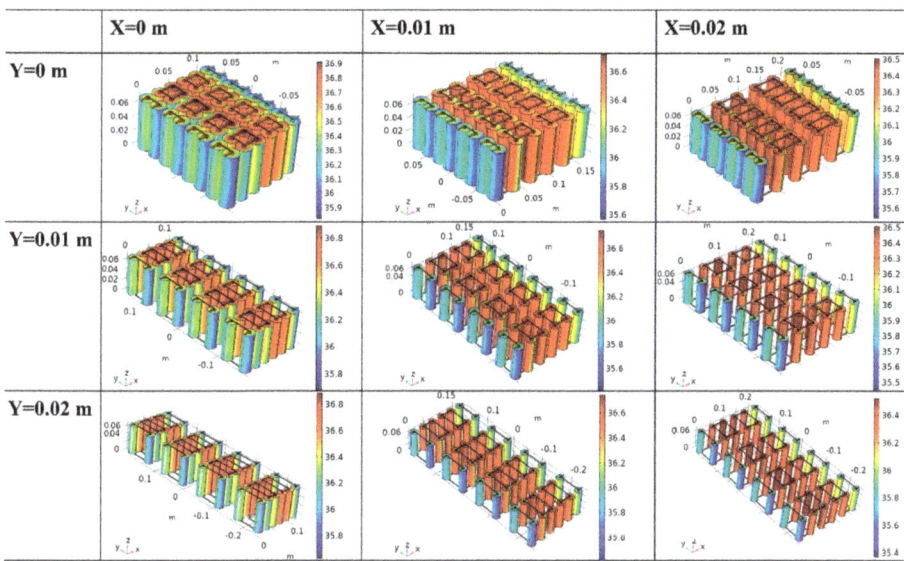

**Figure 4.** Temperature contours of different battery pack configurations.

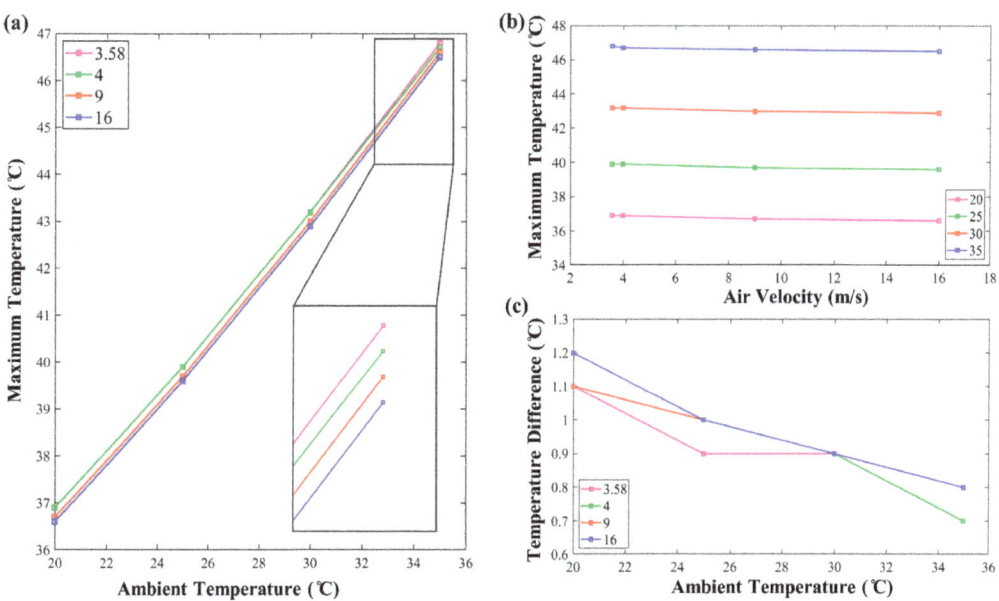

**Figure 5.** The profile of maximum temperature (**a**) and temperature difference (**b**,**c**) under various operation conditions.

## 3.2. Training and Results Analysis

In this study, the multilayer perceptron (MLP) neural network was applied. It is one of the most competitive types of ANN for regression in various research fields. Because this approach shows a considerable capability for universal approximation, it is regularly applied to model quite highly complex and disordered phenomena. As mentioned previously, the ANN utilizes the battery thermal distribution simulation dataset obtained through numerical simulations of a typical battery pack configuration on two cylindrical batteries in parallel and six-coupled battery pairs in series. In summary, it consists of 130 data sets of six parameters (four inputs and two outputs) prepared for the training and testing of the ANN. The detailed inputs and outputs of each dataset are illustrated in Table 3. Note that the heat transfer coefficient was replaced with the air velocity more effectively presentative by the ANN. It is possible to extend the ANN model to predictions on other heat transfer methods not considered in the datasets by analyzing their coolant velocity. The heat transfer coefficient of air can be estimated to:

$$h_t = 10.45 - v - 10v^{1/2} \qquad (23)$$

where $h_t$ represents the heat transfer coefficient, and $v$ is the relative speed between the object exterior and air. This equation is empirical and can be applied to the velocity range from 2 to 20 m s$^{-1}$ [51].

Table 3. Details of the inputs and outputs for the ANN model.

| | Inputs | | | | Outputs | |
|---|---|---|---|---|---|---|
| Parameters | X_Gap | Y_Gap | Air velocity | Ambient temperature | Maximum temperature | Temperature difference |
| Units | m | m | m s$^{-1}$ | °C | °C | °C |
| Range | 0–0.02 | 0–0.02 | 30–39.96 | 20–30 | - | - |

The proposed ANN model has been trained using the Levenberg-Marquardt (LM) optimization technique [52]. The LM method based on Levenberg [53] and Marquardt [54] combines Newton's method and gradient descent. It is one of the most efficient training algorithms for neural network modelling [55]. Generally, this algorithm demands more storage space but less time. The training process will be terminated spontaneously when generalization ends improving, as represented by a growth of the mean square error of the validation samples. In the LM method, the Hessian matrix can be approximated as:

$$Hf = Jf^T Jf \qquad (24)$$

The gradient is given by:

$$\nabla f = Jf^T e \qquad (25)$$

where $Jf$ is the Jacobian matrix and $e$ is the vector of a network error.

The LM working function or the fitness function is based on the mean square error (MSE) between the network output and the target output:

$$F = MSE = \frac{1}{N} \sum_{i=1}^{N} \left( R_{i,network} - R_{i,target} \right)^2 \qquad (26)$$

where $N$ is the number of datasets, $R_{i,network}$ is the network output and $R_{i,target}$ is the target output from the simulation data. The number of hidden neurons has been mentioned in Equation (22).

Figure 6 demonstrates the ANN regression results, and it plots the regression relation between the physical outputs and the targets, which indicates that this ANN model has achieved a good fit with the training datasets. From Figure 7, the error histogram plot shows that most errors reside in the range of −0.03688 to 0.0292. The majority of the predictions had a root mean square (RMS) error of approximately 0.088%, with around 10% of the predictions within ±25% RMS error. The ANN was successfully trained with an overall R (fitness) of 0.999, with most prediction errors within 1% RMS error. In future works, the ANN model can be further refined to achieve an even higher reliability and accuracy. This can be done by adding more simulation results considering a wider range of configurations or applying more advanced ANN training techniques.

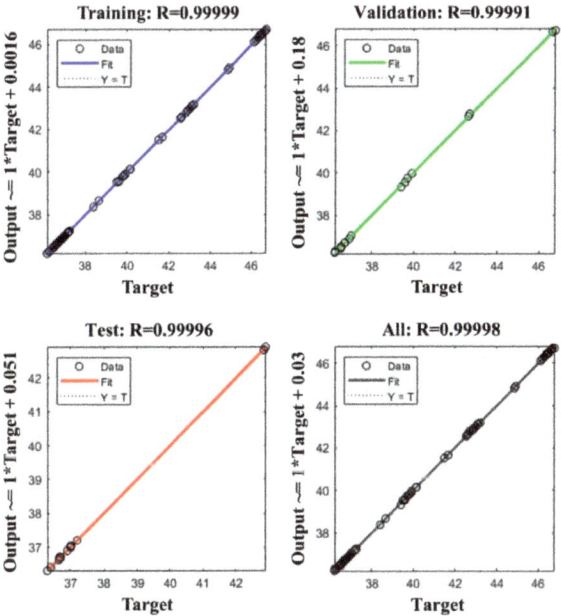

**Figure 6.** Regression results of ANN model.

*3.3. Optimization Analysis and Discussions*

The optimization configuration was proposed further to investigate the designed system with various working temperatures to improve the battery system's fire safety performance and cooling efficiency. These simulation results investigated the air velocity and ambient temperature after successfully training the ANN model. A battery pack comprises many single battery cells, and the operation temperature difference for the single battery cell inside the battery pack will be sourced by the temperature imbalance of the battery pack. This will result in the cell's inconsistency, and over the normal state during the charge and discharge process, harm the battery pack service life. Consequently, an appropriate battery thermal management system should simultaneously lower the maximum temperature and temperature difference of the battery packs to ensure the reliability and consistency of EVs and HEVs battery performance.

The 6,250,000 groups of structure and operation features were created in MATLAB. Inputs 1 and 2 are the configuration features, and Inputs 3 and 4 are the operating conditions. The CFD-ANN model calculates the optimal combination to achieve the perfect battery performance under the existing case arrangement with the current four inputs and two outputs.

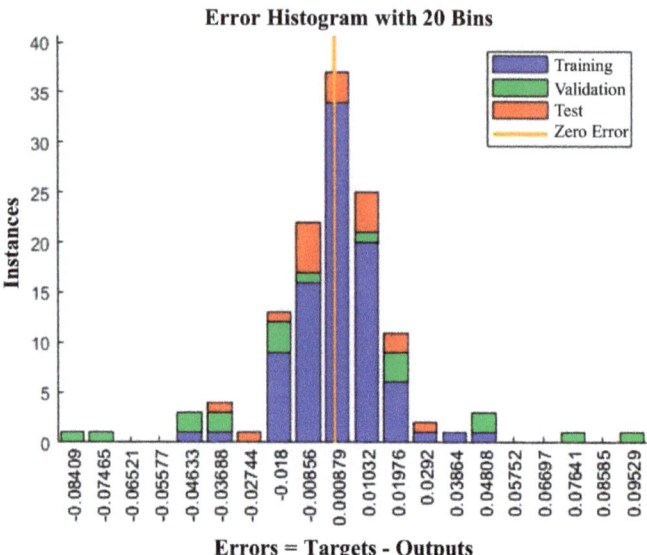

**Figure 7.** RMS error histogram of the ANN model predictions.

After successful training the ANN model, four inputs were divided into 50 intervals separately in a specific range. Then, 6,250,000 groups of structure features were processed in MATLAB, and the optimal result can be selected. According to Figure 8, the batteries' maximum temperature and temperature difference are greatly affected by the battery configuration and operating conditions, with a fluctuation as high as about 7 °C for the maximum temperature and 1.5 °C for the temperature difference. It can be obtained that the instability of the maximum temperature shows a different trend with the fluctuation of the temperature difference under the current range of various inputs. Therefore, the maximum temperature is chiefly influenced by the battery properties. The temperature difference can be treated as an indicator to evaluate the battery cooling performance of the battery pack.

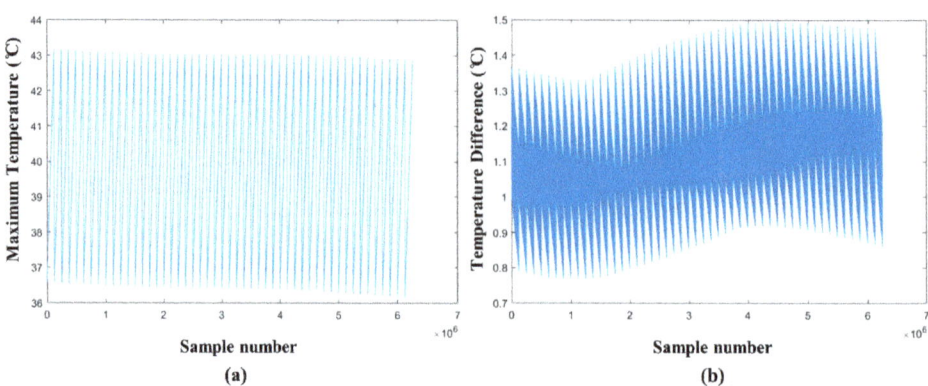

**Figure 8.** (a) Maximum temperature and (b) temperature difference of the battery pack for different input combinations.

From the CFD-ANN model simulation results, the six million sample results were sorted in ascending order of the battery spacing. Regarding achieving a battery cooling performance, it can be concluded that the optimal case is the 0.02 m X-direction gap and 0.01 m Y-direction gap under 20 °C ambient temperature and the air velocity 16 m s$^{-1}$. Compared to the original configuration with the same operating conditions, the maximum temperature decreased by 1.9%, and the temperature difference dropped by 4.5%, which means the CFD-ANN model optimization improved both the cooling efficiency and battery performance. The proposed framework demonstrates an efficient way to improve the thermal performance of the battery pack by optimizing the configuration under different operating conditions.

## 4. Conclusions

The optimal design of battery thermal management systems was achieved by applying a three-dimensional electro-thermal model coupled with the ANN model. Utilizing numerical simulations via CFD, different battery pack configurations were investigated in a simulation environment to positively impact cooling efficiency, battery performance, and battery fire safety. The three-dimensional electro-thermal model was introduced to calculate the temperature distribution and validated with the previous experimental data. The numerical case studies were applied to train the proposed ANN model, demonstrating the relationship among geometric parameters, operating conditions and cooling efficiency. The CFD-ANN model compared 6,500,000 combinations with various configuration and boundary conditions. The optimal design for the current battery setup was the case with a 0.02 m X-direction gap and a 0.01 m Y-direction gap, which lowered the maximum temperature and temperature difference by 1.9% and 4.5%, respectively.

The results highlighted one significant advantage of numerical simulations. All the pertinent information such as structural parameters and operation requirements can be derived from the model. Furthermore, different factors can be simulated and optimized simultaneously in a simulation environment to deliver a constructive perception of the battery performance and thermal behaviour. In the future, more parameters can be introduced with the current CFD-ANN model, such as electrode materials, electrolyte materials, cell numbers, different coolants, and so forth. A universal database of parameters and repercussions for the battery thermal management system can be prepared, which can then be processed by foresight models to create outlooks and predictions of the fire risks of the LIB energy storage system with a set of input variables.

**Author Contributions:** Conceptualization, A.L. and A.C.Y.Y.; methodology, A.L. and W.W. (Wei Wang); software, T.B.Y.C.; validation, A.L. and A.C.Y.Y.; formal analysis, A.L. and C.S.L.; investigation, W.W. (Wei Wang); resources, T.B.Y.C. and W.W. (Wei Wu); data curation, W.Y.; writing—original draft preparation, A.L.; writing—review and editing, Q.N.C. and S.K.; visualization, A.L.; supervision, G.H.Y.; project administration, A.C.Y.Y.; funding acquisition, G.H.Y. All authors have read and agreed to the published version of the manuscript.

**Funding:** This research was funded by the Australian Research Council (ARC Industrial Transformation Training Centre IC170100032) and the Australian Government Research Training Program Scholarship. The authors deeply appreciate all financial and technical supports.

**Conflicts of Interest:** The authors declare no conflict of interest.

## References

1. Yang, Y.; Bremner, S.; Menictas, C.; Kay, M. Battery energy storage system size determination in renewable energy systems: A review. *Renew. Sustain. Energy Rev.* **2018**, *91*, 109–125. [CrossRef]
2. Armand, M.; Axmann, P.; Bresser, D.; Copley, M.; Edström, K.; Ekberg, C.; Guyomard, D.; Lestriez, B.; Novák, P.; Petranikova, M. Lithium-ion batteries–Current state of the art and anticipated developments. *J. Power Sour.* **2020**, *479*, 228708. [CrossRef]
3. Li, Y.; Vilathgamuwa, M.; Xiong, B.; Tang, J.; Su, Y.; Wang, Y. Design of minimum cost degradation-conscious lithium-ion battery energy storage system to achieve renewable power dispatchability. *Appl. Energy* **2020**, *260*, 114282. [CrossRef]
4. Chen, T.; Jin, Y.; Lv, H.; Yang, A.; Liu, M.; Chen, B.; Xie, Y.; Chen, Q. Applications of lithium-ion batteries in grid-scale energy storage systems. *Trans. Tianjin Univ.* **2020**, *26*, 208–217. [CrossRef]
5. Lai, C.S.; Jia, Y.; Xu, Z.; Lai, L.L.; Li, X.; Cao, J.; McCulloch, M.D. Levelized cost of electricity for photovoltaic/biogas power plant hybrid system with electrical energy storage degradation costs. *Energy Convers. Manag.* **2017**, *153*, 34–47. [CrossRef]

6. Zhang, C.; Jiang, J.; Gao, Y.; Zhang, W.; Liu, Q.; Hu, X. Charging optimization in lithium-ion batteries based on temperature rise and charge time. *Appl. Energy* **2017**, *194*, 569–577. [CrossRef]
7. Dong, G.; Wei, J. Determination of the load capability for a lithium-ion battery pack using two time-scale filtering. *J. Power Sour.* **2020**, *480*, 229056. [CrossRef]
8. Xia, Q.; Wang, Z.; Ren, Y.; Yang, D.; Sun, B.; Feng, Q.; Qian, C. Performance reliability analysis and optimization of lithium-ion battery packs based on multiphysics simulation and response surface methodology. *J. Power Sour.* **2021**, *490*, 229567. [CrossRef]
9. Feng, X.; Ouyang, M.; Liu, X.; Lu, L.; Xia, Y.; He, X. Thermal runaway mechanism of lithium ion battery for electric vehicles: A review. *Energy Storage Mater.* **2018**, *10*, 246–267. [CrossRef]
10. Wang, Q.; Mao, B.; Stoliarov, S.I.; Sun, J. A review of lithium ion battery failure mechanisms and fire prevention strategies. *Prog. Energy Combust. Sci.* **2019**, *73*, 95–131. [CrossRef]
11. Sun, P.; Bisschop, R.; Niu, H.; Huang, X. A review of battery fires in electric vehicles. *Fire Technol.* **2020**, *56*, 1–50. [CrossRef]
12. Li, A.; Yuen, A.C.Y.; Wang, W.; Cordeiro, I.M.d.; Wang, C.; Chen, T.B.Y.; Zhang, J.; Chan, Q.N.; Yeoh, G.H. A Review on Lithium-Ion Battery Separators towards Enhanced Safety Performances and Modelling Approaches. *Molecules* **2021**, *26*, 478. [CrossRef] [PubMed]
13. Akinlabi, A.H.; Solyali, D. Configuration, design, and optimization of air-cooled battery thermal management system for electric vehicles: A review. *Renew. Sustain. Energy Rev.* **2020**, *125*, 109815. [CrossRef]
14. Jiang, Z.; Li, H.; Qu, Z.; Zhang, J. Recent progress in lithium-ion battery thermal management for a wide range of temperature and abuse conditions. *Int. J. Hydrog. Energy* **2022**, *47*, 9428–9459. [CrossRef]
15. Al-Zareer, M.; Dincer, I.; Rosen, M.A. Novel thermal management system using boiling cooling for high-powered lithium-ion battery packs for hybrid electric vehicles. *J. Power Sour.* **2017**, *363*, 291–303. [CrossRef]
16. Smith, J.; Singh, R.; Hinterberger, M.; Mochizuki, M. Battery thermal management system for electric vehicle using heat pipes. *Int. J. Therm. Sci.* **2018**, *134*, 517–529. [CrossRef]
17. Al-Zareer, M.; Dincer, I.; Rosen, M.A. Performance assessment of a new hydrogen cooled prismatic battery pack arrangement for hydrogen hybrid electric vehicles. *Energy Convers. Manag.* **2018**, *173*, 303–319.
18. Saw, L.H.; Ye, Y.; Tay, A.A.; Chong, W.T.; Kuan, S.H.; Yew, M.C. Computational fluid dynamic and thermal analysis of Lithium-ion battery pack with air cooling. *Appl. Energy* **2016**, *177*, 783–792.
19. Kirad, K.; Chaudhari, M. Design of cell spacing in lithium-ion battery module for improvement in cooling performance of the battery thermal management system. *J. Power Sour.* **2021**, *481*, 229016. [CrossRef]
20. Shen, M.; Gao, Q. A review on battery management system from the modeling efforts to its multiapplication and integration. *Int. J. Energy Res.* **2019**, *43*, 5042–5075. [CrossRef]
21. Deng, J.; Bae, C.; Marcicki, J.; Masias, A.; Miller, T. Safety modelling and testing of lithium-ion batteries in electrified vehicles. *Nat. Energy* **2018**, *3*, 261–266. [CrossRef]
22. Cordeiro, I.M.D.C.; Liu, H.; Yuen, A.C.Y.; Chen, T.B.Y.; Li, A.; Cao, R.F.; Yeoh, G.H. Numerical investigation of expandable graphite suppression on metal-based fire. *Heat Mass Transf.* **2021**, *58*, 1–17.
23. Xu, M.; Zhang, Z.; Wang, X.; Jia, L.; Yang, L. Two-dimensional electrochemical–thermal coupled modeling of cylindrical LiFePO4 batteries. *J. Power Sour.* **2014**, *256*, 233–243. [CrossRef]
24. Larsson, F.; Anderson, J.; Andersson, P.; Mellander, B.-E. Thermal modelling of cell-to-cell fire propagation and cascading thermal runaway failure effects for lithium-ion battery cells and modules using fire walls. *J. Electrochem. Soc.* **2016**, *163*, A2854. [CrossRef]
25. Jin, X.; Duan, X.; Jiang, W.; Wang, Y.; Zou, Y.; Lei, W.; Sun, L.; Ma, Z. Structural design of a composite board/heat pipe based on the coupled electro-chemical-thermal model in battery thermal management system. *Energy* **2021**, *216*, 119234. [CrossRef]
26. Wang, Y.; Fan, W.; Liu, P. Simulation of temperature field of lithium battery pack based on computational fluid dynamics. *Energy Procedia* **2017**, *105*, 3339–3344. [CrossRef]
27. Wilke, S.; Schweitzer, B.; Khateeb, S.; Al-Hallaj, S. Preventing thermal runaway propagation in lithium ion battery packs using a phase change composite material: An experimental study. *J. Power Sour.* **2017**, *340*, 51–59. [CrossRef]
28. An, Z.; Jia, L.; Li, X.; Ding, Y. Experimental investigation on lithium-ion battery thermal management based on flow boiling in mini-channel. *Appl. Therm. Eng.* **2017**, *117*, 534–543. [CrossRef]
29. Ouyang, D.; Liu, J.; Chen, M.; Weng, J.; Wang, J. An experimental study on the thermal failure propagation in lithium-ion battery pack. *J. Electrochem. Soc.* **2018**, *165*, A2184. [CrossRef]
30. Li, H.; Duan, Q.; Zhao, C.; Huang, Z.; Wang, Q. Experimental investigation on the thermal runaway and its propagation in the large format battery module with Li (Ni1/3Co1/3Mn1/3) O2 as cathode. *J. Hazard. Mater.* **2019**, *375*, 241–254. [CrossRef]
31. Zhang, X.; Wang, Y.; Wu, J.; Chen, Z. A novel method for lithium-ion battery state of energy and state of power estimation based on multi-time-scale filter. *Appl. Energy* **2018**, *216*, 442–451. [CrossRef]
32. Li, K.; Yan, J.; Chen, H.; Wang, Q. Water cooling based strategy for lithium ion battery pack dynamic cycling for thermal management system. *Appl. Therm. Eng.* **2018**, *132*, 575–585. [CrossRef]
33. Zhao, R.; Zhang, S.; Liu, J.; Gu, J. A review of thermal performance improving methods of lithium ion battery: Electrode modification and thermal management system. *J. Power Sour.* **2015**, *299*, 557–577. [CrossRef]
34. Kim, J.; Oh, J.; Lee, H. Review on battery thermal management system for electric vehicles. *Appl. Therm. Eng.* **2019**, *149*, 192–212. [CrossRef]

35. Patel, J.R.; Rathod, M.K. Recent developments in the passive and hybrid thermal management techniques of lithium-ion batteries. *J. Power Sour.* **2020**, *480*, 228820. [CrossRef]
36. Fan, L.; Khodadadi, J.; Pesaran, A. A parametric study on thermal management of an air-cooled lithium-ion battery module for plug-in hybrid electric vehicles. *J. Power Sour.* **2013**, *238*, 301–312. [CrossRef]
37. Wang, T.; Tseng, K.; Zhao, J.; Wei, Z. Thermal investigation of lithium-ion battery module with different cell arrangement structures and forced air-cooling strategies. *Appl. Energy* **2014**, *134*, 229–238. [CrossRef]
38. Choudhari, V.; Dhoble, A.; Panchal, S. Numerical analysis of different fin structures in phase change material module for battery thermal management system and its optimization. *Int. J. Heat Mass Transf.* **2020**, *163*, 120434. [CrossRef]
39. Chen, S.; Peng, X.; Bao, N.; Garg, A. A comprehensive analysis and optimization process for an integrated liquid cooling plate for a prismatic lithium-ion battery module. *Appl. Therm. Eng.* **2019**, *156*, 324–339. [CrossRef]
40. Mitchell, T.M. *Machine Learning*; McGraw-hill: New York, NY, USA, 1997.
41. Jaliliantabar, F.; Mamat, R.; Kumarasamy, S. Prediction of lithium-ion battery temperature in different operating conditions equipped with passive battery thermal management system by artificial neural networks. *Mater. Today Proc.* **2022**, *48*, 1796–1804. [CrossRef]
42. Wu, B.; Han, S.; Shin, K.G.; Lu, W. Application of artificial neural networks in design of lithium-ion batteries. *J. Power Sour.* **2018**, *395*, 128–136. [CrossRef]
43. Qian, X.; Xuan, D.; Zhao, X.; Shi, Z. Heat dissipation optimization of lithium-ion battery pack based on neural networks. *Appl. Therm. Eng.* **2019**, *162*, 114289. [CrossRef]
44. Feng, F.; Teng, S.; Liu, K.; Xie, J.; Xie, Y.; Liu, B.; Li, K. Co-estimation of lithium-ion battery state of charge and state of temperature based on a hybrid electrochemical-thermal-neural-network model. *J. Power Sour.* **2020**, *455*, 227935. [CrossRef]
45. Shi, Y.; Ahmad, S.; Liu, H.; Lau, K.T.; Zhao, J. Optimization of air-cooling technology for LiFePO4 battery pack based on deep learning. *J. Power Sour.* **2021**, *497*, 229894. [CrossRef]
46. Guo, M.; Sikha, G.; White, R.E. Single-particle model for a lithium-ion cell: Thermal behavior. *J. Electrochem. Soc.* **2010**, *158*, A122. [CrossRef]
47. Doyle, M.; Newman, J.; Gozdz, A.S.; Schmutz, C.N.; Tarascon, J.M. Comparison of modeling predictions with experimental data from plastic lithium ion cells. *J. Electrochem. Soc.* **1996**, *143*, 1890. [CrossRef]
48. Chen, S.-C.; Wang, Y.-Y.; Wan, C.-C. Thermal analysis of spirally wound lithium batteries. *J. Electrochem. Soc.* **2006**, *153*, A637. [CrossRef]
49. Chen, M.; Challita, U.; Saad, W.; Yin, C.; Debbah, M. Artificial neural networks-based machine learning for wireless networks: A tutorial. *IEEE Commun. Surv. Tutor.* **2019**, *21*, 3039–3071. [CrossRef]
50. Hunter, D.; Yu, H.; Pukish, M.S., III; Kolbusz, J.; Wilamowski, B.M. Selection of proper neural network sizes and architectures—A comparative study. *IEEE Trans. Ind. Inform.* **2012**, *8*, 228–240. [CrossRef]
51. ToolBox, E. Convective Heat Transfer. Available online: https://www.engineeringtoolbox.com/convective-heat-transfer-d_430.html (accessed on 27 July 2021).
52. Hagan, M.T.; Menhaj, M.B. Training feedforward networks with the Marquardt algorithm. *IEEE Trans. Neural Netw.* **1994**, *5*, 989–993. [CrossRef]
53. Levenberg, K. A method for the solution of certain non-linear problems in least squares. *Q. Appl. Math.* **1944**, *2*, 164–168. [CrossRef]
54. Marquardt, D.W. An algorithm for least-squares estimation of nonlinear parameters. *J. Soc. Ind. Appl. Math.* **1963**, *11*, 431–441. [CrossRef]
55. Yetik, O.; Karakoc, T.H. Estimation of thermal effect of different busbars materials on prismatic Li-ion batteries based on artificial neural networks. *J. Energy Storage* **2021**, *38*, 102543. [CrossRef]

Article

# A Novel Evaluation Criterion for the Rapid Estimation of the Overcharge and Deep Discharge of Lithium-Ion Batteries Using Differential Capacity

Peter Kurzweil [1,*], Bernhard Frenzel [1] and Wolfgang Scheuerpflug [2]

[1] Electrochemistry Laboratory, University of Applied Sciences (OTH), Kaiser-Wilhelm-Ring 23, 92224 Amberg, Germany
[2] Diehl Aerospace GmbH, Donaustraße 120, 90451 Nürnberg, Germany
* Correspondence: p.kurzweil@oth-aw.de; Tel.: +49-9621-482-3317

**Abstract:** Differential capacity $dQ/dU$ (capacitance) can be used for the instant diagnosis of battery performance in common constant current applications. A novel criterion allows state-of-charge (SOC) and state-of-health (SOH) monitoring of lithium-ion batteries during cycling. Peak values indicate impeding overcharge or deep discharge, while $dSOC/dU = dU/dSOC = 1$ is close to "full charge" or "empty" and can be used as a marker for SOC = 1 (and SOC = 0) at the instantaneous SOH of the aging battery. Instructions for simple state-of-charge control and fault diagnosis are given.

**Keywords:** capacitance; state-of-charge estimation; state-of-health; aging; lithium-ion battery

## 1. Introduction

For the safe operation of battery systems in stationary and mobile applications, the reliable indication of the correct state-of-charge (SOC) and state-of-health (SOH) [1] is a growing task and challenge for an efficient battery management system (BMS).

The current state-of-the-art in metrology [2–4] does not provide a universal and rapid method for the diagnosis of a lithium-ion battery without carefully examining the degradation of hundreds of full charge/discharge cycles. The monitoring of batteries by Ampere-hour counting over a few cycles has been used as a quality assurance tool or lithium-ion cells destined for long lifetime applications such as in electric vehicles or aircraft applications.

However, there is no simple and quick method to determine the actual SOC regardless of the age of the battery. Apart from accounting for amp-hours and nominal voltages, it is fundamentally unclear when a battery is sufficiently full during charging without going into overcharge, and when exhaustion is imminent during discharging. Various internal and external faults can occur during the battery operation, leading to performance loss and thermal runaway [5].

### 1.1. Incremental Data Analysis

Originally, "differential voltage analysis" (DVA) was proposed by Bloom et al. [6], who observed the change of $-Q_0 \cdot dU/dQ$ versus battery capacity $Q$ to gain insight into the aging processes of lithium-ion batteries during cycling.

The term "differential capacity analysis" appeared around the year 2000 (for history see [7,8]) for the first derivative of the galvanostatic curve, $U(Q)$. A series of $dQ/dU$ peaks (as a function of electrode potential or cell voltage $U$) corresponds to the potential plateaus (at constant voltage).

"Incremental capacity analysis" (ICA) was described by Dubarry et al. [9] and Dahn et al. [10] who considered the reciprocal quantity $dQ/dU$.

"Differential capacity analysis" [11] using high precision constant current chronopotentiometry and coulometry was employed for a detailed understanding of the aging processes and capacity degradation of lithium-ion batteries. Using the potential-capacity data of the positive and negative electrodes of fresh commercial cells, the differential capacity was calculated as a reference for a theoretical cell. Full $LiMn_2O_4$/graphite cells could be explained by mere relative shifts of the positive and negative potential-capacity curves.

"Delta differential capacity analysis" (Smith et al. [12]) was introduced to study the degradation of lithium-ion cells. By the help of constant-current chronopotentiometry, voltage versus charge data were collected as cells were charged and discharged in subsequent cycles. These $U(Q)$ values were then differentiated, using finite differences, to create differential capacity, $dQ/dU$, for a given measured cycle. For comparison of new and aged batteries, "Delta differential capacity", the difference $\Delta(dQ/dU)$ between the differential capacities of the $n$th and $m$th cycle was calculated. $\Delta dQ/dU$ should be zero for a perfect battery cell that does not degrade from cycle to cycle.

Both ICA and DVA [13–15] are based on the cell terminal voltage. However, the voltage axis may be replaced by the state-of-charge [16]. Differential capacity $dQ/dU$ from charge–discharge curves and pseudo-capacitance [17] at low frequencies from impedance spectra at the same voltage are equivalent [18].

For SOH estimation, the location interval between two inflection points of the differential voltage curve can be evaluated and compared to a new battery [19,20]; the distance between the inflection points is proportional to battery capacity. In recent years, differential voltage analysis has helped to obtain insights into the degradation of lithium-ion cells by capacity loss and resistance increase. An inhomogeneous lithium distribution leads to a flattening of the $dU/dQ$ signals [21]. Metal ions dissolute, migrate and deposit on the counter electrode [22].

Unfortunately, differential curves are very noisy, so that previous smoothing of the data is required. Simple data reduction, moving averages, and FFT smoothing [23] have been described. Fitting of the measured voltage profile with a number of Gaussian curves [24] has been proposed for differential capacity and differential voltage curves of high quality.

*1.2. Scope of This Study*

In aviation, the fast charging of a battery is allowed only in case of emergency. The state-of-charge (SOC) drops due to self-discharge after storing for a long time without power supply. Capacity determination and the recharging of a 2-Ah battery using the constant current discharge method and other diagnosis tools take roughly 1.5 to 2 h according to the regulations in air traffic.

Based on our previous work [25], we measured hundreds of lithium-ion batteries with different cell chemistries and wondered what might be a simple criterion for "full" and "empty" without performing a full charge–discharge cycle each time and risking overcharge or deep discharge of the cell. By the help of differential capacity, we finally found a diagnostic method that did not waste a complete charge–discharge cycle and a subsequent recharge to determine just the available charge, i.e., SOC = 1, while the state-of-health of the battery continuously deteriorates. Differential capacity $dQ/dU$ is the first derivative of the charge–discharge curve $Q(U)$.

In this study, the performance of a novel evaluation criterion for constant current charge–discharge curves is demonstrated based on numerous examples and different cell chemistries. The idea is to characterize the state-of-charge of any lithium-ion battery, whose history and state-of-health is not known, as "empty" or "full" using a simple calculation rule. Ideally, an automated process would determine whether the battery is approaching overcharge or deep discharge based on the small voltage and charge changes during the charging or discharging process, without knowing the actual charge (capacity $Q_0$ of the last charge).

## 2. Materials and Methods

New and old lithium-ion batteries of different manufactures and cell chemistries were charged and discharged at a constant current between the upper and lower cutoff voltage. The batteries investigated in the course of long-time tests under real conditions as in the airplane are compiled in Table 1.

Table 1. Lithium-ion batteries in this study according to the manufacturers' data sheets.

| | Chemistry | Cell | Rated Voltage | Max./Min. Voltage $U$ (V) | Capacity $Q$ (Ah) | Allowed Current (A) | |
|---|---|---|---|---|---|---|---|
| | | | | | | Charge | Discharge |
| 1 | LFP | LithiumWerks ANR26650M1B (LiFePO$_4$) | 3.3 | 3.6 ... 2.0 | 2.6 | 10 (4 C) | 50 (20 C) |
| 2 | NMC | LG ICR18650HE2 | 3.65 | 4.2 ... 2.0 | 2.5 | 4 | 20 |
| 3 | LCO | Sanyo/Panasonic UR18650FK, Li$_{1-x}$CoO$_2$ | 3.7 | 4.2 ... 2.5 | 2.3 | 2.3 | 4.8 |
| 4 | NCA | SONY US18650VTC6 | 3.65 | 4.2 ... 2.0 | 3.0 | 5 | 20 |

The proposed calculation methods work with conventional laboratory equipment and do not require devices from specific manufacturers. In this study, a DC power source (Elektro-Automatik EA-PS 2342-10B, Viersen, Germany), an electronic load (ET Systems ELP/DCM 9712C, Altlußheim, Germany), and a data logger (Agilent 34972A, Meilhaus Electronic, Alling. Germany) were combined in a climatic chamber (Vötsch VT7021, Weiss Technik AG, Altendorf, Switzerland). At a constant ambient temperature, the battery operation between overcharge and deep discharge was considered at slow charge–discharge rates (below 1C). Electric charge (capacity) was determined by coulomb-counting, although this is not strictly required for the proposed method. Measured data can be evaluated using EXCEL, MATLAB or similar software.

### 2.1. Differential Capacity and Resistance

To avoid numerical problems, differential capacity $C = dQ/dU$ (capacitance [17]) is best calculated from charge–discharge curves using the reciprocal of the differential voltage. According to Equation (1), at constant charge and discharge currents, small voltage differences $\Delta U$ in the time interval $\Delta t$ do not cause "division by zero" errors.

$$C = \frac{dQ}{dU} = \left(\frac{dU}{dQ}\right)^{-1} = \left(\frac{dR}{dt}\right)^{-1} \tag{1}$$

The unit of $dQ/dU$ is Farad: F = C/V = As V$^{-1}$. Therefore, the symbol $C$ is used for an electrical capacitance, which must not be confused with the electrical charge (capacity $Q$).

Capacitance $C$ (slope of the $Q(U)$ curve) is small and resistance $R$ (slope of the $U(Q)$ curve) is great when the battery is depleted or overcharged. Equation (2) qualitatively explains this relationship between capacitance and resistance as the state-of-charge changes.

$$\frac{\partial SOC}{\partial U} = \frac{d(Q/Q_0)}{dU} = \frac{d(CU/Q_0)}{dU} = \frac{C}{Q_0} = \frac{I}{Q_0 \frac{dU}{dt}} \sim \frac{1}{R\,I} \tag{2}$$

where $Q_0$ is the maximum capacity, i.e., the actual available (discharged) electrical charge of the battery after the previous full charge. According to Ohm's law, the voltage drop across the battery cell correlates with the internal resistance $R = dU/dI$. At a constant current, $dQ = 0$, "DV" is insensitive to resistance changes. Nevertheless, $dQ/dU$ implicitly represents the change in internal conductance and is therefore a measure of aging.

### 2.2. The Intersection Method: A New Approch

The true state-of-charge SOC($t$) = $Q(t)/Q_0$ (related to the last full charge $Q_0$) is not clearly a simple function of differential capacity. SOC detection using a linear function

$C(SOC)$ works best for flat $U(Q)$ curves, which is true for LFP batteries. However, we are looking for a general empirical method suitable for SOC and SOH monitoring of all battery types.

Indeed, the peaks of differential capacity occur at "almost full" and "almost empty". The novel criterion in Equation (3) could serve as an indication that the battery is virtually fully charged and imminent overcharging is likely. The same is valid for the empty battery short before an undesirable deep discharge.

$$\frac{dQ}{dU} = \frac{dU}{dQ} \tag{3}$$

The criterion also works with dimensionless quantities, if the state-of-charge (SOC) is used instead of the electric charge $Q$.

$$\frac{dSOC}{dU} = \frac{dU}{dSOC} \equiv 1 \tag{4}$$

A descriptive interpretation of the criterion results from the intersection of a straight line (ascending curve $dQ/dU$, in short $x$) and a hyperbola (descending curve $dU/dQ$, in short $1/x$). The equation $x = 1/x$ has the solution $x = \pm 1$, where only +1 is physically meaningful. The differential capacity could also be approximated by $x^2 = (1/x)^2$ or in general $x^n = (1/x)^n$, which leads to the same solution, $x = 1$.

A more complicated approach employs the maximum curvature of the charge–discharge curve. The curvature $K$ of a function $y(x)$ is mathematically defined according to Equation (5) using the first and second derivatives ($y'$ and $y''$). However, it is difficult to determine $K$ with noisy measurement data. The reciprocal of the curvature is the radius of curvature $1/K$.

$$K = \frac{y''(x)}{[1 + (y')^2]^{\frac{3}{2}}} \tag{5}$$

Electric charge $Q(t)$ is calculated by integrating the current $I(t)$ with respect to time $t$, for example using the trapezoid rule. It is advisable to smooth the charge–discharge curve before numerical derivation, for example by a moving average. Charging currents are positive ($I > 0$) and discharging currents have negative signs ($I < 0$). Details are given below.

*2.3. Theoretical Background of the Intersection Criterion*

Simplified, a lithium-ion battery can be modeled by an equivalent circuit consisting of a series combination of the electrolyte resistance $R_1$ and the charge transfer impedance $R_2 \mid\mid C$, which is a parallel combination of the charge transfer resistance $R_2$ and the double layer capacitance $C$. For charging and discharging with a constant current, the state-of-charge as a function of battery voltage is shown in Figure 1.

Differential capacitance $dQ/dU$ (incremental capacity "IC") and the "incremental voltage" $dU/dQ$ (differential voltage "DV") intersect below the upper cut-off voltage at a point corresponding to the kink point of the charge curve near full charge. A similar intersection point is obtained for the discharge above the lower cut-off voltage. The distance between the intersection points defines the voltage window in which the battery can be operated without long-term damage.

The model does not consider phase changes that cause further steps in the charge–discharge curve. The idea of the intersection criterion Equation (3) is to detect phase changes in the charge–discharge curve at an early stage to prevent overcharging and deep discharging and to switch from a constant current to a constant voltage operation.

Differential capacity turns to a local maximum where the cell voltage reaches a constant value. $dQ/dU$ peaks occur where the $U(Q)$ curve is flat (or the $Q(U)$ curve is steep), e.g., when the battery reaches a phase equilibrium ($\Delta U \to 0$).

Incremental voltage d$U$/d$Q$ rises abruptly as soon a constant current can no longer be fed into or drawn from the cell ($\Delta I \to 0$). d$U$/d$Q$ peaks show the steepest decent where phase changes, overcharge or deep discharge occur.

Peaks in the d$Q$/d$U$ curve indicate coexisting equilibrium phases with different lithium concentrations (d$U$ = 0), whereas d$U$/d$Q$ peaks reflect phase transitions (Bloom [6]).

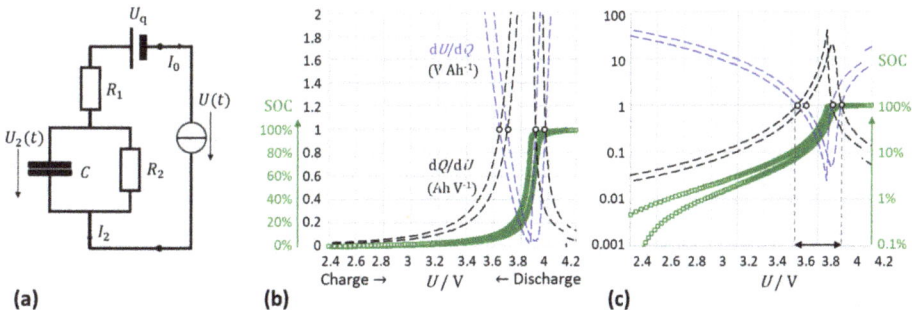

**Figure 1.** General model of a lithium-ion battery: (**a**) Network elements: $U_q$ = 4.2 V, $Q_0$ = 2 Ah, $C$ = 12500 F, $R_1$ = 0.004 Ω, $R_2$ = 0.02 Ω. (**b**) Calculated battery voltage $U$ (green), differential capacity d$Q$/d$U$ ("IC", dashed black) and "incremental voltage" d$U$/d$Q$ ("DV", V/Ah, dashed blue) versus state-of-charge for constant current charge and discharge ($I$ = 0.5 A) on a linear scale. (**c**) Calculated values on the logarithmic scale. (○) Intersection points: Charge: 3.70 V and 3.98 V; discharge 3.64 V and 3.91 V.

## 3. Results and Discussion

### 3.1. Battery Monitoring Using the Intersection Method

The suitability of differential capacity d$Q$/d$U$ (incremental capacity "IC", first derivative of the voltage–charge curve $U(Q)$) as a quality criterion for the state-of-health (SOH) is now shown for the experimental charge–discharge curves of new and aged lithium-iron phosphate cells. Figure 2 displays d$Q$/d$U$ and its reciprocal d$U$/d$Q$. The scaling on the $y$-axis is Ah/V or V/Ah. However, drawing dSOC instead of d$Q$ gave the same numerical results.

(a) Differential capacity

Differential capacity indicates small changes in the battery during aging far better than the $U(Q)$ curve. The d$Q$/d$U$ signals were sharper on the voltage scale than against SOC. The voltage positions of the peaks were well reproducible and differed only slightly for individual new batteries of the same type. For lithium-iron phosphate chemistry, the central peak of d$Q$/d$U$ at about 3.25 V indicated roughly the "almost empty" state (SOC $\to$ 0). The third peak showed the "almost full" battery at 3.3 V (SOC $\to$ 1).

Due to the internal resistance of the battery, charge and discharge had capacity peaks at different voltages (see Figure 1b). At a constant temperature, old and new cells almost did not differ in the voltage positions. Both deep discharge and overcharge did not significantly shift the d$Q$/d$U$ peaks on the voltage axis. However, the height of the d$Q$/d$U$ peak became smaller with old cells. With aging, the oxidation peaks shifted toward higher voltages (with charging) and the reduction peaks toward lower voltages (with discharging).

(b) Incremental voltage

This rise of signal can serve as a criterion that the battery is now fully charged and not yet overcharged. Due to phase transitions, often a sharp rise in temperature was observed shortly before the overcharge or exhaustion of the battery took place. The d$U$/d$Q$ peak ("DV") was an early warning of upcoming exhaust heat events and thermal runaway.

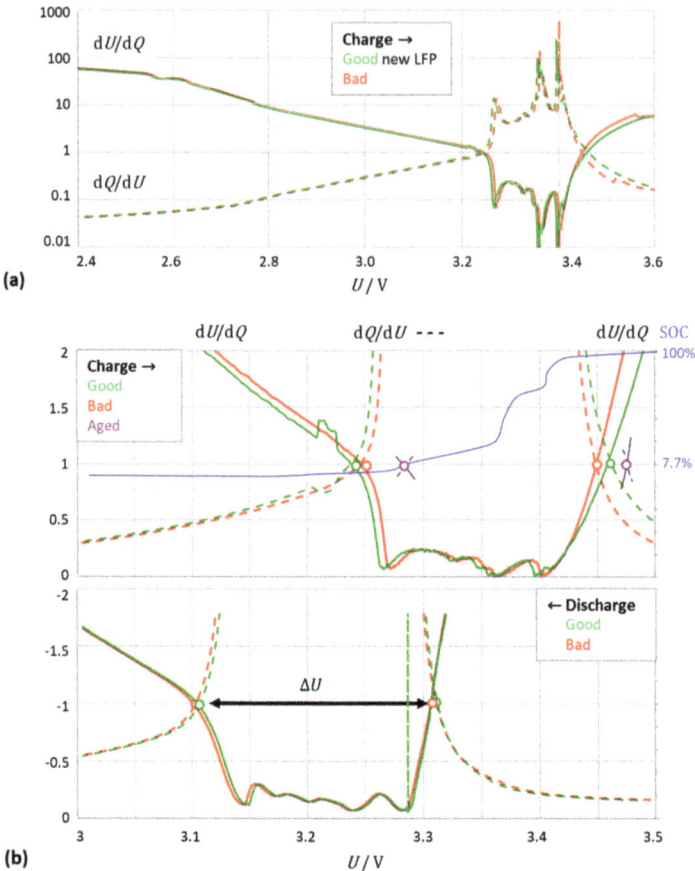

**Figure 2.** LFP batteries of the same type (LithiumWerks). (**a**) First derivative of the charge curve on the logarithmic scale: the best and the worst battery of 100 samples. (**b**) The intersection points (○) of differential capacity $dQ/dU$ ("IC" in Ah V$^{-1}$) and "incremental voltage" $dU/dQ$ ("DV" in V Ah$^{-1}$) define the voltage window $\Delta U$ between almost full charge and almost full discharge. The cross × indicates the intersection criterium of a battery after long-term aging (900 cycles, DOD 20, 0.3 A). Quantitative data see below.

(c) Intersection method

The intersection of $dQ/dU$ and $dU/dQ$ indicated the almost full charge (SOC → 1) at a high voltage and an almost empty state (SOC → 0) at a low voltage. As the current increased, the intersection points shifted to higher voltages (charging) or lower voltages (discharging) because the voltage dropped across the internal resistance of the battery increases. The voltage window (distance between the upper and lower intersection points) depended on the applied current (see Section 3.3).

Table 2 compiles the quantitative results of different batteries. The new LFP battery was 98% fully charged at 3.45 V (2.49 Ah of 2.54 Ah, cut-off voltage 3.61 V). However, only 2.36 Ah could be discharged until the intersection criterion was reached. All experiments were performed at the same current. The bad battery consumed 2.46 Ah (of 2.51 Ah) until the overcharge warning was reached; 2.33 Ah could be discharged until the deep discharge warning. This means that the small additional overcharge capacity (0.05 Ah) and the small residual capacity (0.19 Ah) which were lost using the intersection criterion, were negligible in relation to the risk of repeated overcharge and deep discharge.

We conclude that the intersection method allowed the economical operation of LFP batteries, with about 10% of the theoretical charge remaining unused.

**Table 2.** Intersection criterion for lithium-ion batteries at the beginning of life and end of life: Cell voltage $U$, state-of-charge (SOC = $Q/Q_0$ in %) and available electric charge $Q$ ($Q_0$ = capacity at full charge) at the upper and lower intersection: $dQ/dU = dU/dQ \approx 1$.

| Battery and Cell Chemistry | | State-of-Health: Capacity $Q_0$ and Electrolyte Resistance | Charge Window | | | Discharge Window | | |
|---|---|---|---|---|---|---|---|---|
| | | | V | SOC | mAh | V | SOC | mAh |
| LFP | LithiumWerks See Figure 1 | New #0 2.5 Ah, 5.3 mΩ | 3.45 ... 3.25 | 98 ... 7.6 | 2493 ... 192 | 3.29 ... 3.10 | 98 ... 7.6 | 2356 ... 43 |
| | | Good #74 2.5 Ah, 5.0 mΩ | 3.46 ... 3.24 | 98 ... 7.5 | 2497 ... 192 | 3.30 ... 3.11 | 98 ... 7.6 | 2365 ... 40 |
| | | Bad #32 2.5 Ah, 6.4 mΩ | 3.45 ... 3.25 | 98 ... 7.7 | 2463 ... 192 | 3.30 ... 3.10 | 98 ... 7.7 | 2329 ... 45 |
| | | Aged #3 2.0 Ah | 3.47 ... 3.28 | 98 ... 8.9 | 2088 ... 189 | 3.25 ... 3.06 | 98 ... 9.2 | 1957 ... 42 |
| NCM | LG See Figure 2 | New 2.2 Ah | 4.2 ... 3.46 | 100 ... 7.5 | 2170 ... 164 | 4.11 ... 3.35 | 99.6 ... 14 | 1871 ... 9 |
| | | DoD 20 1.9 Ah (0.3 A) | 4.2 ... 3.49 | 100 ... 7.6 | 1908 ... 145 | 4.07 ... 3.36 | 99.4 ... 12 | 1674 ... 11 |
| | | DoD 100 1.6 Ah (0.3 A) | 4.2 ... 3.55 | 100 ... 11 | 1590 ... 176 | 4.00 ... 3.35 | 98.5 ... 15 | 1365 ... 24 |
| NCA | Sony See Figure 3 See Figure 4 | New 3.1 Ah, 31 mΩ | – | – | – | 4.15 ... 3.05 | 99.9 ... 7.0 | 2850 ... 0 |
| | | DoD 20 2.3 Ah, 74 mΩ | 4.21 ... 3.43 | 100 ... 6.5 | 2268 ... 148 | 4.04 ... 3.31 | 99.5 ... 16 | 1921 ... 10 |
| | | DoD 100 1.7 Ah, 70 mΩ | 4.13 ... 3.47 | 92 ... 8.1 | 1607 ... 142 | 3.97 ... 3.36 | 98.7 ... 19 | 1415 ... 23 |
| LCO | Panasonic | New #8 2.26 Ah | 4.09 ... 3.73 | 86 ... 5.3 | 1942 ... 120 | 4.09 ... 3.58 | 99.2 ... 5.3 | 2128 ... 18 |
| | | DoD 20 #4 2.21 Ah | 4.21 ... 3.70 | 100 ... 4.8 | 2200 ... 106 | 4.11 ... 3.56 | 99.5 ... 9.2 | 1702 ... 9 |
| | | DoD 100 #6 2.26 Ah | 4.21 ... 3.73 | 99 ... 5.2 | 2237 ... 118 | 4.09 ... 3.75 | 94.8 ... 1.3 | 2152 ... 29 |

### 3.2. The Intersection Method Indicates the Degree of Aging

Figure 3 shows the first derivative of the discharge–voltage curve of NMC batteries that reached their end-of-lives after extended long-term cycle tests. Again, differential capacity and its reciprocal intersected at 1. The distance between the intersection points at low and high voltage defined the useful working range. It can be clearly seen that the new battery covered 0.76 V between "almost full" (1.9 Ah) and "almost empty" (0.009 Ah). The old battery offered only 0.65 V between 1.4 Ah and 0.024 Ah.

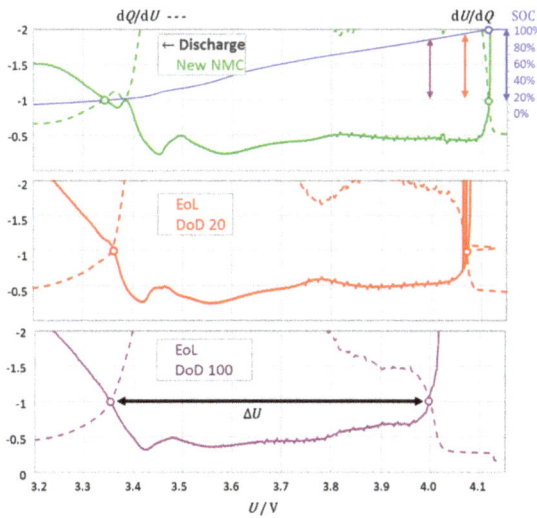

**Figure 3.** NMC batteries of the same type (LG ICR18650HE): Intersection method for brand new condition and end-of-life after flat and deep cycling test at constant current (900 cycles, 0.3 A = C/10). The usable voltage range = distance between the intersections (○) becomes smaller with forced aging. Differential capacity $dQ/dU$ ("IC", in Ah V$^{-1}$), incremental voltage $dU/dQ$ ("DV", in V Ah$^{-1}$).

| Charge | | | | | | | | | | | Discharge | | | | | | | | | |
|---|---|---|---|---|---|---|---|---|---|---|---|---|---|---|---|---|---|---|---|---|
| $\frac{U}{V}$ | $\frac{I}{A}$ | $\frac{\Delta U}{V}$ | $\frac{\Delta U}{V}$ | $\frac{\Delta t}{s}$ | $\frac{\Delta Q}{Ah}$ | $x = \frac{dU}{dQ}$ | $x^{-1} = \frac{dQ}{dU}$ | $\frac{Q}{Ah}$ | SOC % | $S \to 0$ | $S' > 1$ | $\frac{U}{V}$ | $\frac{\Delta U}{V}$ | $\frac{\Delta U}{V}$ | $\frac{\Delta Q}{Ah}$ | $x$ | $x^{-1}$ | $\frac{Q}{Ah}$ | SOC % | $S \to 0$ | $S' > 1$ |
| | | meas. | smooth | fix | | calc. | smooth | sum | | criterion | | | Meas. smooth | | | | smooth | sum | | criterion | |
| 2,579 | 0,001 | | 0,012 | 10 | 0,000 | 5685 | | 0,027 | 0,000 | 0,0 | 36,34 | -2,560 | 4,136 | | -0,002 0,000 | -758 | -0,001 | 0,000 | 100,0 | 758 | -3,880 |
| 2,596 | 0,001 | 0,017 | 0,011 | | 0,000 | 4139 | 0,000 | 0,027 | 0,000 | 0,0 | 36,34 | -2,560 | 4,132 | -0,004 | -0,002 0,000 | -512 | -0,002 | 0,000 | 100,0 | 512 | -3,710 |
| ... | | | | | | | | | | | | | | | | | | | | | |
| 2,949 | 0,295 | 0,013 | 0,015 | | 0,001 | 18,68 | 0,054 | 0,070 | 0,010 | 0,4 | 14,25 | -2,154 | 4,035 | -0,001 | -0,001 0,001 | -1,231 | -0,812 | 0,010 | 99,6 | 0,419 | -0,623 |
| 2,964 | 0,295 | 0,015 | 0,013 | | 0,001 | 15,67 | 0,064 | 0,079 | 0,011 | 0,5 | 12,53 | -2,098 | 4,035 | -0,001 | -0,001 0,001 | -1,005 | -0,995 | 0,010 | 99,5 | 0,010 | 1,000 |
| 2,975 | 0,295 | 0,011 | 0,012 | | 0,001 | 14,35 | 0,070 | 0,084 | 0,012 | 0,5 | 11,79 | -2,072 | 4,034 | -0,001 | -0,001 0,001 | -0,972 | -1,029 | 0,011 | 99,5 | 0,058 | 0,240 |
| ... | | | | | | | | | | | | | | | | | | | | | |
| 3,424 | 0,295 | 0,001 | 0,001 | | 0,001 | 1,035 | 0,967 | 0,978 | 0,146 | 6,5 | 0,044 | 0,353 | 4,000 | 0,000 | 0,000 0,001 | -0,114 | -8,748 | 0,153 | 93,3 | 8,633 | -1,936 |
| 3,425 | 0,295 | 0,001 | 0,001 | | 0,001 | 1,012 | 0,988 | 0,996 | 0,147 | 6,5 | 0,009 | 1,068 | 4,000 | 0,000 | 0,000 0,001 | -0,115 | -8,688 | 0,154 | 93,3 | 8,573 | -1,933 |
| 3,426 | 0,295 | 0,001 | 0,001 | | 0,001 | 0,997 | 1,003 | 1,014 | 0,148 | 6,5 | 0,027 | 0,561 | 4,000 | 0,000 | 0,000 0,001 | -0,118 | -8,499 | 0,155 | 93,2 | 8,381 | -1,923 |
| ... | | | | | | | | | | | | | | | | | | | | | |
| 3,579 | 0,298 | 0,000 | 0,000 | | 0,001 | 0,272 | 3,672 | 3,633 | 0,455 | 20,1 | 3,358 | -1,526 | 3,889 | 0,000 | -0,001 0,001 | -0,625 | -1,600 | 0,475 | 79,3 | 0,975 | -0,989 |
| ... | | | | | | | | | | | | | | | | | | | | | |
| 3,722 | 0,291 | 0,000 | 0,000 | | 0,001 | 0,470 | 2,126 | 2,134 | 0,801 | 35,3 | 1,665 | -1,221 | 3,786 | 0,000 | 0,000 0,001 | -0,219 | -4,563 | 0,838 | 63,4 | 4,344 | -1,638 |
| ... | | | | | | | | | | | | | | | | | | | | | |
| 3,904 | 0,290 | 0,000 | 0,000 | | 0,001 | 0,238 | 4,198 | 4,175 | 1,374 | 60,6 | 3,935 | -1,595 | 3,565 | -0,001 | 0,000 0,001 | -0,569 | -1,756 | 1,447 | 36,8 | 1,187 | -1,074 |
| ... | | | | | | | | | | | | | | | | | | | | | |
| 3,944 | 0,290 | 0,000 | 0,000 | | 0,001 | 0,248 | 4,037 | 4,054 | 1,548 | 68,2 | 3,807 | -1,581 | 3,486 | 0,000 | 0,000 0,001 | -0,276 | -3,620 | 1,633 | 28,7 | 3,343 | -1,524 |
| ... | | | | | | | | | | | | | | | | | | | | | |
| 4,056 | 0,294 | 0,000 | 0,000 | | 0,001 | 0,459 | 2,179 | 2,178 | 1,820 | 80,3 | 1,719 | -1,235 | 3,313 | -0,001 | -0,001 0,001 | -0,997 | -1,003 | 1,921 | 16,1 | 0,006 | 1,214 |
| 4,056 | 0,294 | 0,000 | 0,000 | | 0,001 | 0,458 | 2,182 | 2,178 | 1,821 | 80,3 | 1,719 | -1,235 | 3,312 | -0,001 | -0,001 0,001 | -0,994 | -1,007 | 1,922 | 16,1 | 0,013 | 0,887 |
| ... | | | | | | | | | | | | | | | | | | | | | |
| 4,204 | 0,297 | 0,000 | 0,000 | | 0,001 | 0,441 | 2,269 | 2,255 | 2,259 | 99,6 | 1,812 | -1,258 | 3,154 | -0,001 | -0,001 0,001 | -1,040 | -0,962 | 2,066 | 9,8 | 0,078 | 0,110 |
| 4,207 | 0,278 | -0,001 | 0,000 | | 0,001 | 0,207 | 4,836 | 4,167 | 2,268 | 100,0 | 3,927 | -1,594 | 2,504 | -0,007 | -0,007 0,001 | -7,684 | -0,130 | 2,290 | 0,0 | 7,554 | -1,878 |

**Figure 4.** Calculation recipe for the intersection method. Example data from this work. The criterion $S = |x - x^{-1}| \to 0$, or $S' = -\lg(10 \cdot S) > 1$ shows the cell voltage at the intersection point, $dQ/dU = dU/dQ \approx 1$, in the charge–discharge curve (below in green). Electric charge $Q$, available capacity $Q_0$ (in bold), and state-of-charge (SOC = $Q/Q_0$) are given for information only.

In Figure 5, the intersection method was applied to lithium NCA batteries. Qualitatively the same results were obtained as with the other cell chemistries. For NMC and NCA chemistries, the upper intercept around 4 V reflected "virtually full" (SOC > 99%).

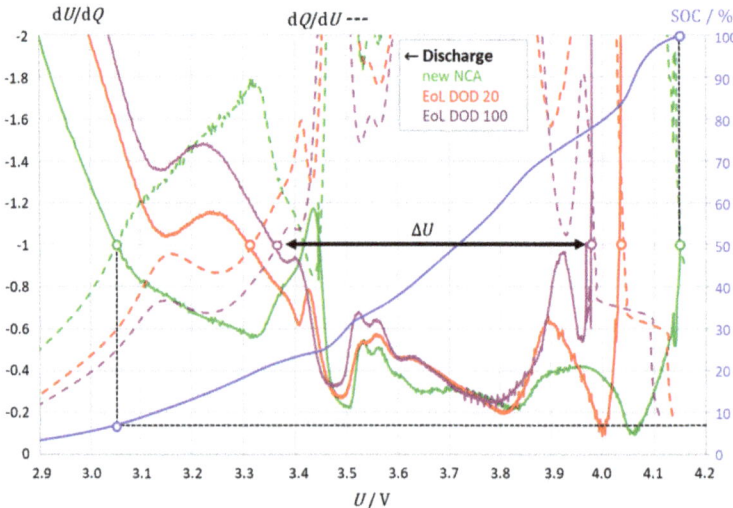

**Figure 5.** NCA batteries of the same type (SONY): First derivative of the constant current discharge characteristics for new samples and end-of-life parts after constant current cycling test (900 cycles, 0.3 A). The usable voltage range and SOC window between the intersection points (○) becomes smaller with forced aging. For comparison: SOC curve (blue) of the new battery.

Table 2 adds an LCO battery; the discharge profiles (not shown here) were not qualitatively different from the examples discussed.

## 3.3. The Intersection Criterion Reflects Kinetic Inhibitions

As a mathematical tool, the intersection method is nothing more than a mirror of the charge–discharge curve. It is designed to provide a timely indication of impending overcharge or deep discharge. Figure 6 shows exemplary measurements at different currents and temperatures.

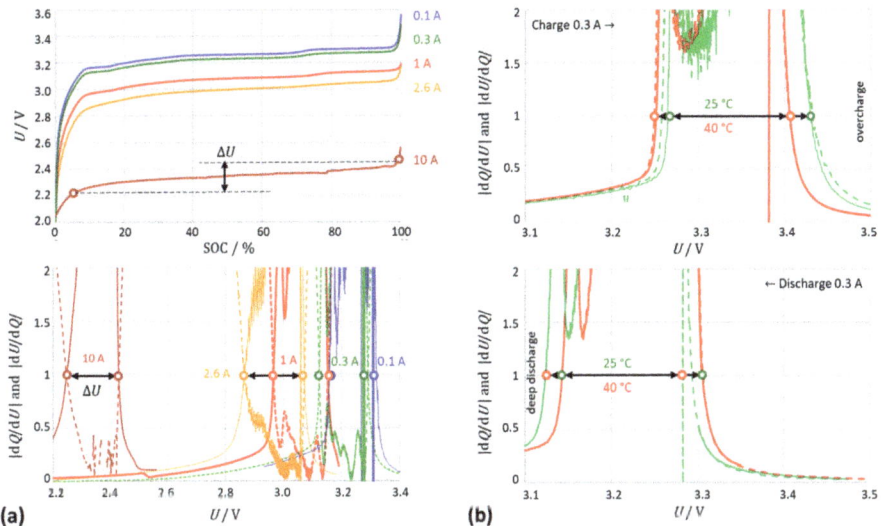

**Figure 6.** Application of the intersection method to measurements on a lithium iron phosphate battery: (**a**) Variation of current: discharge characteristics (above) and change of voltage window based on differential capacity and its reciprocal (below). $\Delta U$ is the 'safe' voltage window and a measure of the battery's internal resistance. (**b**) Variation of temperature during charging (**above**) and discharging (**below**) the battery. The scatter at small currents is a consequence of the used measurement technique.

For different currents, the intersection points reliably indicated the "safe" voltage window. As the discharge current increased, the intersection points moved to a lower cell voltage, as expected, because the voltage drop across the internal resistance of the battery increased, $U(I) = U_0 - I R$. As the temperature increased and the kinetic inhibitions decreased, the intersection points shifted to lower cell voltages.

## 3.4. Practical Implementation

The voltage values in a common charge–discharge curve are not equidistant; therefore, the usual formulas for numerical differentiation including smoothing over multiple data points are not applicable. Small differences between adjacent values cause outliers and spikes. Therefore, for noisy signals, we successfully used derivatives with central differences.

$$\frac{dy}{dx} \approx \frac{y_{i+1} - y_{i-1}}{x_{i+1} - x_{i-1}} \tag{6}$$

The intersection method can be easily carried out by machine according to the calculation recipe in Figure 4.

In Section 3.5, the method was applied to synthetic data without prior smoothing of the measurement values. However, the method worked reliably even with slightly noisy data.

When the current signal was very noisy and the voltage changes were very small, outliers (spikes) occurred in the derivative which, in case of doubt, may be deleted point by point. If noisy measurement data are available, the voltage signal is smoothed by a moving average. With millivolt resolution, 8 or 13 data points are helpful. Then the quotient, $S = \Delta U / \Delta Q$, and its reciprocal are formed. Comparing how $dQ/dU$ and $dU/dQ$ approach each other and ideally reach the value of one, the operating range of the battery between "almost empty" and "almost full" was found.

For acceptable results with highly noisy measurement data, as shown in the figures above and below, the step-by-step calculation scheme is as follows:

1. Measurement of voltage $U$ and electric current $I$ during charging and discharging, e.g., every 10 s;
2. Calculation of the differences $\Delta U = U_{i+1} - U_i$ for all voltage values ($i = 1, \ldots, n$);
3. Calculation of charge differences $\Delta Q = I \cdot \Delta t$ using the average value of constant current $I$ and time interval $\Delta t$. Informatively, SOC = $Q/Q_{\max}$ can be added for each voltage point;
4. Smoothing of voltage differences by averaging over 15 data points (total curve has 3000 data points);
5. Calculation of incremental voltage ("DV"), $dU/dQ \approx \Delta U / \Delta Q$, using the smoothed voltage vector;
6. Calculation of the reciprocal $dQ/dU \approx (\Delta U / \Delta Q)^{-1}$;
7. Smoothing of $dQ/dU$ by averaging over 17 data points.

*3.5. Limitations of the Method: Application to Different Cell Chemistries*

Figure 7 shows the constant current charge–discharge curves of various lithium-ion batteries and the application of the intersection method. The latter worked best when the $Q(U)$ curve was S-shaped and steep.

(a) Lithium–iron phosphate (LFP)

Curvatures that are easy to read by eye cause considerable difficulties in computer-aided analysis of data curves. The derivative $dQ/dU$ gave the steepest slope. In contrast to that, the intercept criterion Equation (3) provided the voltage at the greatest curvature of the curve, i.e., close to the kink point, before a constant voltage prevails. The S-shaped charge curve exhibited the lower and upper kink point at about 3.27 V (SOC 0.1) and 3.45 V (SOC 0.98); the discharge curve had kink points at 3.14 V (SOC 0.1) and 3.28 V (SOC 0.9). The LFP discharge curve reached the radius of curvature one at 3.1 V and 3.4 V. The voltage range where the radius of curvature was less than 1 again marked the operating range between "almost full" and "almost empty". At the end of discharge, the radius of the curvature increased steeply. The benefit of curvature and radius of curvature was small, because one could also observe the almost constant voltage. However, it was not clear from the voltage when overcharging and when deep discharging began. This is the advantage of the intersection criterion as demonstrated above.

(b) Lithium–cobalt oxide (LCO)

This battery reached the "radius of curvature one" at 4.0 V and 3.5 V. Due to the small change in voltage, the radius of curvature increased at the beginning and end of the discharge. From the first derivative of the discharge curve, the values $dSOC/dU = 1$ were basically evident, but the noise of the measured values was disturbing.

(c) Lithium–manganese oxide (LMO)

The radius of curvature $1/K = 1$ at 4.1 V and 3.4 V again showed the beginning rise of the discharge curve. The operating range could be read more easily from the voltage difference between the horizontal at SOC = 1 and SOC = 0.

(d) Lithium nickel manganese cobalt oxide (NMC)

The charge–discharge curve was less steep than that of LFP and LCO. The kink points at 3.4 V and 4.0 V (discharge) were approximately represented by the intersection of the first derivative dSOC/dU = 1 (and the intersection criterion). The maxima of the derivative indicated the steepest slope of the discharge curve.

(e) Lithium nickel cobalt aluminum oxide (NCA)

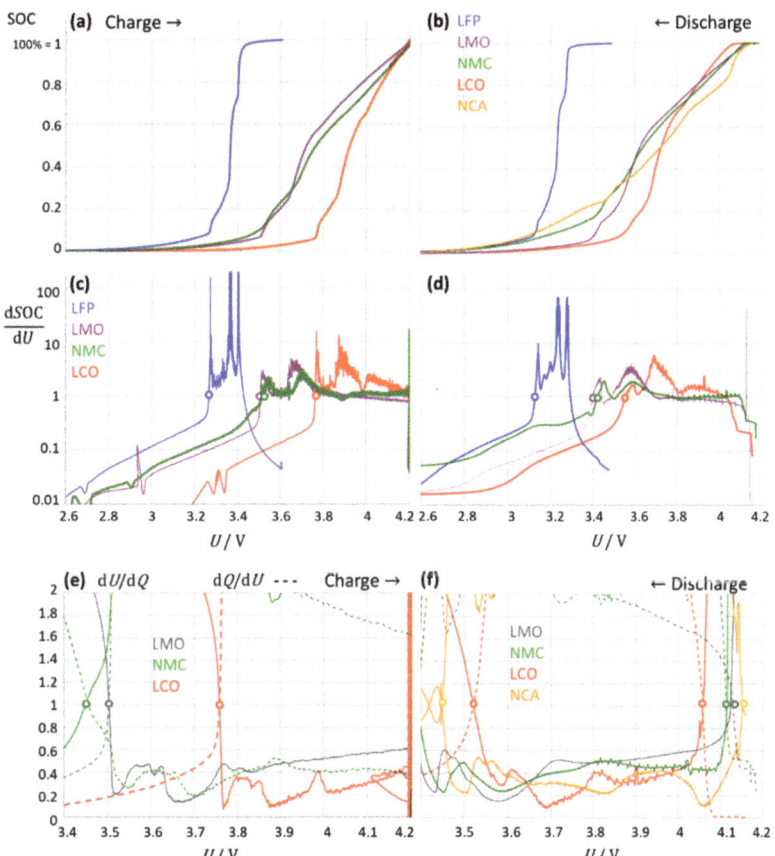

**Figure 7.** Test of the method with highly noisy data. Complete charge –discharge curves of different lithium-ion cell chemistries. (**a**,**b**) State-of-charge (SOC = $Q/Q_0$) versus cell voltage $U$. (**c**) Differential capacity dSOC/dU as measured without smoothing. (**d**) Differential capacity dSOC/dU with smoothing by moving average over 13 data points. (**e**) Intersection method for the charge curves: differential capacity dQ/dU ("IC", dashed), incremental voltage ("DV", solid). (**f**) Intersection method for the discharge curves.

The discharge curve was relatively flat and less S-shaped than that of other battery chemistries. Noisy measurement values led to large scatter and numerical errors. The derivative dQ/dU showed several passes through one (see Figure 7d). The radius of curvature was one at the kink point (4.15 V) and at other points; the numerical evaluation was unsatisfactory despite strong smoothing.

### 3.6. Verification of the Intersection Criterion Using Synthetic Data

In this section, the proposed method is applied to synthetic data of LFP batteries provided by Dubarry et al. [26]. The intersection method was applied to the unsmoothed data set of 705,638 charge-discharge curves, $U(SOC)$. The data were evaluated using MATLAB (code see Supplementary Materials).

Figure 8 compares the first derivative of the charge–discharge curves with respect to the absolute state-of-charge, $S = dSOC/dU$. Here, normalized capacity refers to the initial state to compare a "new" and an "aged" system on the same scale.

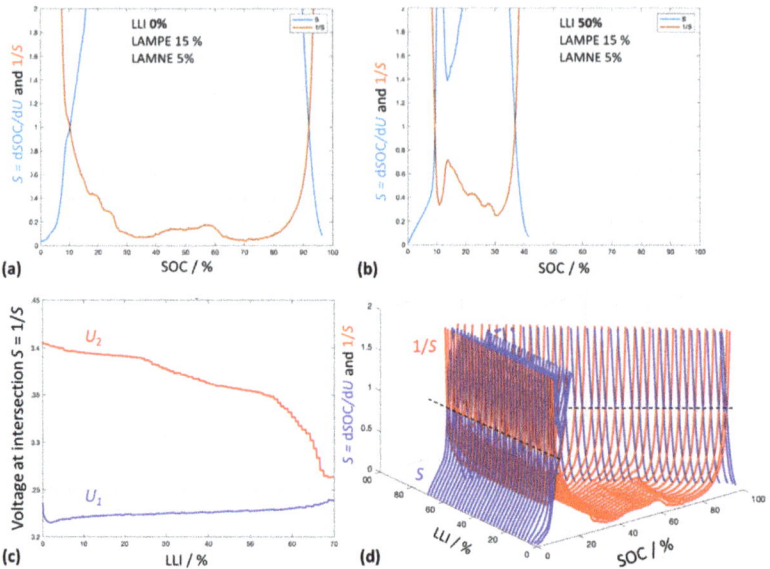

**Figure 8.** Application of the intersection method (this paper) to synthetic data from [26] for lithium iron phosphate chemistry. Here: loss of lithium-ion inventory (LLI) with constant loss of active material at the positive electrode (LAMPE) and the negative electrode (LAMNE). (**a**) Example curve "new" at the beginning of lithium loss, (**b**) "old" at 50% lithium loss versus the absolute state-of-charge with respect to the initial system (here: normalized capacity). The unit of $S$ is $V^{-1}$, the unit of $1/S$ is V. (**c**) Change of voltage window during aging: $U_1$ and $U_2$ are the lower and the upper intersection points at $S = 1/S = 1$. (**d**) Intersection criterion for all states of lithium loss between 0% and 70%.

The intersection points were found by searching for the smallest value of the difference, $|S - 1/S|^2 \to$ min, on the "left" and on the "right" side of the data set.

The intersection points at $S = 1/S = 1$ occurred reliably, so that the associated voltage could be determined fairly accurately. Figure 8c clearly shows how the usable voltage window (distance between die intersection points $U_1$ and $U_2$) became smaller and smaller during aging. For lithium iron phosphate and other cell chemistries, intermediate peaks below $S = 1/S = 1$ do not matter. For diagnostic purposes, such peaks would be interpreted as phase changes.

In the Supplementary Materials, a video shows the method in action for the special case of increasing the loss of the lithium inventory.

## 4. Conclusions

For all common lithium-ion chemistries, the empirical intersection criterium proposed in this work, $dSOC/dU = dU/dSOC = 1$, allows the operation of a lithium-ion battery to be monitored between "almost full" and "almost empty leaving some percent of the available capacity unused to avoid overcharge and deep discharge:

- The intersection points corresponded to the kink points in the charge–voltage curve. At the upper intersection, the battery was "virtually full" and was considered a

warning of impending overcharge or phase change. The small deviation to full charge depended on the cell chemistry;
- At the lower intersection point, the battery was "virtually empty" with a small amount of residual charge remaining in the battery; this was considered as a warning of impending deep discharge or phase change. For "completely empty", including deep discharge, the criterion diverged from one: $dQ/dU \to 0$ and $dU/dQ \to -\infty$;
- With an increasing current, the intersection points shifted to higher voltages (charging) or lower voltages (discharging); the distance between the intersections slightly increased with the current due to the voltage drop across the internal resistance of the cell.

The small deviation of the intersection method compared to ampere-hour counting contained a useful "reserve" to protect the battery against overcharging and deep discharge. The intersection method is suitable for simple intelligent battery monitoring without the need for Ah counting.

Starting from any battery condition, it is sufficient to record the voltage and current over time, calculate the charge and voltage differences, and evaluate the quotient $\Delta U/\Delta Q$. The criterion $dSOC/dU = dU/dSOC = 1$ will stop an automated full charge or discharge in time and prevent thermal runaway as far as possible. To avoid zero differences and rounding errors, the calculation rule in Equation (6) is recommended as a practical implementation of the intersection criterion.

$$S = \left| \frac{\Delta U}{\Delta Q} - \left( \frac{\Delta U}{\Delta Q} \right)^{-1} \right|^2 \to 0 \quad \text{or} \quad -\lg(10 \cdot S) \gg 1 \qquad (7)$$

For constant current charge and discharge, one current–voltage point per 10 s is sufficient. The voltage position of the upper intersection can be improved by a shorter measuring interval.

*Relevance to Battery Management Systems*

In practice, the intersection method is useful for determining when to switch from constant current (CC) to constant voltage charging and when to stop CC discharging. The criterion is not intended as a replacement for monitoring cut-off voltages, but as a tool for the evaluation of conventional charge–discharge curves.

The benefit of the intersection method is a mathematical one; it does not repeat the lower and upper cut-off voltage (defined by the manufacturer for a particular battery) but does provide empirical kink points of the charge–discharge curve just before reaching "full" and "empty" (for any battery). The physical basis of the criterion is solely the course of the charge–discharge curve, which changes over time due to aging. Cut-off voltages need not be known in advance.

The criterion shows a snapshot of the charge–discharge curve. It is irrelevant for the method at which point of the charge–discharge curve, in which direction or under which operating conditions (SOC, SOH, current, temperature, load change) it is used. The result $1/S = S = 1$ means "Attention, almost full, overcharge is imminent" or "Attention, almost empty, deep discharge is imminent". At operating voltages to the left and right of the intersection points, little additional charge flows into or out of the battery. The intersection criterion is intended as an early warning of impending heat events and phase changes.

The criterion does not claim to extend battery life. However, the relationship between shortened service life due to overcharge and deep discharge is known, e.g., [27]. The voltage range between the upper and lower intersection point becomes smaller during aging. It is a mirror image of the charge–discharge curve and thus of electrode kinetics. The intersections shift to lower cell voltages as the internal resistance of the battery increases. Cut-off voltages listed in data sheets do not depend on the actual state-of-health but are specifications of the manufacturer. The conventional diagram also shows that the initial curvature of the upper branch of the curve shifts more and more towards low voltages (see Video S3). The intercept criterion makes this visual observation machine-readable (see Supplementary Video S2).

The proposed mathematical procedure helps a digital machine to find the inflection points of the charge/discharge curve, which cannot always be detected accurately and quickly even by the human eye. Such an automated process would determine if the battery is approaching overcharge or deep discharge without the need to know the actual state-of-charge or state-of-health. The method is suitable for implementation on microcontrollers of battery management systems. It is simple and consumes few resources (memory, computing power).

**Supplementary Materials:** The following supporting information can be downloaded at: https://www.mdpi.com/article/10.3390/batteries8080086/s1, Video S1: Visualization of the intersection method using synthetic data (Dubarry [26]) for lithium-iron phosphate chemistry: Loss of lithium inventory 0% to 70% at constant loss of active material. Narrowing of the usable SOC range; Video S2: Intersection method on the voltage scale showing the narrowing of the 'safe' voltage window during aging; Video S3: Conventional discharge curves with pre-defined cut-off voltages; Figure S1: MATLAB CODE: MATHLAB code fragment for evaluating LFP chemistry.

**Author Contributions:** Writing—Original Draft Preparation, Review and Editing, all authors. All authors have read and agreed to the published version of the manuscript.

**Funding:** This work was supported by Diehl Aerospace GmbH.

**Conflicts of Interest:** The authors declare no conflict of interest.

**List of Symbols and Abbreviations**

| | | | |
|---|---|---|---|
| $C$ | Capacitance, $dQ/dU = 1/x$ (F) | LCO | lithium cobalt oxide |
| $Q$ | electric charge, battery capacity (Ah) | LMO | lithium manganese spinel |
| $Q_0$ | capacity of a fully charged battery (Ah) | LFP | lithium iron phosphate |
| $R$ | ohmic resistance, real part of impedance ($\Omega$) | NCA | nickel cobalt aluminum |
| $R_e$ | electrolyte resistance ($\Omega$) | NMC | nickel manganese cobalt |
| $U$ | cell voltage (V) | SOC | state-of-charge |
| $x$ | incremental voltage: $dU/dQ$ ($\Omega$/s) | SOH | state-of-health |

## References

1. Waag, W.; Sauer, D.U. State-of-Charge/Health. In *Encyclopedia of Electrochemical Power Sources*; Garche, J., Dyer, C., Moseley, P.T., Ogumi, Z., Rand, D., Scrosati, B., Eds.; Elsevier: Amsterdam, The Netherlands, 2009; Volume 4, pp. 793–804.
2. Piller, S.; Perrin, M.; Jossen, A. Methods for state-of-charge determination and their applications. *J. Power Sources* **2001**, *96*, 113–120. [CrossRef]
3. Xiong, R.; Pan, Y.; Shen, W.; Li, H.; Sun, F. Lithium-ion battery aging mechanisms and diagnosis method for automotive applications: Recent advances and perspectives. *Renew. Sustain. Energy Rev.* **2020**, *131*, 110048. [CrossRef]
4. Gauthier, R.; Luscombe, A.; Bond, T.; Bauer, M.; Johnson, M.; Harlow, J.; Louli, A.J.; Dahn, J.R. How do Depth of Discharge, C-rate and Calendar Age Affect Capacity Retention, Impedance Growth, the Electrodes, and the Electrolyte in Li-Ion Cells? *J. Electrochem. Soc.* **2022**, *169*, 020518. [CrossRef]
5. Tran, M.-K.; Fowler, M. A Review of Lithium-Ion Battery Fault Diagnostic Algorithms: Current Progress and Future Challenges. *Algorithms* **2020**, *13*, 62. [CrossRef]
6. Bloom, I.; Jansen, A.N.; Abraham, D.P.; Knuth, J.; Jones, S.A.; Battaglia, V.S.; Henriksen, G.L. Differential voltage analyses of high-power, lithium-ion cells: 1. Technique and application. *J. Power Sources* **2005**, *139*, 295–303. [CrossRef]
7. Barai, A.; Uddin, K.; Dubarry, M.; Somerville, L.; McGordon, A.; Jennings, P.; Bloom, I. A comparison of methodologies for the non-invasive characterization of commercial Li-ion cells. *Prog. Energy Combust. Sci.* **2019**, *72*, 1–31. [CrossRef]
8. Hatzikraniotis, E.; Mitsas, C.L.; Siapkas, D.I. Differential Capacity Analysis, a Tool to Examine the Performance of Graphites for Li-Ion Cells. In *Materials for Lithium-Ion Batteries*; NATO Science Series; Julien, C., Stoynov, Z., Eds.; Springer: Dordrecht, The Netherlands, 2020; Volume 85. [CrossRef]
9. Dubarry, M.; Svoboda, V.; Hwu, R.; Liaw, B.Y. Incremental capacity analysis and close-to-equilibrium OCV measurements to quantify capacity fade in commercial rechargeable lithium batteries. *Electrochem. Solid State Lett.* **2006**, *9*, A454. [CrossRef]
10. Dahn, H.M.; Smith, A.J.; Burns, J.C.; Stevens, D.A.; Dahn, J.R. User-Friendly Differential Voltage Analysis Freeware for the Analysis of Degradation Mechanisms in Li-Ion Batteries. *J. Electrochem. Soc.* **2012**, *159*, A1405. [CrossRef]

11. Smith, A.J.; Burns, J.C.; Dahn, J.R. High-Precision Differential Capacity Analysis of LiMn$_2$O$_4$/graphite Cells. *Electrochem. Solid-State Lett.* **2011**, *14*, A39. [CrossRef]
12. Smith, A.J.; Dahn, J.R. Delta Differential Capacity Analysis. *J. Electrochem. Soc.* **2012**, *159*, A290. [CrossRef]
13. Zheng, L.; Zhu, J.; Lu, D.D.C.; Wang, G.; He, T. Incremental capacity analysis and differential voltage analysis based state of charge and capacity estimation for lithium-ion batteries. *Energy* **2018**, *150*, 759–769. [CrossRef]
14. Jehnichen, P.; Wedlich, K.; Korte, C. Degradation of high-voltage cathodes for advanced lithium-ion batteries—differential capacity study on differently balanced cells. *Sci. Technol. Adv. Mater.* **2019**, *20*, 1–9. [CrossRef]
15. Krupp, A.; Ferg, E.; Schuldt, F.; Derendorf, K.; Agert, C. Incremental capacity analysis as a state of health estimation method for lithium-ion battery modules with series-connected cells. *Batteries* **2021**, *7*, 2. [CrossRef]
16. Zhang, S.; Guo, X.; Dou, X.; Zhang, X. A rapid online calculation method for state of health of lithium-ion battery based on coulomb counting method and differential voltage analysis. *J. Power Sources* **2020**, *479*, 228740. [CrossRef]
17. Kurzweil, P.; Ober, J.; Wabner, D.W. Method for extracting kinetic parameters from measured impedance spectra. *Electrochim. Acta* **1989**, *34*, 1179–1185. [CrossRef]
18. Guo, D.; Yang, G.; Zhao, G.; Yi, M.; Feng, X.; Han, X.; Lu, L.; Ouyang, M. Determination of the Differential Capacity of Lithium-Ion Batteries by the Deconvolution of Electrochemical Impedance Spectra. *Energies* **2020**, *13*, 915. [CrossRef]
19. Wang, L.; Pan, C.; Liu, L.; Cheng, Y.; Zhao, X. On-board state of health estimation of LiFePO$_4$ battery pack through differential voltage analysis. *Appl. Energy* **2016**, *168*, 465–472. [CrossRef]
20. Wang, L.; Zhao, X.; Liu, L.; Pan, C. State of health estimation of battery modules via differential voltage analysis with local data symmetry method. *Electrochim. Acta* **2017**, *256*, 81–89. [CrossRef]
21. Sieg, J.; Storch, M.; Fath, J.; Nuhic, A.; Bandlow, J.; Spier, B.; Sauer, D.U. Local degradation and differential voltage analysis of aged lithium-ion pouch cells. *J. Energy Storage* **2020**, *30*, 101582. [CrossRef]
22. Zhan, C.; Wu, T.; Lu, J.; Amine, K. Dissolution, migration, and deposition of transition metal ions in Li-ion batteries exemplified by Mn-based cathodes–a critical review. *Energy Environ. Sci.* **2018**, *11*, 243–257. [CrossRef]
23. Press, W.H.; Teukolsky, S.A.; Vetterling, W.T.; Flannery, B.F. *The Art of Scientific Computing*, 3rd ed.; Cambridge University Press: Cambridge, UK, 2007.
24. Christophersen, J.P.; Shaw, S.R. Using radial basis functions to approximate battery differential capacity and differential voltage. *J. Power Sources* **2010**, *195*, 1225–1234. [CrossRef]
25. Kurzweil, P.; Scheuerpflug, W. State-of-charge monitoring and battery diagnosis of different lithium-ion chemistries using impedance spectroscopy. *Batteries* **2021**, *7*, 17. [CrossRef]
26. Dubarry, M.; Beck, D. Analysis of Synthetic Voltage vs. Capacity Datasets for Big Data Li-ion Diagnosis and Prognosis. *Energies* **2021**, *14*, 2371. [CrossRef]
27. Ouyang, D.; Chen, M.; Liu, J.; Wei, R.; Weng, J.; Wang, J. Investigation of a commercial lithium-ion battery under overcharge/over-discharge failure conditions. *RSC Adv.* **2018**, *8*, 33414. [CrossRef]

Article

# State of Health Trajectory Prediction Based on Multi-Output Gaussian Process Regression for Lithium-Ion Battery

Jiwei Wang [1], Zhongwei Deng [2,*], Jinwen Li [3], Kaile Peng [3], Lijun Xu [4], Guoqing Guan [1] and Abuliti Abudula [1,*]

[1] Graduate School of Science and Technology, Hirosaki University, Hirosaki 036-8560, Japan
[2] School of Mechanical and Electrical Engineering, University of Electronic Science and Technology of China, Chengdu 611731, China
[3] College of Mechanical and Vehicle Engineering, Chongqing University, Chongqing 400044, China
[4] Xinjiang Coal Mine Electromechanical Engineering Technology Research Center, Xinjiang Institute of Engineering, Urumqi 830023, China
* Correspondence: dengzw1127@gmail.com (Z.D.); abuliti@hirosaki-u.ac.jp (A.A.)

**Abstract:** Lithium-ion battery state of health (SOH) accurate prediction is of great significance to ensure the safe reliable operation of electric vehicles and energy storage systems. However, safety issues arising from the inaccurate estimation and prediction of battery SOH have caused widespread concern in academic and industrial communities. In this paper, a method is proposed to build an accurate SOH prediction model for battery packs based on multi-output Gaussian process regression (MOGPR) by employing the initial cycle data of the battery pack and the entire life cycling data of battery cells. Firstly, a battery aging experimental platform is constructed to collect battery aging data, and health indicators (HIs) that characterize battery aging are extracted. Then, the correlation between the HIs and the battery capacity is evaluated by the Pearson correlation analysis method, and the HIs that own a strong correlation to the battery capacity are screened. Finally, two MOGPR models are constructed to predict the HIs and SOH of the battery pack. Based on the first MOGPR model and the early HIs of the battery pack, the future cycle HIs can be predicted. In addition, the predicted HIs and the second MOGPR model are used to predict the SOH of the battery pack. The experimental results verify that the approach has a competitive performance; the mean and maximum values of the mean absolute error (MAE) and root mean square error (RMSE) are 1.07% and 1.42%, and 1.77% and 2.45%, respectively.

**Keywords:** lithium-ion battery; health indicators; state of health; multi-output Gaussian process regression; health prediction

## 1. Introduction

### 1.1. Literature Review

For numerous advantages, lithium-ion batteries have been widely used in electric vehicles, consumer electronics devices, and energy storage systems [1,2]. Like many electrochemical systems, the repeated charging and discharging during the application of batteries inevitably cause gradual aging, resulting in an increase in internal resistance and a decrease in capacity. In the later stage of battery aging, it can more easily cause failure or fire, and the method to enhance the performance and safety of lithium-ion battery systems is a critical research hotspot [3]. Prognostic and Health Management has been one of the indispensable functions of the battery management system. Generally, the aging degree of the battery is characterized by State of health (SOH). SOH is defined as the ratio of the maximum available capacity to the rated capacity. According to the different application fields of lithium-ion batteries, the battery is usually considered to approach its end of life when its capacity reaches 80% of the normal value or its internal resistance increases to twice the initial value [4]. Existing studies have shown that lithium-ion battery

SOH prediction methods can be classified into electrochemical-based methods (EM) [5–7], equivalent circuit-based methods (ECM) [8,9], and data-driven methods (DDM) [10,11].

The relevant mathematical techniques can simplify the EM, which is established through the mathematical modeling of the internal mechanism of the battery and can be used to predict battery SOH. Nonetheless, the aging models based on different side reactions inside the battery are usually coupled with some partial differential equations [12], leading to some deficiencies such as high computational costs and hindering their applicability. Unlike EM, ECM is the most commonly used model that combines voltage sources, resistors, capacitors, and other components. The computational cost of this model is small, and its parameters are easy to identify. As a complex, nonlinear time-varying system, it is difficult to establish a model for the battery to characterize dynamic properties accurately. The data-driven approach does not consider the battery operating mechanism and aging mechanism compared with the model-based approach. It does not require a specific physical model, meaning it is more flexible. In addition, data-driven approaches typically treat the battery as a "black box" that maps external measurements such as voltage, current, and temperature to capacity based on machine learning algorithms.

With the continuous progress of computer technology and artificial intelligence technology, data-driven methods in battery health research receive more attention than physical methods due to their flexibility and model-free characteristics [13]. The following are data-driven methods for battery health prognostics: an artificial neural network (ANN) [14], a support vector machine (SVM) [6,15], a relevance vector machine (RVM) [16], and a Gaussian process regression (GPR) [17,18]; the advantages and disadvantages of the four approaches are listed in Table 1 [10].

Table 1. Comparison of the advantages and disadvantages of the four data-driven approaches.

| Approach | Advantages | Disadvantages |
|---|---|---|
| ANN | support for multidimensional spaces; high prediction accuracy; ability to learn independently; | high computational complexity large-scale samples; poor uncertainty expression; complex structure; |
| SVM | support for multidimensional spaces; strong generalization ability; better performance in the nonlinear system; | kernel function satisfying the Mercer criterion; more computing resources are required; sensitive to missing data; |
| RVM | high sparsity; not subject to Mercer restrictions; | depends on kernel function selection; susceptibility to falling into a local optimum; |
| GPR | availability of uncertainty expressions; applicable to high-dimensional and small sample data; | poor long-term forecasting; high cost of computing large samples of data; |

Compared with ANN, RVM, and SVM, the GPR model has better adaptability to deal with complex issues such as high-dimensional data, small samples, non-parameters, uncertainty expression, and nonlinearity. At the same time, the GPR model has received widespread attention in battery SOH research. Liu et al. [19] successfully prognosticated the cycle capacity of lithium-ion batteries based on the improved single-output Gaussian process regression (SOGPR) model. Li et al. [20] used the SOGPR to prognosticate the SOH of a battery with satisfactory accuracy. However, the SOGPR model cannot fully utilize the information of other batteries as prior knowledge to accurately predict the SOH of the target battery. To cope with this shortcoming, Zheng et al. [21] proposed a multiple SOGPR model to achieve the prediction of battery SOH based on different weights, but this method requires a large amount of computation. In order to further overcome the shortcomings of SOGPR, Boyle et al. [22] regarded the Gaussian process as a convolution between a smoothing kernel and Gaussian white noise.

Multi-output Gaussian process regression (MOGPR) [23,24] utilizes a covariance matrix for each channel to model their possible dependencies and each channel can use the information of other channels to enhance performance. Richardson et al. [25] employed the SOGPR to predict battery SOH directly and predicted the capacity based on the MOGPR and the correlation between different battery cells. However, research on the SOH prediction of battery packs is still rare. In addition, most of the above studies mainly use the NASA public data set to conduct the battery SOH estimation, but the data set was collected in 2008, and the cycle life of batteries is less than 200 cycles. Due to the rapid development of battery technology, this dataset is not suitable for the current commercial batteries, which usually have a larger capacity and a longer cycling life.

*1.2. The Thought of This Paper*

It is still challenging to achieve an accurate SOH prediction of battery packs only using early aging data of the pack. To this end, this paper proposes a method to build an accurate SOH prediction model for battery packs based on the MOGPR by employing the entire life cycling data of battery cells and the initial cycle data of the battery pack. Firstly, a battery aging experimental platform is constructed to collect battery aging data, and health indicators (HIs) that characterize battery aging are extracted. Then, we use correlation coefficients to evaluate the correlation between the HIs and the capacity, and the HIs that have a strong correlation to the battery capacity are screened. Finally, two MOGPR prediction models are constructed, namely, an MOGPR HIs prediction model and an MOGPR SOH prediction model. The MOGPR HIs prediction model is trained by employing the first 20% cycle HIs of the battery pack and the entire life cycle HIs of the battery cell. Based on this model, the future cycle HIs of the battery pack can be obtained. The MOGPR SOH prediction model is established by using the early HIs and SOH data of the battery pack as a training sample. Based on this model and the predicted HIs, the future SOH of the battery pack can be predicted.

The method consists of three main parts: data acquisition, model construction, and health prognostics, as shown in Figure 1. First, the battery aging experiment platform is built to collect the battery aging data such as voltage, current, capacity, and temperature. Secondly, the HIs are extracted from the charge/discharge aging experimental data and filtered by using the correlation analysis method. Additionally, they are combined with the MOGPR to construct the battery SOH prediction model. Finally, the SOH-predicted results of the battery cells and packs are evaluated by three metrics.

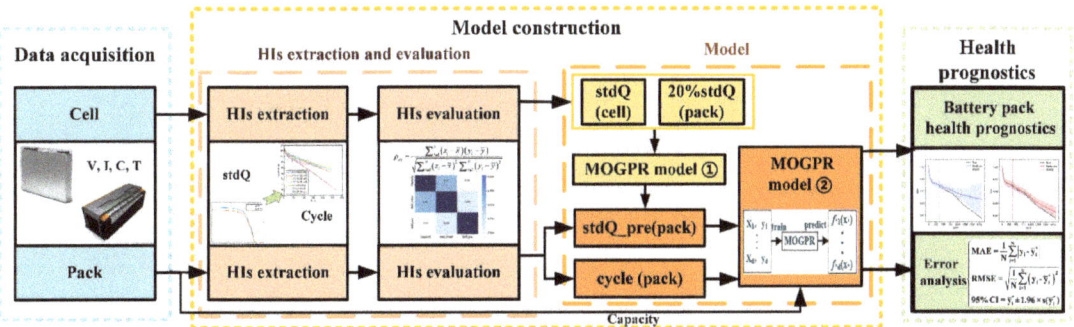

**Figure 1.** The composition and implementation principle of the battery SOH prediction method.

The contributions of this paper are as follows:

(1) Two HIs, namely, cycle number and standard deviation of discharge capacity (*stdQ*) are combined to achieve a highly accurate SOH prediction for battery packs.
(2) The proposed MOGPR model can maintain a high-precision SOH prediction of battery cells and battery packs under different working conditions.
(3) Only 20% early aging data of battery packs are employed to achieve an accurate SOH trajectory prediction for the battery pack, which saves lots of time and energy in whole-life aging tests of battery packs.

The remaining sections of this paper are as follows: Section 2 analyzes the results; Section 3 is the experimental testing; Section 4 extracts and evaluates the HIs from the experimental aging data; the methodology is described in Section 5; the conclusions of this paper are given in the end.

## 2. Results and Discussion

In this section, the HIs prediction model and the SOH prediction model based on the MOGPR are validated by using the aging experimental data of battery cells and battery packs. The battery pack stdQ prediction results based on the MOGPR model are presented in Section 2.1. In Section 2.2, the SOH prediction results of battery cells under different models and working conditions are presented, while the SOH prediction results of the battery pack under different models are illustrated in Section 2.3.

### 2.1. The HIs Prediction of Battery Pack

Under the working condition of 35 °C_0.5C0.5C (35 °C: ambient temperature, 0.5C0.5C: charge–discharge rate), based on the MOGPR model, the future cycle $stdQ$ ($stdQ\_pre$) of the battery pack can be obtained by learning the entire life data of the battery cell $stdQ$ and the initial 20% life data of the battery pack $stdQ\_mean$, and the results are presented in Figure 2. As shown in Figure 2a, the vertical dashed line represents the 20% cycle data of the battery pack, with the left side representing the observed value and the right side representing the predicted value. The predicted HIs not only have the same trend as the observed HIs, but also have a small error, and their MAE and RMSE are 0.36% and 0.496%, respectively. Figure 2b illustrates the correlation between the $stdQ$ and the capacity in the battery pack, where the $stdQ\_mean$ represents the observed value of the $stdQ_{1-15}$ of the battery pack. The results show that the $stdQ\_pre$ still has a strong correlation with the capacity, and thus the above $stdQ$ can be used for the health prediction of the battery pack. Although the $stdQ$ has a good correlation with capacity, there is still a certain degree of deviation. In order to improve prediction accuracy, the next section will fuse the two HIs, namely the cycle number and the $stdQ$, for the SOH prediction of the battery pack.

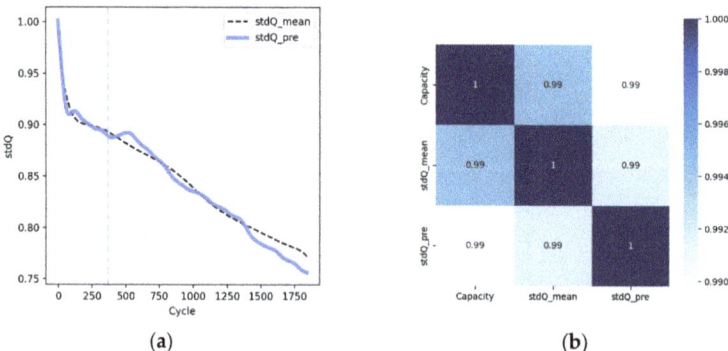

**Figure 2.** The prediction results of the $stdQ$ for the battery pack: (**a**) the predicted future $stdQ$ of the battery pack; (**b**) the correlation coefficient between observed and predicted HIs and the capacity.

### 2.2. SOH Prediction of Battery Cells

#### 2.2.1. Prediction Results of Two Different Models

The SOH prediction results of the battery cell based on the SOGPR model and the MOGPR model are compared in this section. Under the working condition of 35 °C_0.5C0.5C, the SOGPR prediction model is trained by using the first 20% aging data of the 1# cell and the entire life aging data of the 2# cell, and the results are shown in Figure 3a. The bold solid line represents the observed value used to train the SOGPR model, the dotted line is the future SOH of the battery cell, and the solid blue line is the predicted SOH. The light blue area displays the 95% confidence interval (CI), which is used to assess the prediction results. The narrower the 95% CI, the more reliable. It can be seen from the picture that the SOGPR prediction model cannot achieve accurate SOH prediction of the battery cell.

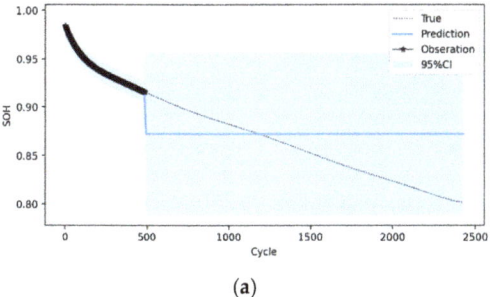
(a)

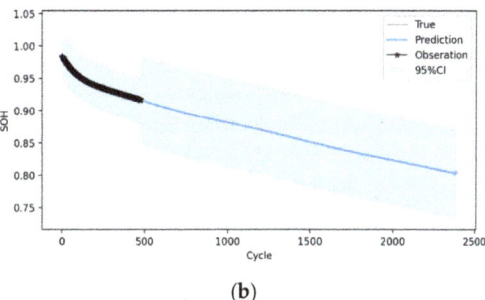
(b)

**Figure 3.** SOH prediction results for battery cell (35 °C_0.5C0.5C): (**a**) SOH prediction result based on the SOGPR; (**b**) SOH prediction result based on the MOGPR.

Compared with the SOGPR model, the MOGPR model accounts for the disadvantage of the SOGPR model and can take use of the prior SOH information of the 2# cell to obtain higher prediction accuracy. Under the working condition of 35 °C_0.5C0.5C, the MOGPR prediction model is trained by employing the entire life cycling data of the 2# cell and the first 20% aging data of the 1# cell, and the prediction result is shown in Figure 3b. Their MAE and RMSE are 0.278% and 0.337%, respectively.

2.2.2. Prediction Results for Two Different Conditions

In order to analyze the influence of temperature and the charge–discharge rate on the MOGPR model, the future SOH of battery cell under 35 °C_0.5C0.5C is predicted based on the battery cell aging data of two different operating conditions: 25 °C_0.5C0.5C and 35 °C_0.3C1C, and the predicted results are shown in Figure 4.

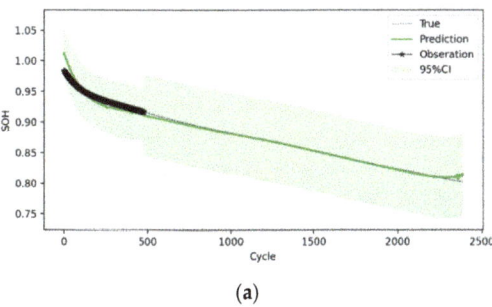
(a)

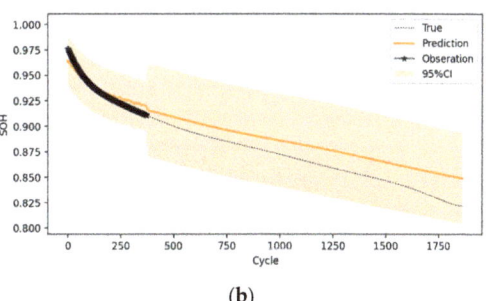
(b)

**Figure 4.** SOH prediction results of battery cell based on the MOGPR under different conditions: (**a**) results of MOGPR-based battery cell SOH prediction (25 °C_0.5C0.5C); (**b**) results of MOGPR-based battery cell SOH prediction (35 °C_0.3C1C).

At first, in order to analyze the influence of temperature on the prediction accuracy of the SOH of the battery cell, the entire life aging data of the 3# cell under the working condition of 25 °C_0.5C0.5C and the first 20% cycle data of the 1# cell under 35 °C_0.5C0.5C working conditions are employed to train the MOGPR model. The SOH prediction results are shown in Figure 4a, the MAE and RMSE are 0.31%, and 0.99%, respectively. The result shows that temperature has little effect on the SOH prediction of the battery cell. The model can obtain satisfactory SOH prediction accuracy under different working temperatures by using the aging data in other working temperatures.

Then, to analyze the influence of the charge–discharge rate on the prediction accuracy of battery cell SOH, the entire life aging data of the 5# cell under 35 °C_0.3C1C operating conditions and the first 20% cycle data of the 1# cell under 35 °C_0.5C0.5C working conditions are utilized to train the MOGPR model. The deviation of the predicted value from the true value gradually increases with the increasing number of cycles; the SOH prediction result is shown in Figure 4b. Their MAE and RMSE are 1.71% and 1.89%, respectively.

By verifying the MOGPR model based on two different working conditions, it can be seen that, compared with temperature, the impact of the charge–discharge rate on the prediction accuracy of the battery cell SOH is more obvious.

*2.3. SOH Prediction of Battery Pack*

In this section, firstly, the prediction results of two different HIs are validated separately based on the SOGPR model. Secondly, based on the MOGPR model, the prediction results of two different HIs are verified separately. The black dotted line is the actual value of the SOH. the solid colorful line represents the predicted value of the SOH, and the corresponding area is the 95% CI.

2.3.1. Prediction Results Based on the SOGPR Model

Under 35 °C_0.5C0.5C operating conditions, the first 20% of a battery pack is used to train the SOGPR model. The prediction result is shown in Figure 5. The battery used in the experiments in this paper is affected by polarization, and the SOH of the first 100 cycles of the pack shows a rapid decline. In Figure 5a, the SOGPR model is trained using only the cycle number as the HI, and the SOH prediction error of the battery pack is large. The MAE and RMSE are 2.554% and 3.64%, respectively. While in Figure 5b, the cycle number and *stdQ_pre* are used as the input HIs. Although the future cycling SOH prediction of the battery pack can be achieved by training two sets of HIs, the predicted values deviate significantly from the actual values and are located in the unreliable region, and its MAE and RMSE are 2.79% and 3.74%, respectively. The results show the SOGPR model cannot obtain accurate SOH predictions of the battery pack.

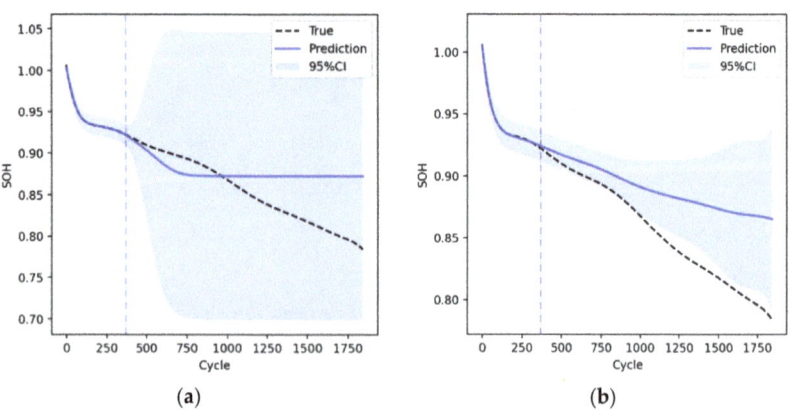

**Figure 5.** SOH prediction of the battery pack based on the SOGPR with different HIs: (**a**) only using the cycle number as the HI; (**b**) using the cycle number and *stdQ* as the HIs.

2.3.2. Prediction Results Based on the MOGPR Model

In this section, the MOGPR prediction model is trained by the data of a single HI (cycle number) and two HIs (cycle number and *stdQ_pre*), and the SOH prediction results of the battery pack are analyzed. Under the working condition of 35 °C_0.5C0.5C, the whole life aging data of the 1# cell and the first 20% of the battery pack are used to train the MOGPR prediction model of the battery pack. In Figure 6a, only the cycle number is used as the

HI to train the model. The results show that the capacity drops rapidly in the early cycle. In the first 400 cycles, it still has high prediction accuracy and can effectively capture the overall trend of battery decline, with an MAE and RMSE of 1.87% and 2.69%, respectively. Compared with SOGPR, the prediction accuracy of SOH has been significantly improved. Although the deviation of the predicted future SOH of the battery pack from the actual value is small, its 95% CI is still much larger than the normal threshold. The results show that, based on the MOGPR SOH prediction model, satisfactory reliability prediction results cannot be obtained by training using only the cycle number.

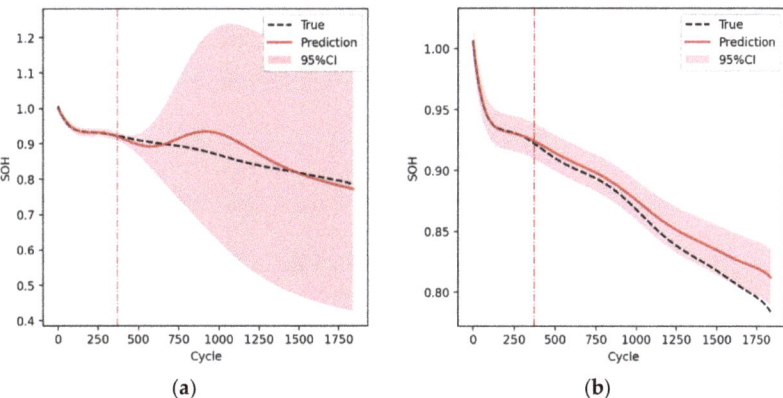

**Figure 6.** SOH prediction of the battery pack based on the MOGPR with different HIs: (**a**) only using the cycle number as the HI; (**b**) using the cycle number and $stdQ$ as the HIs.

The SOH prediction results of the battery pack based on two HIs (cycle number and $stdQ\_pre$) are shown in Figure 6b. It can be observed that the model can not only capture the general trend of battery aging but also has better results, with an MAE and RMSE of 0.91% and 1.18%, respectively. Based on the fusion feature to account for the deficiency of a single feature, the prediction accuracy has been significantly improved.

## 3. Experiment

As an electrochemical system, the battery inevitably leads to the gradual degradation of its performance during constant use and long-term storage. To study its aging characteristics, an aging experiment platform is set up to conduct different charging and discharging tests. The battery aging test platform mainly includes a battery tester, a thermal chamber, a computer, a data logger, etc., as shown in Figure 7. First, set the experimental steps and parameters through the computer. Then, use the battery tester to run the battery cells and pack in the thermal chamber according to the preset steps. Finally, save the experimental data to the computer through the data logger.

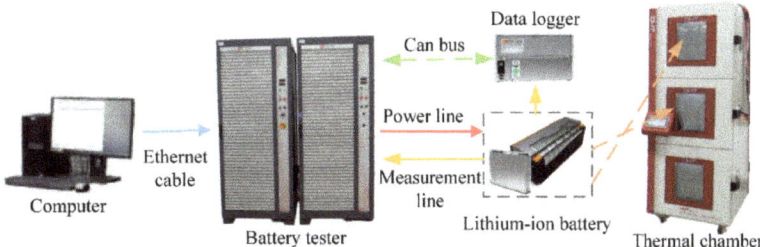

**Figure 7.** The platform for the battery aging experiment.

The battery used in this aging test is a prismatic battery cell with a LiFePO$_4$ cathode and a graphite anode. A batch of batteries with a rated capacity of 110Ah is applied to the aging experiment. The battery pack is composed of fifteen battery cells in series. Both battery cells and battery packs are subjected to aging tests at 35 °C_0.5C0.5C. In order to analyze the influence of temperature and the current rate on aging characteristics, aging experiments are carried out on battery cells under two operating conditions, namely 25 °C_0.5C0.5C and 35 °C_0.3C1C, respectively. The aging test conditions of battery cells and battery packs are shown in Table 2.

**Table 2.** Battery aging test conditions.

| Conditions | | Cell | | Pack |
|---|---|---|---|---|
| Temperature (°C) | 35 | 25 | 35 | 35 |
| Charge rate (C) | 0.5 | 0.5 | 0.3 | 0.5 |
| Discharge rate (C) | 0.5 | 0.5 | 1 | 0.5 |

In this work, a test battery cell (35 °C_0.5C0.5C_cell, 1#) (35 °C: ambient temperature, 0.5C0.5C: charge–discharge rate, cell 1#: battery number) is an example for the description. The battery tester conducts the aging experiment under preset working conditions, where the ambient temperature is set to 35 °C. Consistently use the 0.5 C rate to complete charging until 3.65 V, and the current becomes 0.05 C. In the discharge process, the discharge rate is set to 0.5 C to discharge until 2.5 V, and then the discharge step is terminated. The voltage and current curves in a charge–discharge cycle are shown in Figure 8a, then the above process is repeated until the capacity reaches a preset value of the initial capacity of the battery. Compared with the battery cells, the cycling life of the battery pack is usually much shorter, as shown in Figure 8b. Specifically, since the battery pack is affected by the inconsistency of the battery cells and multiple factors, the aging rate of the battery pack is increased, and its cycle life is shortened.

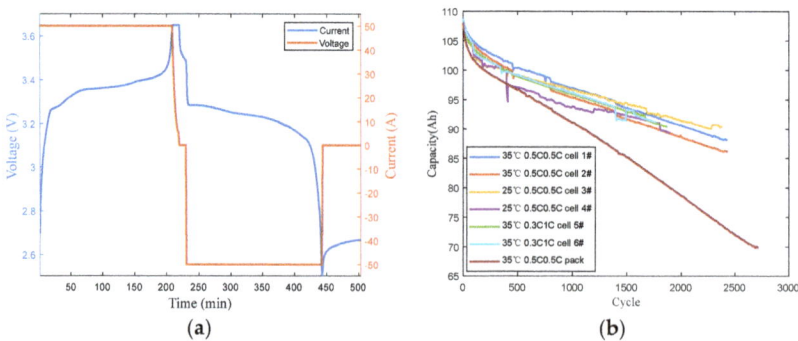

**Figure 8.** Battery cycling profile and capacity aging curves: (**a**) voltage and current curves in a charge–discharge cycle; (**b**) capacity decay curves of battery cells and packs.

## 4. Health Indicators Extraction and Evaluation

### 4.1. Health Indicators Extraction

As a battery is repeatedly charged and discharged, its active material will gradually decrease. The gradual thickening of the solid electrolyte interphase (SEI) eventually leads to its capacity fading and a power drop. Experiments show differences in the charge and discharge capacity under different cycles, so capacity would usually be used as an indicator to evaluate battery aging. The discharge capacity of the lithium-ion battery in this experiment is completed between the lower and upper cut-off voltage. This work chooses to extract the voltage segment in this voltage interval, then obtains the corresponding discharge capacity sequence ($Q$) through the ampere-hour integration. In order to further improve the reliability and operability of this indicator, the standard deviation of the $Q$ is adopted, which is named as $stdQ$ [26]. The specific implementation principle of $stdQ$ is as follows:

Assuming that the voltage discharge curve is divided into the same $N$ sub-intervals, the voltage interval $\Delta V$ can be calculated by Equation (1):

$$\Delta V = \frac{V_{\max} - V_{\min}}{N} \tag{1}$$

Based on the voltage interval $\Delta V$, the $i$-th voltage interval sequence $V_{s,i}$ can be obtained by Equation (2):

$$V_{s,i} = [V_{\max}, V_{\max} - \Delta V, \cdots, V_{\min}] \tag{2}$$

The accumulated charge sequence $Q$ corresponding to the voltage sequence can be obtained according to the ampere-hour integration method, as shown in Equation (3):

$$Q^i(V) = [Q_1, Q_2, \cdots, Q_N] \tag{3}$$

The $dQ$ sequence of Equation (4) can be obtained as the difference between the sequence of Equation (3) and the element $Q_1$ which corresponds to the first voltage interval. Then, the variance of the $dQ$ sequence can be obtained to obtain the health indicator of the $i$-th cycle and named as $stdQ$.

$$dQ^i(V) = \left[Q_1^i - Q_1^i, Q_2^i - Q_1^i, \cdots, Q_N^i - Q_1^i\right] \tag{4}$$

Through the analysis of the experimental data, it can be seen that as the number of battery cycles increases, the battery capacity decreases continuously, so the cycle number can also be used as an HI to capture the aging status of the battery.

*4.2. Health Indicators Evaluation*

The selection of HIs is critical for the prediction of battery SOH based on machine learning, which can not only effectively eliminate a large number of unimportant and redundant features but also help to reduce the computational cost and obtain reliable prediction results. The Pearson correlation analysis method is suitable for the quantitative analysis of the linear relationship between the extracted HIs and the battery capacity. The Pearson correlation coefficient can be expressed by Equation (5) [27]:

$$\rho = \frac{\sum (x_i - \bar{x}_i)(y - \bar{y})}{\sqrt{\sum (x_i - \bar{x}_i)^2 \sum (y - \bar{y})^2}} \tag{5}$$

where $x_i$ represents the HIs, and $y$ represents the capacity observations. $\bar{x}_i$ and $\bar{y}$ represent their mean values, respectively. $\rho$ represents the correlation coefficient between the HIs and the capacity.

Based on the above analysis, the features of fifteen battery cells in the battery pack are extracted. The correlation between the HIs and the battery capacity is calculated by using Pearson correlation analysis. Figure 9 shows the correlation between the sixteen HIs ($stdQ_{1-15}$, $stdQ\_mean$) and capacity, respectively, in the battery pack, where the $stdQ\_mean$ represents the average value of the $stdQ$ for the fifteen battery cells. The Pearson correlation analysis shows that the correlation coefficients are greater than 0.99, indicating that there is a strong correlation between the $stdQ$ and the capacity. Therefore, the $stdQ\_mean$ can be used as an HI to represent all changes in the $stdQ_{1-15}$ of the battery pack.

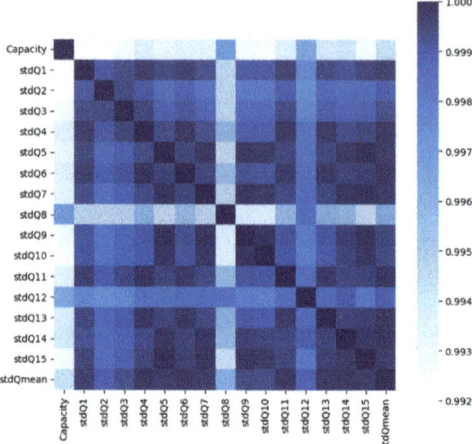

**Figure 9.** The correlation coefficient between *stdQ* and the capacity of each battery cell.

## 5. Methodology

The Gaussian process regression model, as a machine learning method based on Bayesian framework, has the advantages of non-parametric and uncertainty expression. According to the number of model outputs, the model can be divided into single-output Gaussian process regression (SOGPR) models and multiple-output Gaussian process regression (MOGPR) models. Since the traditional machine learning methods cannot fit well for heterogeneous data, we use the MOGPR model to predict battery pack health. A comparison of the implementation principles of the two different Gaussian process regression models can be found in the literature [28], and an intuitive illustration is shown in Figure 10. For the multiple-input multiple-output prediction problem, the traditional method often uses multiple SOGPR models to build models separately, where the input is $\{X_i, y_i\}$, and the output is $\{f_i\}$. However, this method ignores the correlation during multiple outputs; in contrast, the MOGPR model accounts for the deficiencies of the SOGPR model.

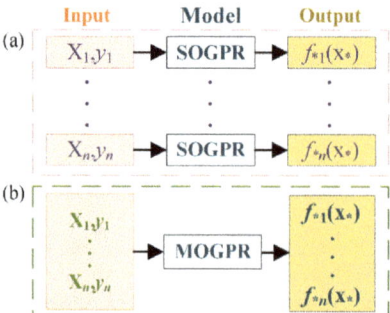

**Figure 10.** Basic principles of two GPR models: (**a**) basic principles of the SOGPR model; (**b**) basic principles of the MOGPR model.

### 5.1. Single-Output Gaussian Process Regression Model

A typical single-output Gaussian process is a collection of any finite number of random variables with a joint Gaussian distribution, whose properties are completely determined by their means and covariance functions. The Gaussian process definition is shown in Equation (6):

$$f(x) \sim GP(m(x), k(x, x')) \tag{6}$$

where $x$ and $x'$ represent two different input samples, $m(x)$ is the mean function, and its value usually takes zero (the assumption does not affect the generalization and learning performance of the Gaussian process), and $k(x, x')$ is the covariance function of the Gaussian process, which characterizes the correlation between random variables.

According to Bayesian theory, the posterior distribution of the predicted value $y_*$ can be obtained, as shown in Equation (7):

$$p(y_*|x, y, x_*) = N\left(\mu_*, \sigma_*^2\right) \tag{7}$$

where $x$ and $y$ are the input and output of the training set, respectively. $x_*$ and $y_*$ are the input and prediction output of the testing set, respectively. $\mu_*$ is the prediction mean, and $\sigma_*^2$ is the prediction covariance.

### 5.2. Multi-Output Gaussian Process Regression Model

The MOGPR model is obtained by extending the SOGPR model. Compared with the SOGPR model, the MOGPR model accounts for the deficiency of the SOGPR model in that each output needs to be modeled separately and cannot capture the potential correlation between multiple outputs. The MOGPR establishes a covariance matrix for each output, so as to learn the correlation between each output. It assumes that the multiple outputs are related to some extent, and employs the mutual information to obtain more accurate prediction results than the SOGPR model [28].

The MOGPR assumes that the set containing $D$ functions, $\{f_d(x)\}_{d=1}^{D}$, where any function can be expressed as the convolution of the smooth kernel function $\{G_d(x)\}_{d=1}^{D}$, with the implicit function $\mu(x)$, as shown in Equation (8):

$$f_d(x) = \int_x G_d(x - z)\mu(z)dz \tag{8}$$

Similar to the SOGPR, the MOGPR multi-output random variable $f(x)$ is assumed to obey a Gaussian distribution, as shown in Equation (9):

$$f(x) \sim GP(m(x), K_{MOGP}) \tag{9}$$

where $m(x)$ is the mean function, predicted by the mean of the test aging data series, and the multi-output covariance $K_{MOGP}$ is defined as Equation (10):

$$K_{MOGP}(x, x') = k_{f_d, f_{d'}}(x, x') = \begin{bmatrix} k_{11}(x, x') & \cdots & k_{1D}(x, x') \\ \vdots & \ddots & \vdots \\ k_{D1}(x, x') & \cdots & k_{DD}(x, x') \end{bmatrix} \tag{10}$$

The multiple output regression problem can be defined in Equation (11):

$$y_d(x) = f_d(x) + \varepsilon_d \tag{11}$$

where $f(x)$ is the multi-output function, $\varepsilon_d$ is a Gaussian noise $\varepsilon_d \sim N(0, \sigma_n^2)$, and $y_d$ is the multi-output observation.

Based on Bayesian theory, the posterior distribution of MOGP predicted values $y_{d*}$ can be expressed in Equation (12):

$$p(y_{d*}|x, y, x_*) = N\left(\mu_{d*}, \sigma_{d*}^2\right) \tag{12}$$

where $x$ and $y$ are the input and output of the training set, $x_*$ and $y_{d*}$ are the input and predicted output of the test set, $\mu_{d*}$ is the predicted mean, and $\sigma_{d*}^2$ is the predicted covariance.

In this work, for battery SOH prediction, the HIs and capacity data of battery cells and packs are used as the input of the MOGPR model, and the corresponding SOHs are taken as the output, respectively. Firstly, the battery cell entire-life aging data and the pack early cycle data are selected and loaded into the model for training. Then, the MOGPR prediction model is used to complete the prediction of the battery pack SOH. We use three metrics to evaluate the accuracy of the prediction results, namely, MAE, RMSE, and 95% CI, as shown in Equation (13):

$$\begin{cases} MAE = \frac{1}{N}\sum_{i=1}^{N}|y_i - \overline{y}_i^*| \\ RMSE = \sqrt{\frac{1}{N}\sum_{i=1}^{N}(y_i - \overline{y}_i^*)^2} \\ 95\%CI = \overline{y}_i^* \pm 1.96 \times \sigma(\overline{y}_i^*) \end{cases} \quad (13)$$

where $y_i$ and $\overline{y}_i^*$ represent the actual and predicted values of the battery SOH, respectively, and $\sigma(\overline{y}_i^*)$ is the variance of the predicted capacity. The 95% CI represents the confidence interval of the predicted value of the battery SOH.

## 6. Conclusions

In this paper, a battery pack SOH prediction method based on the MOGPR model is proposed with satisfactory accuracy. Firstly, two HIs are proposed from the battery cells and the battery pack by analyzing the battery aging characteristics. Then, the Pearson correlation analysis method is used to quantify the correlation between the HIs and the capacity. At last, the SOH prediction result based on the MOGPR is verified by employing the entire life cycling data of the battery cell and the initial cycle data of the battery pack. Based on the *stdQ* of the battery cell, the prediction of the future *stdQ* of the battery pack is realized through the MOGPR model. Then, the cycle number and *stdQ_pre* are combined to form the HIs set, and the MOGPR model is employed again to achieve the prediction of the future SOH of the battery pack. The results show that its MAE and RMSE are 0.91% and 1.18%, respectively. The results of this paper show that the prediction effect based on two features is better than that of a single feature, and the performance of the MOGPR model is better than that of the SOGPR model. By comparison, the MOGPR model based on the two features has better reliability and accuracy. Only the basic RBF kernel function is used in this model, and the performance of other kernel functions has not been compared. In addition, this paper only used the LFP battery to verify the MOGPR model, and whether this method is applicable to other material batteries needs further research. In the future, we will try to use different kernel functions in our scenario and combine MOGPR with ANN. Furthermore, our proposed model will be validated on different materials of batteries.

**Author Contributions:** Conceptualization, J.W., Z.D. and A.A.; methodology, J.W., Z.D., J.L. and A.A.; software, J.W. and J.L.; validation, J.W. and J.L.; formal analysis, J.W. and J.L.; investigation, J.W. and J.L.; resources, J.W. and K.P.; data curation, J.W. and K.P.; writing—original draft preparation, J.W.; writing—review and editing, J.W., Z.D., J.L., K.P., L.X., G.G. and A.A.; visualization, J.W. and J.L.; supervision, Z.D. and A.A; project administration, Z.D. All authors have read and agreed to the published version of the manuscript.

**Funding:** This research received no external funding.

**Institutional Review Board Statement:** Not applicable.

**Informed Consent Statement:** Not applicable.

**Data Availability Statement:** The data presented in this study are available on request from the corresponding author.

**Conflicts of Interest:** The authors declare no conflict of interest.

## References

1. Farmann, A.; Waag, W.; Marongiu, A.; Sauer, D.U. Critical review of on-board capacity estimation techniques for lithium-ion batteries in electric and hybrid electric vehicles. *J. Power Sources* **2015**, *281*, 114–130. [CrossRef]
2. Liu, J.; Bao, Z.; Cui, Y.; Dufek, E.J.; Goodenough, J.B.; Khalifah, P.; Li, Q.; Liaw, B.Y.; Liu, P.; Manthiram, A. Pathways for practical high-energy long-cycling lithium metal batteries. *Nat. Energy* **2019**, *4*, 180–186. [CrossRef]
3. Khumprom, P.; Yodo, N. A data-driven predictive prognostic model for lithium-ion batteries based on a deep learning algorithm. *Energies* **2019**, *12*, 660. [CrossRef]
4. Zubi, G.; Dufo-López, R.; Carvalho, M.; Pasaoglu, G.; Reviews, S.E. The lithium-ion battery: State of the art and future perspectives. *Renew. Sustain. Energy Rev.* **2018**, *89*, 292–308. [CrossRef]
5. Zhang, Q.; Wang, D.; Yang, B.; Cui, X.; Li, X. Electrochemical model of lithium-ion battery for wide frequency range applications. *Electrochim. Acta* **2020**, *343*, 136094. [CrossRef]
6. Wang, Y.; Tian, J.; Sun, Z.; Wang, L.; Xu, R.; Li, M.; Chen, Z. A comprehensive review of battery modeling and state estimation approaches for advanced battery management systems. *Renew. Sustain. Energy Rev.* **2020**, *131*, 110015. [CrossRef]
7. Xiong, R.; Li, L.; Li, Z.; Yu, Q.; Mu, H. An electrochemical model based degradation state identification method of Lithium-ion battery for all-climate electric vehicles application. *Appl. Energy* **2018**, *219*, 264–275. [CrossRef]
8. Plett, G.L. *Battery Management Systems, Volume II: Equivalent-Circuit Methods*; Artech House: Norwood, MA, USA, 2015.
9. Hu, X.; Li, S.; Peng, H. A comparative study of equivalent circuit models for Li-ion batteries. *J. Power Sources* **2012**, *198*, 359–367. [CrossRef]
10. Hu, X.; Xu, L.; Lin, X.; Pecht, M. Battery lifetime prognostics. *Joule* **2020**, *4*, 310–346. [CrossRef]
11. Ge, M.-F.; Liu, Y.; Jiang, X.; Liu, J. A review on state of health estimations and remaining useful life prognostics of lithium-ion batteries. *Measurement* **2021**, *174*, 109057. [CrossRef]
12. Deng, Z.; Hu, X.; Lin, X.; Xu, L.; Li, J.; Guo, W. A reduced-order electrochemical model for all-solid-state batteries. *IEEE Trans. Transp. Electrif.* **2020**, *7*, 464–473. [CrossRef]
13. Hu, X.; Jiang, J.; Cao, D.; Egardt, B. Battery health prognosis for electric vehicles using sample entropy and sparse Bayesian predictive modeling. *IEEE Trans. Ind. Electron.* **2015**, *63*, 2645–2656. [CrossRef]
14. You, G.-w.; Park, S.; Oh, D. Real-time state-of-health estimation for electric vehicle batteries: A data-driven approach. *Appl. Energy* **2016**, *176*, 92–103. [CrossRef]
15. Klass, V.; Behm, M.; Lindbergh, G. A support vector machine-based state-of-health estimation method for lithium-ion batteries under electric vehicle operation. *J. Power Sources* **2014**, *270*, 262–272. [CrossRef]
16. Hu, C.; Jain, G.; Schmidt, C.; Strief, C.; Sullivan, M. Online estimation of lithium-ion battery capacity using sparse Bayesian learning. *J. Power Sources* **2015**, *289*, 105–113. [CrossRef]
17. Deng, Z.; Hu, X.; Lin, X.; Xu, L.; Che, Y.; Hu, L. General discharge voltage information enabled health evaluation for lithium-ion batteries. *IEEE/ASME Trans. Mechatron.* **2020**, *26*, 1295–1306. [CrossRef]
18. Wang, J.; Deng, Z.; Yu, T.; Yoshida, A.; Xu, L.; Guan, G.; Abudula, A. State of health estimation based on modified Gaussian process regression for lithium-ion batteries. *J. Energy Storage* **2022**, *51*, 104512. [CrossRef]
19. Liu, K.; Hu, X.; Wei, Z.; Li, Y.; Jiang, Y. Modified Gaussian process regression models for cyclic capacity prediction of lithium-ion batteries. *IEEE Trans. Transp. Electrif.* **2019**, *5*, 1225–1236. [CrossRef]
20. Li, X.; Wang, Z.; Yan, J. Prognostic health condition for lithium battery using the partial incremental capacity and Gaussian process regression. *J. Power Sources* **2019**, *421*, 56–67. [CrossRef]
21. Zheng, X.; Deng, X. State-of-health prediction for lithium-ion batteries with multiple gaussian process regression model. *IEEE Access* **2019**, *7*, 150383–150394. [CrossRef]
22. Boyle, P.; Frean, M. Dependent gaussian processes. *Adv. Neural Inf. Process. Syst.* **2004**, *17*, 17,217–224.
23. Li, Y.; Sheng, H.; Cheng, Y.; Stroe, D.-I.; Teodorescu, R. State-of-health estimation of lithium-ion batteries based on semi-supervised transfer component analysis. *Appl. Energy* **2020**, *277*, 115504. [CrossRef]
24. Li, J.; Deng, Z.; Liu, H.; Xie, Y.; Liu, C.; Lu, C. Battery capacity trajectory prediction by capturing the correlation between different vehicles. *Energy* **2022**, *260*, 125123. [CrossRef]
25. Richardson, R.R.; Osborne, M.A.; Howey, D.A. Gaussian process regression for forecasting battery state of health. *J. Power Sources* **2017**, *357*, 209–219. [CrossRef]
26. Deng, Z.; Hu, X.; Li, P.; Lin, X.; Bian, X. Data-driven battery state of health estimation based on random partial charging data. *IEEE Trans. Power Electron* **2021**, *37*, 5021–5031. [CrossRef]
27. Lee Rodgers, J.; Nicewander, W.A. Thirteen ways to look at the correlation coefficient. *Am. Stat.* **1988**, *42*, 59–66. [CrossRef]
28. Liu, H.; Cai, J.; Ong, Y.-S. Remarks on multi-output Gaussian process regression. *Knowl. Based Syst.* **2018**, *144*, 102–121. [CrossRef]

Article

# State of Charge Estimation of Lithium-Ion Battery for Electric Vehicles under Extreme Operating Temperatures Based on an Adaptive Temporal Convolutional Network

Jiazhi Miao [1,2], Zheming Tong [1,2,*], Shuiguang Tong [1,2], Jun Zhang [2] and Jiale Mao [3]

1 State Key Laboratory of Fluid Power and Mechatronic Systems, Zhejiang University, Hangzhou 310027, China
2 School of Mechanical Engineering, Zhejiang University, Hangzhou 310027, China
3 College of Chemical and Biological Engineering, Zhejiang University, Hangzhou 310027, China
* Correspondence: tzm@zju.edu.cn

**Citation:** Miao, J.; Tong, Z.; Tong, S.; Zhang, J.; Mao, J. State of Charge Estimation of Lithium-Ion Battery for Electric Vehicles under Extreme Operating Temperatures Based on an Adaptive Temporal Convolutional Network. *Batteries* **2022**, *8*, 145.
https://doi.org/10.3390/batteries8100145

Academic Editor: Matthieu Dubarry

Received: 23 August 2022
Accepted: 23 September 2022
Published: 27 September 2022

**Publisher's Note:** MDPI stays neutral with regard to jurisdictional claims in published maps and institutional affiliations.

**Copyright:** © 2022 by the authors. Licensee MDPI, Basel, Switzerland. This article is an open access article distributed under the terms and conditions of the Creative Commons Attribution (CC BY) license (https://creativecommons.org/licenses/by/4.0/).

**Abstract:** The accurate estimation of state of charge (SOC) under various conditions is critical to the research and application of batteries, especially at extreme temperatures. However, few studies have examined the SOC estimation performance of estimation algorithms for several types of batteries under such conditions. In this study, a new method was derived for SOC estimation and a series of experiments were conducted covering five types of lithium-ion batteries with three kinds of cathode materials (i.e., LiFePO$_4$, Li(Ni$_{0.5}$Co$_{0.2}$Mn$_{0.3}$)O$_2$, and LiCoO$_2$), three test temperatures, and four real driving cycles to verify the proposed method. The test temperatures for battery operation ranges from −20 to 60 °C. Then, an adaptive machine learning (ML) framework based on the deep temporal convolutional network (TCN) and Coulomb counting method was proposed, and the structure of the estimation model was designed through the Taguchi method. The accuracy and generalizability of the proposed method were evaluated by calculating the estimation errors and their standard deviations (SDs), its average errors showed a decline of at least 49.66%, and its SDs showed a decline of at least 45.88% when compared to four popular ML methods. These traditional ML methods performed poor accuracy and stability at extreme temperatures (−20 and 60 °C) when compared to 25 °C, while the proposed adaptive method exhibited stable and high performances at different temperatures.

**Keywords:** machine learning; state of charge estimation; temporal convolutional network; extreme temperature

## 1. Introduction

Recently, environmental pollution has become increasingly serious, with greenhouse gas emissions rising rapidly due to large-scale fossil fuel consumption [1,2]. Therefore, it is important to develop green and clean energy sources, which are efficient and convenient [3,4]. Lithium-ion batteries (LIBs) exhibit high mobility, a long lifespan and high energy-density, which are widely used in clean transportation systems, smart grids, and renewable energy sources [1,5,6]. There are two key methods to accelerate the application of LIBs: (1) Develop new materials with better electrochemical properties, which makes batteries safer, more efficient, and have longer lifespans [7,8]; (2) Higher-accuracy quantification of the internal electrochemical changes of the LIBs from the macroscopic point of view, such that the appropriate operations (e.g., charge, discharge, or maintain) can be decided at any time to ensure that the battery works in a safe and healthy condition [9]. This prompts the real-time estimation of battery health state. The state of charge (SOC) of LIB, as one of the most important indicators of internal battery health state, can show the remaining battery energy and help to formulate an appropriate charge and discharge strategy [10–13].

Since the battery health states including SOC cannot be measured directly through sensors, there has been a tremendous amount of research to develop estimation algorithms for higher accuracy [14–17]. The SOC estimation methods can be divided into three categories: conventional methods, model-based methods, and machine learning (ML) methods [18–21]. The conventional methods include the open circuit voltage (OCV) method [22], the Coulomb counting method [23], and the electrochemical impedance spectroscopy (EIS) method [24,25]. The OCV method estimates the SOC through a one-to-one relationship between the OCV and the SOC [26]; however, it cannot be used for the LiFePO4 battery because of a flat plateau in the OCV–SOC curve [27]. In addition, a long rest time is needed to make the battery reach an equilibrium condition before measuring the OCV, which limits its online application [28]. The key characteristics of the Coulomb counting method are low computational costs and poor accuracy [18]. The EIS method utilizes the battery impedance and internal resistance to describe the electrical characteristic, but it is expensive [29].

For the model-based approaches, the Kalman filter (KF) and the particle filter (PF) are the most widely used methods [30,31]. Since the KF is more likely to be used for a linear system, while the LIB is a highly nonlinear system, some extensions were proposed to address the adaptivity problem [32]. Xiong et al. [33] used an extended Kalman filter (EKF) to estimate the SOC of the vanadium redox flow battery, and the results showed that the maximum estimation error was within 5.5%. Wang et al. [34] proposed a robust adaptive unscented Kalman filter (UKF) method for unbiased SOC estimation. Their method was applied on a $LiNi_xCo_yMn_zO_2$ battery with a rated capacity of 40 Ah under 25 °C, and the maximum absolute error exceeded 5%. Yang et al. [35] applied a novel fuzzy adaptive cubature Kalman filter (CKF) to estimate the SOC of LIB under 35 °C. The results revealed that this method has a faster convergence speed compared with the traditional CKF method. Wang et al. [36] utilized the PF method to estimate the SOC of the LiFePO4 battery under a dynamic temperature condition of $-3.5$ to 45 °C, which showed better performance with less than 1% error compared to EKF.

In contrast, ML methods have become highly attractive in many fields in recent years. For example, they have been used for the prediction of material properties and characteristics [37,38], estimation of battery health states [39,40], and energy storage applications [41]. For SOC estimation, the ML methods treat the battery system as a black box and build the model by fitting the collected data. The commonly used ML methods include support vector machine for regression (SVR) [42], long-short term memory (LSTM) network [43], and convolutional neural network (CNN) [44], as well as the combination of CNN and LSTM. Lu et al. [45] developed a novel deep operator network that has better performance compared to traditional algorithms. In recent years, the temporal convolutional network (TCN) has proved to have excellent performance for the estimation of time series data due to its characteristics of casual convolution, dilation convolution, and residual block [46]. Liu et al. [47] utilized the TCN to estimate the SOC of the LiFePO4 battery and applied it to the LiPO battery through transfer learning. However, the maximum estimation errors were large at 4.14% under the constant temperature and 10% under rising ambient temperature. In addition, the adaptive activations have proved that they are very helpful to train the neural networks [48]. Jagtap et al. [48–50] have proposed many kinds of adaptive activation functions for regression problems. They provide excellent learning capacity compared to the traditional method.

Despite the above advancements, there are two aspects in the research of SOC estimation that need to be further improved: (1) The generalizability of the algorithm to different types of batteries. This means that improved methods can have a broader range of applications. Many studies only utilize one type of battery to validate their algorithms [26,27,29,30,51], which is not sufficient to prove their generalizability; (2) The adaptability of the algorithm to extreme temperatures. It will be helpful to accelerate the applications of LIB in special situations or extreme environments with high safety and reliability, such as ultra-low temperature cold storage, cold area, and high-temperature workshop. In fact, few studies have validated the estimation performance of the algorithm with the battery working at extreme temperatures, as the testing temperatures in most studies are distributed from $-10$ to 50 °C [26,27,29,30].

In order to further improve the accuracy and generalizability of estimation algorithms to different types of batteries and extreme temperatures, in this paper, the temporal convolutional network (TCN) is utilized to estimate the SOC of LIBs. Its hyperparameters are optimized by the Taguchi method, and the Coulomb counting method is used to produce the input parameter that is highly related to the SOC for the TCN. To sufficiently prove the generalizability of the proposed method, five types of commonly used batteries are employed, which have three kinds of cathode materials (i.e., LiFePO$_4$, Li(Ni$_{0.5}$Co$_{0.2}$Mn$_{0.3}$)O$_2$, and LiCoO$_2$), and they are tested under a constant ambient temperature of 25 °C and two extreme temperatures (−20 and 60 °C), as well as four real driving cycles, including the dynamic stress test (DST), the federal urban driving schedule (FUDS), the urban dynamometer driving schedule (UDDS), and a supplemental federal test procedure driving schedule called US06.

## 2. Experimental and Methodology

### 2.1. Experimental Procedure and Dataset

Five types of batteries made from three kinds of cathode materials were utilized to better prove the generalizability of the proposed method in this paper. The detailed battery information is shown in Table 1. The recommended operating temperatures for the five types of batteries were distributed from −20 to 60 °C. To verify the performance of the proposed method under extreme temperatures and its generalizability to different working temperatures, all batteries were tested separately at −20, 25 and 60 °C. To evaluate the performance of the proposed method in practical applications and its generalizability to different conditions, tests with four real driving cycles (i.e., DST, FUDS, UDDS, and US06) were developed. Before performing these tests, peak power tests and OCV tests were performed under different temperatures, with the test strategy shown in Figure 1. The parameters used in peak power test are defined as follows [52]:

$$P_r = I_1 \times V_2 \qquad (1)$$

$$I_2 = 80\% \times \frac{P_r}{V_1}, \qquad (2)$$

$$V_{limit} = \max(V_1, V_2), \qquad (3)$$

$$I_{high} = \min(I_1, I_2), \qquad (4)$$

$$I_{base} = \frac{\left(12 \times Q_n - I_{high}\right)}{35}, \qquad (5)$$

where $P_r$ is the rated peak power, $I_1$ represents the recommended maximum discharge current, $V_2$ represents the cut-off voltage, $I_2$ represents the calculated maximum discharge current, $V_1$ is the 2/3 open circuit voltage at 80% depth of discharge, $V_{limit}$ is the cut-off voltage of the peak power test, $I_{high}$ is the maximum current in the peak power test, $Q_n$ is the rated capacity, and $I_{base}$ is the minimum current in the peak power test.

The detailed information of peak power tests for five types of batteries under different temperatures is shown in Tables S1–S3, and the results are depicted in Figure 2. The peak power declined from 20% to 80% when the battery worked under extreme temperatures (−20 or 60 °C) compared with 25 °C. The current, voltage and temperature were collected every second through the sensors. In summary, the total number of tests reached 90 groups, including 15 groups of OCV tests, 15 groups of peak power tests, and 60 groups of real driving cycle tests. The partial test data were depicted in Figure 3. Importantly, while the temperature increased, the voltage became more stable.

Table 1. Details of five types of battery information.

| Battery Types | Electrode Material | Nominal Capacity (Ah) | Cut-off Voltage (V) | Charging Voltage (V) | Recommended Operating Temperatures |
|---|---|---|---|---|---|
| LR1865EH | $LiFePO_4$/graphite | 1.7 | 2.0 | 3.6 | 0~45 °C (charge); −20~60 °C (discharge) |
| LR1865SK | $LiFePO_4$/graphite | 2.6 | 2.75 | 4.2 | |
| LR1865SZ | $LiFePO_4$/graphite | 2.5 | 3.0 | 4.2 | |
| LR2170SA | $Li(Ni_{0.5}Co_{0.2}Mn_{0.3})O_2$/graphite | 4.0 | 2.75 | 4.2 | |
| ICR18650 | $LiCoO_2$/graphite | 2.55 | 2.5 | 4.2 | |

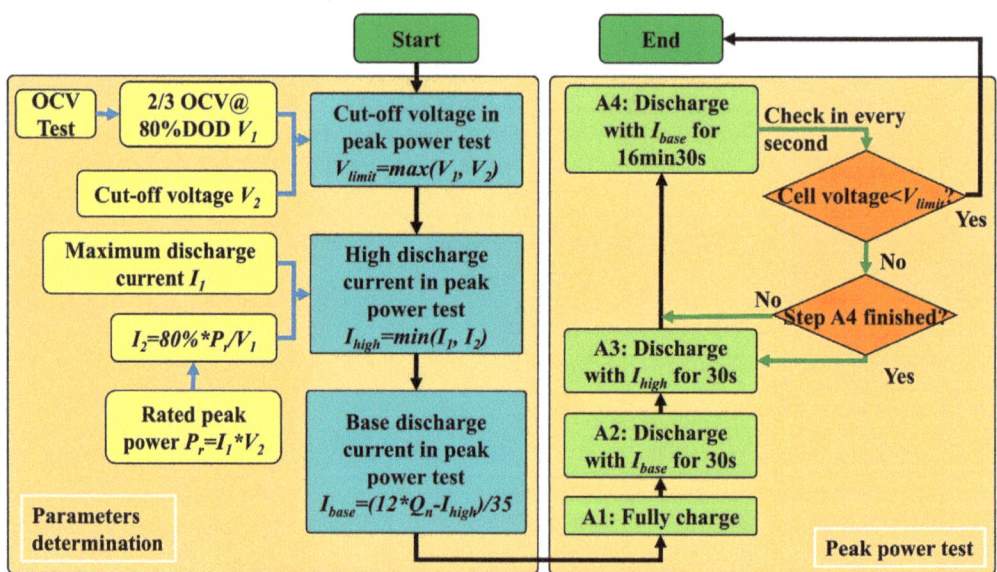

Figure 1. Schematic diagram of the peak power test.

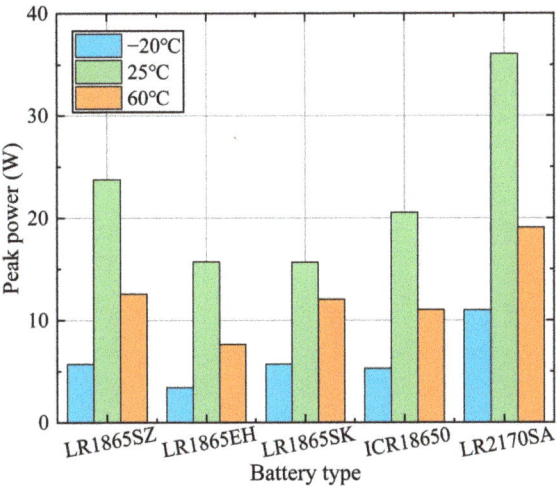

**Figure 2.** Peak power of different types of batteries (LR1865SZ, LR1865EH, LR1865SK, ICR18650, and LR2170SA) under various temperatures (−20, 25 and 60 °C).

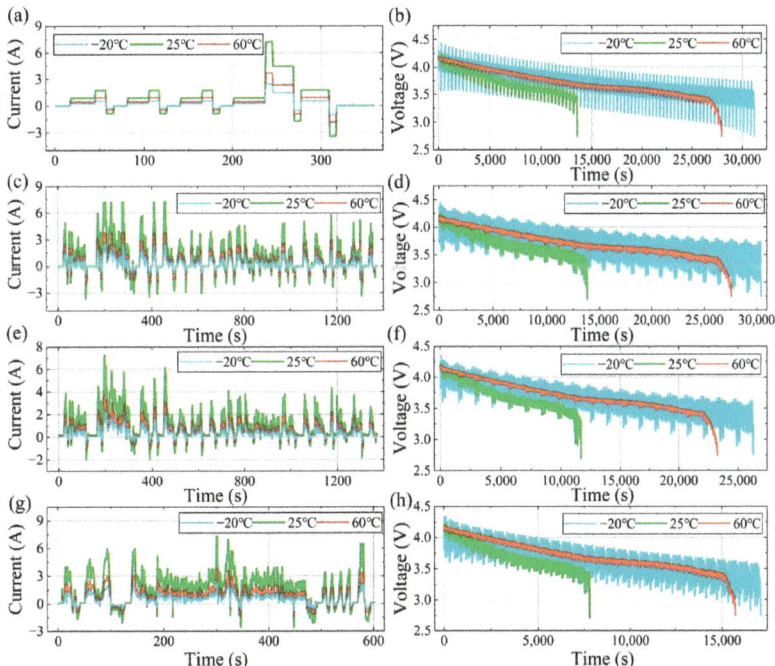

**Figure 3.** Voltage and current measurement under (**a**,**b**) DST, (**c**,**d**) FUDS, (**e**,**f**) UDDS, and (**g**,**h**) US06 experiments for the battery LR2170SA at different temperatures: −20, 25, and 60 °C.

*2.2. Machine Learning*

In this paper, the TCN was utilized to estimate the SOC of batteries. The architectural elements in a TCN were shown in Figure 4. Causal convolution makes the method suitable for sequence modeling, which the traditional CNN cannot deal with. Dilated convolution confers the TCN a larger receptive field size with fewer network layers compared to the traditional CNN; thus, more input data can be considered for every step of SOC estimation. In addition, the utilization of residual blocks can address the vanishing gradient problem of the deep neural network. MAE and RMSE were used to evaluate the performance of the estimation algorithm, which are calculated as:

$$MAE = \frac{1}{n}\sum_{i=1}^{n}\left|y_i - \hat{y_i}\right| \qquad (6)$$

$$RMSE = \sqrt{\frac{1}{n}\sum_{i=1}^{n}\left(y_i - \hat{y_i}\right)^2}, \qquad (7)$$

where $n$ denotes the total time steps of the real driving cycles, and $\hat{y_i}$ and $y_i$ are the estimated SOC value and the experimental SOC value, respectively, for the $i$-th time step.

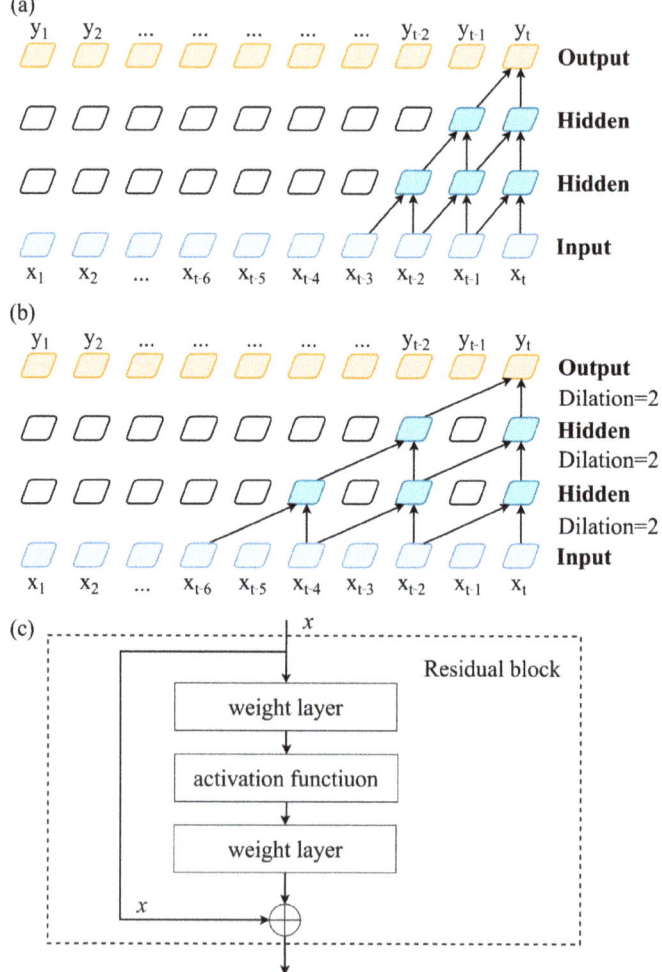

**Figure 4.** Architectural elements in a TCN: (**a**) causal convolution; (**b**) dilation convolution; (**c**) residual block.

## 2.3. Feature Selection Used in the TCN Method

Feature selection is critical to the performance of ML methods. Traditional methods usually utilize the current, voltage and temperature as the input data [26]. However, the estimation results fluctuate wildly and need to be further processed by other algorithms. The reason is that there is a poor correlation between the input data and SOC. In this section, the correlation between three parameters and SOC was evaluated based on the ICR18650 battery that operates at the FUDS driving cycle under 25 °C. In addition, to improve the estimation performance, a new parameter called observed SOC was introduced based on the Coulomb counting method. First, the initial state function was simplified to reduce the calculation amount [26]. Then, a degradation factor f was introduced to reduce the errors caused by temperature. Since the DST data will be used to train the ML method in this paper, they could also be utilized to calculate the degradation factor f. Finally, the observed SOC was defined as:

$$S\hat{O}C_k = S\hat{O}C_{k-1} - \frac{I_{k-1} * \Delta t}{Q_n} * f, \tag{8}$$

where $I_{k-1}$ denotes the current at the k−1 time step, $\Delta t$ is the change of time, and $Q_n$ represents the nominal capacity of the battery.

The results are shown in Figure 5. The $R^2$ was defined through Equation (9). Among the four parameters, the observed SOC had the best correlation with the SOC. The results also prove that the temperature and voltage are highly correlated to the SOC value.

$$R^2 = 1 - \frac{\sum_{i=1}^{n}(y_i - x_i)^2}{\sum_{i=1}^{n}(y_i - \overline{x_i})} \tag{9}$$

where $y_i$ represents the SOC, $x_i$ represents the input parameter, and $\overline{x_i}$ is the average value of $x_i$.

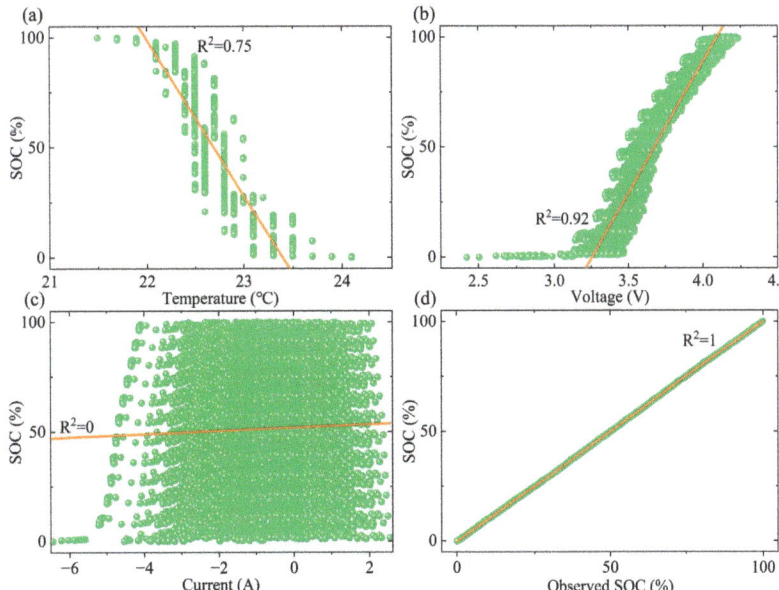

**Figure 5.** Correlations between four input parameters and SOC based on the ICR18650 battery when working in FUDS under 25 °C: (**a**) temperature, (**b**) voltage, (**c**) current, and (**d**) observed SOC calculated by the optimized state function.

The eight strategies of input features were evaluated by the TCN model based on the ICR18650 battery when working in the FUDS driving cycle under 25 °C (see Figure 6).

Figure 6a presents the traditional strategy, utilized current, voltage and temperature as input features, and it shows a poor estimation performance. In addition, the other strategies that use the current as an input feature also exhibit poor estimation accuracy, as shown in Figure 6c,d,g. The reason is that the current has a poor correlation to the SOC and affected the convergence of the model. In Figure 6f,h, the strategies with a single feature perform the best performance. Finally, the observed SOC was selected as the input feature according to the results in Figure 6.

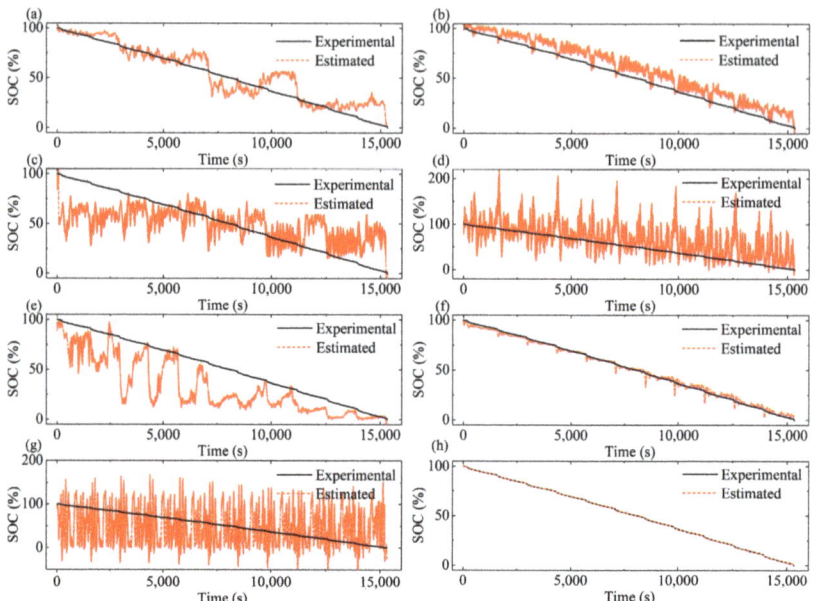

**Figure 6.** The SOC estimation results of the TCN model with different input parameters based on the ICR18650 battery (works at the FUDS condition under 25 °C): (**a**) current, voltage, and temperature; (**b**) voltage and temperature; (**c**) current and temperature; (**d**) current and voltage; (**e**) temperature; (**f**) voltage; (**g**) current; (**h**) observed SOC calculated by the optimized state function.

*2.4. Network Structure Optimization Using the Taguchi Method*

In this section, we first propose three TCNs with different architectures (see Figure 7). To find out the best architecture, the three TCNs with the relevant optimal hyperparameters were compared to each other. The DST data of ICR18650 under 25 °C were utilized to train each TCN, and the estimation performances for FUDS, UDDS and US06 data were used as the evaluation basis. Traditional methods usually employ the trial-and-error method to obtain the optimal topology; however, this approach is time-consuming and inefficient. In this paper, three design factors were settled, each of them with five levels (see Tables S4 and S5). This means that 375 groups of tests would need to be performed for all TCNs with the trial-and-error method. In contrast, the Taguchi method has proved to have a lower cost and higher efficiency for parameter optimization [53,54]; therefore, it was introduced and the number of total tests declined to 75 groups. The orthogonal array L25($5^3$) and experimental results for three TCNs can be seen in Tables S6–S8. The best estimation results of the TCNs are concluded in Table 2. The TCN-v1 model had the minimum MAE, and the TCN-v2 model had the minimum RMSE. Since the RMSE is more sensitive to outliers, it is considered more important than the MAE. Therefore, the adopted TCN-v2 model featured 128 filters, the kernel size of 10, and the list of the dilation of {2, 2, 2, 2, 2, 2} according to the results in Tables S6–S8.

**Table 2.** The SOC estimation performance for ICR18650 under 25 °C based on the three kinds of TCN (TCN-v1, TCN-v2, TCN-v3) with the relevant optimal hyperparameters.

| Neural Networks | FUDS | | UDDS | | US06 | |
|---|---|---|---|---|---|---|
| | MAE (%) | RMSE (%) | MAE (%) | RMSE (%) | MAE (%) | RMSE (%) |
| TCN-v1 | 0.059 | 0.102 | 0.089 | 0.122 | 0.116 | 0.159 |
| TCN-v2 | 0.062 | 0.091 | 0.099 | 0.119 | 0.128 | 0.157 |
| TCN-v3 | 0.062 | 0.102 | 0.099 | 0.128 | 0.129 | 0.167 |

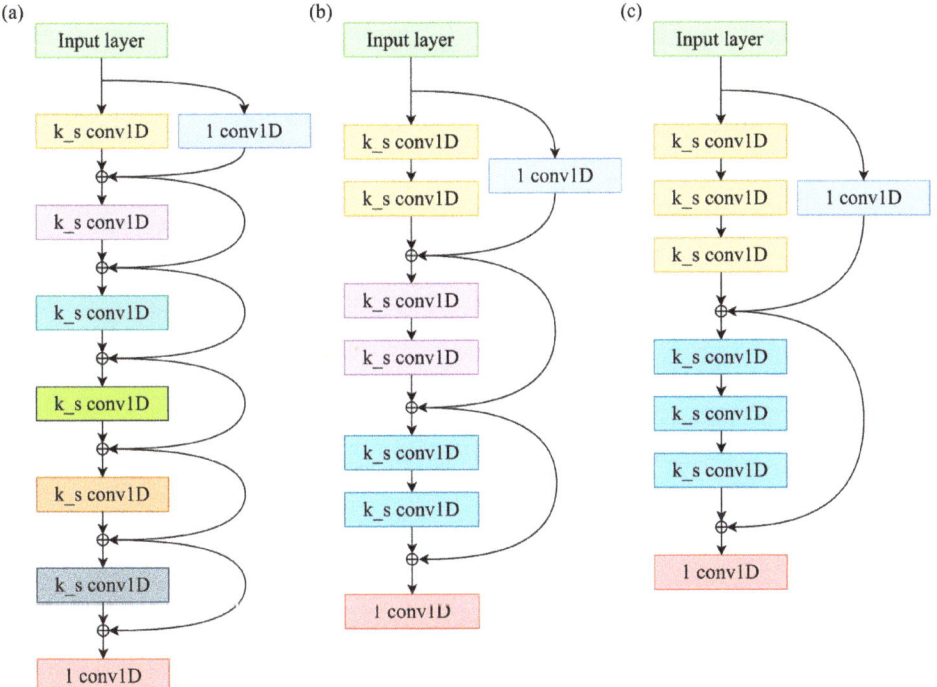

**Figure 7.** Architectures of TCN for SOC estimation: (**a**) TCN-v1; (**b**) TCN-v2; (**c**) TCN-v3.

## 2.5. Adaptive SOC Estimation Method

In this section, the adaptive TCN-v2 (ATCN-v2) model is proposed to estimate the battery SOC. As shown in Figure 8, the ATCN-v2 model can be divided into two parts: offline training and online estimation. Firstly, the DST data are utilized to train the optimized state function to obtain the degradation factor $f$. Secondly, the experimental SOC and the observed SOC calculated by the optimized state function are utilized to train the TCN model. Thirdly, the structure of the TCN is optimized based on the Taguchi method, and the TCN-v2 model is established. For the part of the online estimation, the FUDS, UDDS, and US06 data are utilized to test the model. Since the TCN-v2 has the architectural element of causal convolution, the estimation results of the first 108 s will not be accurate due to the padding operation (see Figure S1). In contrast, the optimized state function has a better performance at the initial stage of real driving cycles. Therefore, the observed SOC calculated by the optimized state function is utilized to correct the estimation results of the initial 108 s based on the TCN-v2 model. Finally, the estimated SOC is output by the ATCN-v2 model. In this study, the proposed model is developed based on the PyCharm software and TensorFlow package. The detailed training information is given in Table S9.

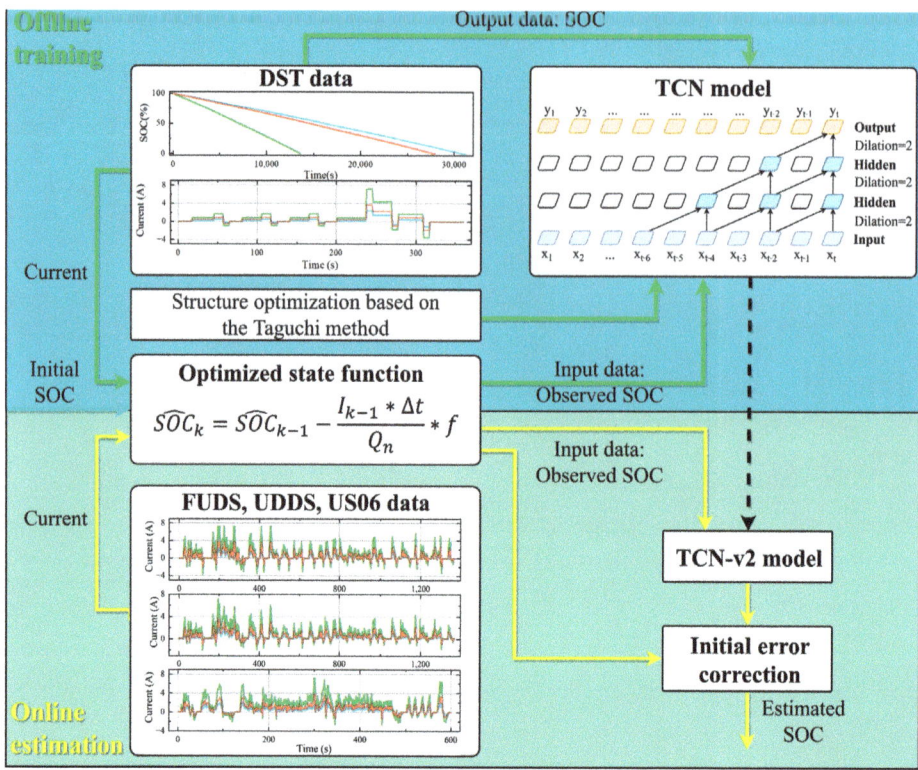

**Figure 8.** Graphical description of the proposed ATCN-v2 model.

## 3. Results and Discussion

### 3.1. SOC Estimation Results Based on the ATCN-v2 Model

In order to verify the performance of the ATCN-v2 model, 45 samples were utilized under different conditions, including three temperatures, three real driving cycles, and five types of batteries. The SOC estimation results of the ATCN-v2 model are shown in Figures S2–S6. They have almost no error as compared with the experimental values, which demonstrates an excellent estimation performance of the proposed method for all samples. As depicted in Figure 9, the estimation errors for all samples exhibit periodical changes. They are within ±0.2% for the temperature of −20 °C, ±0.6% for the temperature of 25 °C, and ±0.3% for the temperature of 60 °C. The results further indicate that the proposed method has a good estimation performance under extreme temperatures (−20 and 60 °C). Finally, the MAEs and RMSEs of the ATCN-v2 model for different types of batteries under different real driving cycles and temperatures were calculated (see Table 3). The MAEs were distributed from 0.021% to 0.185%, and the RMSEs were distributed from 0.026% to 0.277%, indicating that the proposed method has good generalizability.

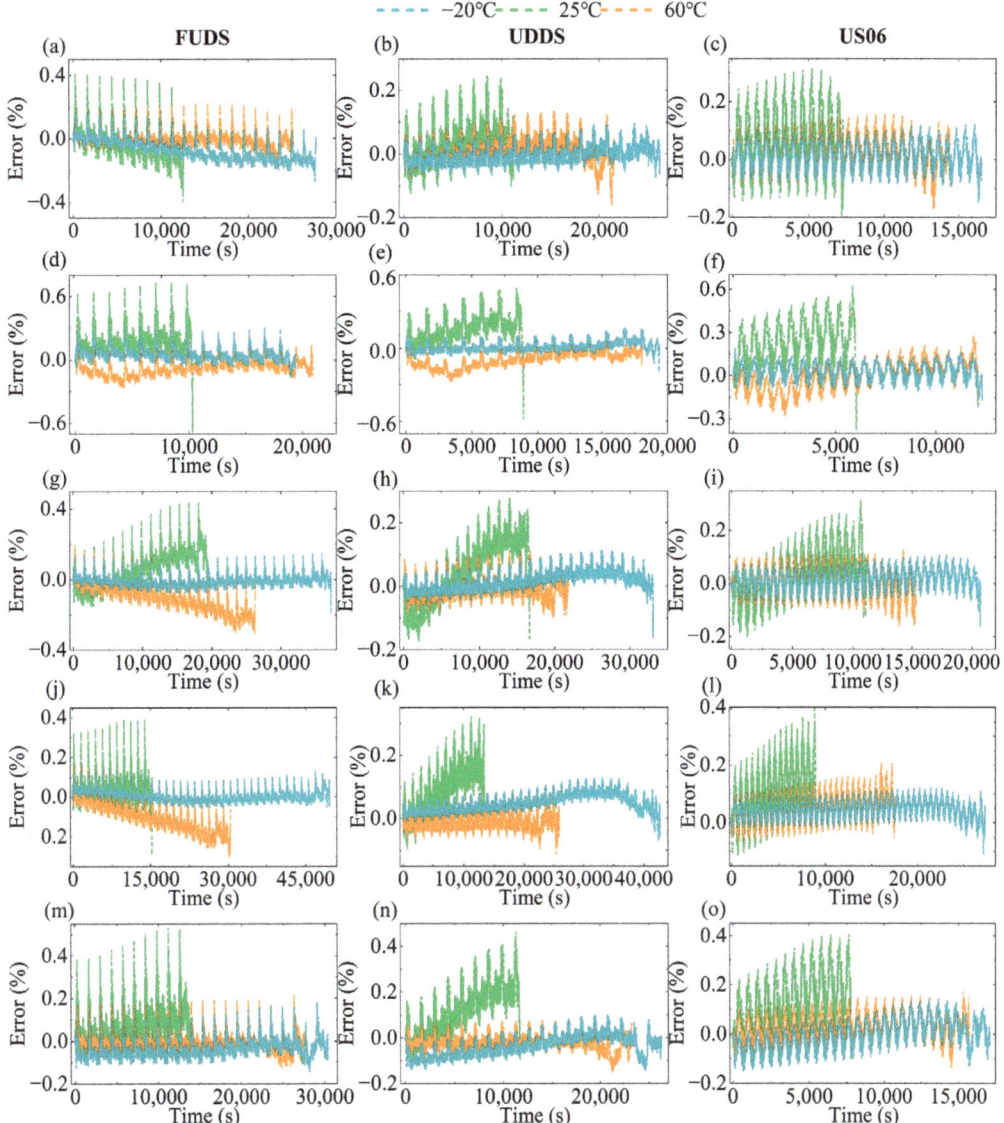

**Figure 9.** The estimation errors of ATCN-v2 model for different types of batteries under different conditions and temperatures: (**a–c**) LR1865SZ; (**d–f**) LR1865EH; (**g–i**) LR1865SK; (**j–l**) ICR18650; (**m–o**) LR2170SA.

Table 3. SOC estimation performance of the ATCN-v2 model for different types of batteries under different real driving cycles and temperatures.

| Temperature | Cases | FUDS | | UDDS | | US06 | |
|---|---|---|---|---|---|---|---|
| | | MAE (%) | RMSE (%) | MAE (%) | RMSE (%) | MAE (%) | RMSE (%) |
| −20 °C | LR1865EH | 0.060 | 0.080 | 0.028 | 0.038 | 0.055 | 0.066 |
| | LR1865SK | 0.027 | 0.034 | 0.026 | 0.033 | 0.034 | 0.041 |
| | LR1865SZ | 0.083 | 0.097 | 0.021 | 0.026 | 0.041 | 0.048 |
| | LR2170SA | 0.049 | 0.056 | 0.047 | 0.058 | 0.049 | 0.060 |
| | ICR18650 | 0.022 | 0.029 | 0.050 | 0.056 | 0.047 | 0.055 |
| 25 °C | LR1865EH | 0.185 | 0.226 | 0.156 | 0.182 | 0.237 | 0.277 |
| | LR1865SK | 0.099 | 0.123 | 0.094 | 0.109 | 0.086 | 0.107 |
| | LR1865SZ | 0.088 | 0.113 | 0.048 | 0.064 | 0.107 | 0.132 |
| | LR2170SA | 0.094 | 0.128 | 0.127 | 0.154 | 0.145 | 0.177 |
| | ICR18650 | 0.061 | 0.090 | 0.099 | 0.118 | 0.127 | 0.156 |
| 60 °C | LR1865EH | 0.092 | 0.105 | 0.097 | 0.114 | 0.083 | 0.103 |
| | LR1865SK | 0.114 | 0.132 | 0.028 | 0.034 | 0.049 | 0.058 |
| | LR1865SZ | 0.036 | 0.051 | 0.031 | 0.040 | 0.052 | 0.064 |
| | LR2170SA | 0.034 | 0.047 | 0.029 | 0.037 | 0.051 | 0.063 |
| | ICR18650 | 0.106 | 0.123 | 0.024 | 0.029 | 0.059 | 0.073 |

*3.2. Performance Evaluation*

In this section, the widely used algorithms, including the LSTM, gated recurrent unit (GRU) neural network, CNN, and CNN-LSTM, were selected for comparison with the proposed ATCN-v2 model. To make the results credible, all algorithms in this paper were settled with the same random seed to avoid randomness. In addition, the algorithms used for comparison were developed with the same structure and hyperparameters as the proposed ATCN-v2 model. The criteria of algorithm evaluation involved the generalizability to temperature, conditions, battery type, and estimation accuracy. First, the MAEs and RMSEs of the LSTM, GRU, CNN, and CNN-LSTM for all 45 samples were calculated and summarized in Tables S9–S12. It can be seen that, for the LSTM, the MAEs were distributed from 0.077% to 9.020%, and the RMSEs were distributed from 0.101% to 10.483%. For the GRU, the MAEs were distributed from 0.087% to 9.196%, and the RMSEs were distributed from 0.104% to 10.599%. For the CNN, the MAEs were distributed from 0.037% to 0.391%, and the RMSEs were distributed from 0.211% to 0.829%. For the CNN-LSTM, the MAEs were distributed from 0.062% to 9.006%, and the RMSEs were distributed from 0.068% to 10.475%. When compared with these methods, the proposed method has the best accuracy. Besides, the average MAEs and average RMSEs of all algorithms were also depicted in Figure 10 for visual comparison. The LSTM, GRU and CNN-LSTM showed poor adaptation to battery type and temperature. For example, they produced abnormally large errors for the battery LR1865EH working under an extreme temperature of −20 °C, for the battery LR2170SA working under 25 °C, and for the battery LR1865SZ working under an extreme temperature of 60 °C. Among them, the proposed method obtained the highest accuracy and the best adaptation to battery type and temperature. Finally, the average MAEs and average RMSEs of five algorithms for the total 45 samples were calculated as shown in Table 4. The performances of LSTM, GRU and CNN-LSTM are very similar. When compared with these methods, the CNN exhibited better performance. Among them, the optimal algorithm was still the ATCN-v2, and its average MAE showed a decline of at least 49.66%, and the average RMSE exhibited a decline of at least 79.95%.

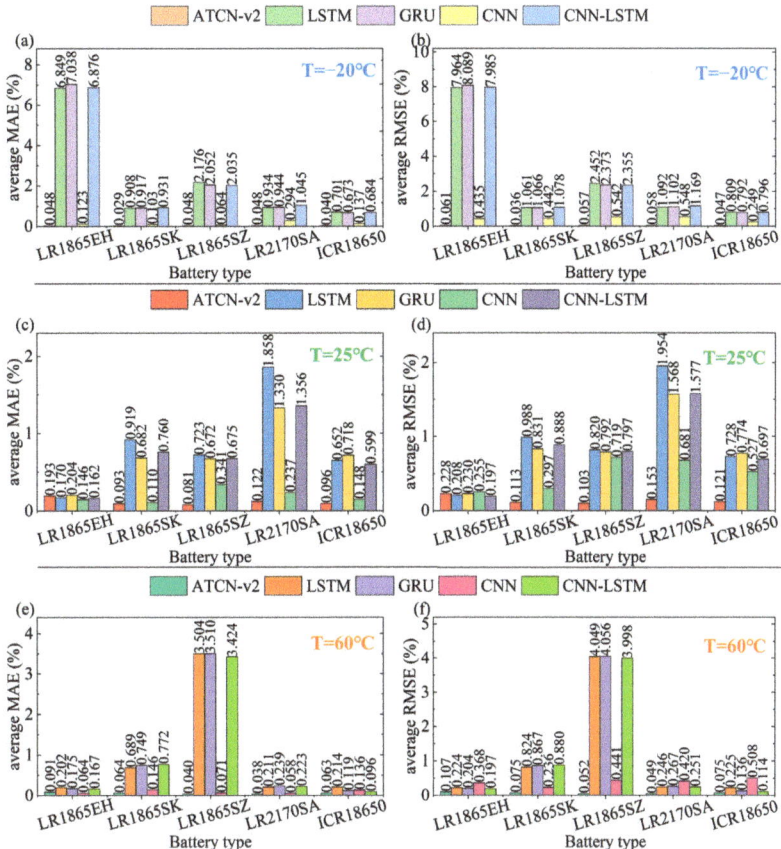

**Figure 10.** SOC estimation errors of the different algorithms for five types of batteries (LR1865EH, LR1865SK, LR1865SZ, LR2170SA, and ICR18650): (**a**,**b**) average MAE and average RMSE under −20 °C; (**c**,**d**) average MAE and average RMSE under 25 °C; (**e**,**f**) average MAE and average RMSE under 60 °C.

**Table 4.** SOC estimation performance of the different algorithms.

| Algorithm | Average MAE (%) | Average RMSE (%) |
|---|---|---|
| ATCN-v2 | 0.073 | 0.089 |
| LSTM | 1.381 | 1.576 |
| GRU | 1.335 | 1.543 |
| CNN | 0.145 | 0.444 |
| CNN-LSTM | 1.320 | 1.532 |

In order to further quantify the generalizability of algorithms, the standard deviations (SDs) of estimation errors (i.e., MAEs and RMSEs) were calculated, that is, to evaluate if the algorithms performed stably when only one parameter (i.e., temperature, driving cycle, or battery type) was changed. The standard deviation was defined as:

$$\sigma = \sqrt{\frac{1}{n}\sum_{i=1}^{n}\left(X_i - \overline{X}\right)^2} \tag{10}$$

where $X_i$ represents the SOC estimation error, and $\overline{X}_i$ represents the average error.

The final results are summarized in Table 5. The adaptation was evaluated for three kinds of parameters, namely, temperature, driving cycle, and battery type. Among them, five algorithms proved to be more adaptable to the different driving cycles, and their performances for different battery types or different temperatures were highly unstable. The traditional methods including LSTM, GRU and CNN-LSTM provided the worst generalizability compared with the proposed method and CNN. The SDs of all the MAEs and RMSEs were calculated as the comprehensive index to evaluate the generalizability of algorithms. The SD of MAEs can be utilized to measure the change of entire errors, and the SD of RMSEs can be utilized to measure the change of the discrete degree of errors. Hence, the smaller the SD of MAEs and SD of RMSEs, the better the generalizability of the algorithm. The SDs of MAEs and RMSEs were 0.046 and 0.055 for the ATCN-v2, 1.862 and 2.160 for LSTM, 1.897 and 2.185 for GRU, 0.085 and 0.152 for CNN, and 1.857 and 2.159 for CNN-LSTM, respectively. Thus, the ATCN-v2 model was indicated to have the best generalizability as compared to other methods.

**Table 5.** SOC estimation performance of the ATCN-v2 model for different types of batteries under different real driving cycles and temperatures.

| Parameter | Algorithm | The SD of MAE (%) | The SD of RMSE (%) |
|---|---|---|---|
| Temperature | ATCN-v2 | 0.037 | 0.044 |
| | LSTM | 1.245 | 1.430 |
| | GRU | 1.237 | 1.427 |
| | CNN | 0.064 | 0.102 |
| | CNN-LSTM | 1.215 | 1.413 |
| Driving cycle | ATCN-v2 | 0.018 | 0.019 |
| | LSTM | 0.548 | 0.630 |
| | GRU | 0.543 | 0.627 |
| | CNN | 0.022 | 0.048 |
| | CNN-LSTM | 0.535 | 0.622 |
| Battery type | ATCN-v2 | 0.027 | 0.031 |
| | LSTM | 1.433 | 1.640 |
| | GRU | 1.407 | 1.620 |
| | CNN | 0.068 | 0.131 |

The effects of temperature on the performance of algorithms were also analyzed. In Table 6, the stability of algorithms under different temperatures without considering the battery type and driving cycle was evaluated by the standard deviation of estimation errors (i.e., MAE and RMSE). The traditional methods, including LSTM, GRU and CNN-LSTM, obtained the largest standard deviation of MAE and RMSE for the battery working at an extremely low temperature ($-20$ °C), and also had a larger standard deviation of MAE and RMSE for the battery working at an extremely high temperature (60 °C), which declined by more than 50% compared to that when working at 25 °C. In other words, these methods result in large errors when used for estimating the SOC of batteries operating at extreme temperatures ($-20$ and 60 °C). Among the five algorithms, the ATCN-v2 model had the minimum standard deviation of MAE and RMSE at three temperatures ($-20$, 25, 60 °C). The results also indicate that the proposed method is suitable for SOC estimation at extreme temperatures ($-20$ and 60 °C). We also noticed that the ATCN-v2 model had a faster convergence velocity compared to the LSTM, GRU and CNN-LSTM. In the same computing environment, the ATCN-v2 method needed about 300 epochs to achieve convergence; meanwhile, the LSTM, GRU and CNN-LSTM required about 500 epochs.

Table 6. The stability of algorithms under different temperatures.

| Algorithms | −20 °C | | 25 °C | | 60 °C | |
|---|---|---|---|---|---|---|
| | The SD of MAE (%) | The SD of RMSE (%) | The SD of MAE (%) | The SD of RMSE (%) | The SD of MAE (%) | The SD of RMSE (%) |
| ATCN-v2 | 0.092 | 0.105 | 0.097 | 0.114 | 0.083 | 0.103 |
| LSTM | 0.114 | 0.132 | 0.028 | 0.034 | 0.049 | 0.058 |
| GRU | 0.036 | 0.051 | 0.031 | 0.040 | 0.052 | 0.064 |
| CNN | 0.034 | 0.047 | 0.029 | 0.037 | 0.051 | 0.063 |
| CNN-LSTM | 0.106 | 0.123 | 0.024 | 0.029 | 0.059 | 0.073 |

## 4. Conclusions

This work proposed a new algorithm based on the TCN and Coulomb counting method for estimating the SOC of five types of LIBs. First, different parameters, including current, voltage, temperature, and observed SOC, were evaluated for suitability as the input data of the model. Eight strategies for input data were designed, and the strategies with a single and highly SOC-related parameter showed the best performance. Therefore, the observed SOC calculated by optimized state function was selected as the input data. Then, three TCNs with different structures were compared to find the best scheme, whose structure was optimized by the Taguchi method. Considering that there are initial errors caused by the padding operation, the estimation results of the first 108 time steps were corrected. Finally, the ATCN-v2 model was established.

In order to better verify the accuracy and generalizability of the proposed algorithm, five types of widely used batteries composed of three kinds of cathode materials (i.e., $LiFePO_4$, $Li(Ni_{0.5}Co_{0.2}Mn_{0.3})O_2$, and $LiCoO_2$) were tested under four real driving cycles and three temperatures. The real driving cycles included DST, FUDS, UDDS, and US06. The test temperatures include an extremely low temperature (−20 °C), an extremely high temperature (60 °C) for the mentioned batteries above, and a constant temperature of 25 °C. Finally, 60 groups of real driving cycle tests were developed. The DST data were utilized for model training, and FUDS data, UDDS data and US06 data were utilized for model testing. The SOC estimation errors of the ATCN-v2 model were within ±0.6%, and the MAEs and RMSEs were less than 0.185% and 0.277%, respectively. Subsequently, the proposed method was compared with four popular algorithms, including LSTM, GRU, CNN and CNN-LSTM. The algorithms adopted the same structure, input data and hyperparameters. The average MAE of the ATCN-v2 model declined by at least 49.66%, and the average RMSE declined by at least 79.95% when compared to other methods. The results indicate that the proposed method has the highest estimation accuracy.

In addition, the SDs of MAEs and RMSEs were calculated to quantify the generalizability of the algorithms. Three experimental variables could be considered to test the generalization ability, which comprise battery type, real driving cycle and temperature. First, the generalization ability to an individual variable of the algorithms was evaluated. The results show that it is hard to adapt to different types of batteries and different temperatures for the LSTM, GRU and LSTM-CNN. The proposed model exhibited the best generalization ability to any of these variables with the minimum SDs of MAEs and RMSEs. The effects of different temperatures on the algorithm performance were also studied in this way. The results suggest that extremely low temperature (−20 °C) and extremely high temperature (60 °C) for the batteries in this paper have a significant impact on the performance of LSTM, GRU and CNN-LSTM, while the ATCN-v2 method is highly adaptable to these two extreme temperatures. Finally, the comprehensive generalization ability to three experimental variables was obtained by computing the SDs of all of the MAEs and RMSEs. Compared to other methods, the SDs of MAEs and RMSEs of the ATCN-v2 method dropped by at least 45.88% and 63.82%, respectively. In summary, the proposed ATCN-v2 method demonstrated excellent SOC estimation accuracy and good generalization ability, and had better performance and stability of battery operation for all battery types, and also real driving cycles under extreme temperatures, as compared with LSTM, GRU, CNN and CNN-LSTM. In further research, we will consider applying the ATCN-v2 to a larger dataset. The proposed method can be helpful to accelerate the research and application process in battery energy storage and green transport.

**Supplementary Materials:** The following supporting information can be downloaded at: https://www.mdpi.com/article/10.3390/batteries8100145/s1.

**Author Contributions:** Conceptualization, J.M. (Jiazhi Miao) and Z.T.; methodology, J.M. (Jiazhi Miao); software, Z.T. and S.T.; validation, J.M. (Jiazhi Miao) and J.Z.; investigation, Z.T. and J.M. (Jiale Mao); resources, S.T.; data curation, J.Z.; writing—original draft preparation, J.M. (Jiazhi Miao); writing—review and editing, Z.T. and J.M. (Jiazhi Miao); project administration, S.T.; funding acquisition, Z.T. All authors have read and agreed to the published version of the manuscript.

**Funding:** This research was funded by the National Natural Science Foundation of China (52075481) and Zhejiang Provincial Natural Science Foundation (LR19E050002).

**Data Availability Statement:** The inventory used in this study is available upon request from the corresponding author.

**Conflicts of Interest:** The authors declare no conflict of interest.

## References

1. Hu, X.; Xu, L.; Lin, X.; Pechet, M. Battery lifetime prognostics. *Joule* **2020**, *4*, 310. [CrossRef]
2. Sulaiman, N.; Hannan, M.A.; Mohamed, A.; Majlan, E.H.; Wan Daud, W.R. A review on energy management system for fuel cell hybrid electric vehicle: Issues and challenges. *Renew. Sustain. Energy Rev.* **2015**, *52*, 802. [CrossRef]
3. Zhang, Q.; Tong, Z.; Tong, S.; Cheng, Z. Modeling and dynamic performance research on proton exchange membrane fuel cell system with hydrogen cycle and dead-ended anode. *Energy* **2021**, *218*, 119476. [CrossRef]
4. Zhang, Q.; Tong, Z.; Tong, S. Effect of cathode recirculation on high potential limitation and self-humidification of hydrogen fuel cell system. *J. Power Sources* **2020**, *468*, 228388. [CrossRef]
5. Rietmann, N.; Hügler, B.; Lieven, T. Forecasting the trajectory of electric vehicle sales and the consequences for $CO_2$ emissions. *J. Clean. Prod.* **2020**, *261*, 121038. [CrossRef]
6. Li, Y.; Liu, K.; Foley, A.M.; Zülke, A.; Berecibar, M.; Nanini-Maury, E.; Mierlo, J.V.; Hoster, H.E. Data-driven health estimation and lifetime prediction of lithium-ion batteries: A review. *Renew. Sust. Energ. Rev.* **2019**, *113*, 109254. [CrossRef]
7. Zhang, W.; Fan, L.; Tong, Z.; Miao, J.; Shen, Z.; Li, S.; Chen, F.; Qiu, Y.; Lu, Y. Stable Li-metal deposition via a 3D nanodiamond matrix with ultrahigh young's modulus. *Small Methods* **2019**, *3*, 1900325. [CrossRef]
8. Tong, Z.; Miao, J.; Li, Y.; Tong, S.; Zhang, Q.; Tan, G. Development of electric construction machinery in China: A review of key technologies and future directions. *J. Zhejiang Univ.-SCI. A* **2021**, *22*, 245–264. [CrossRef]
9. Shen, J.; Ma, W.; Xiong, J.; Shu, X.; Zhang, Y.; Chen, Z.; Liu, Y. Alternative combined co-estimation of state of charge and capacity for lithium-ion batteries in wide temperature scope. *Energy* **2022**, *244*, 123236. [CrossRef]
10. Rahman, A.; Lin, X. Li-ion battery individual electrode state of charge and degradation monitoring using battery casing through auto curve matching for standard CCCV charging profile. *Appl. Energy* **2022**, *321*, 119367. [CrossRef]
11. Tian, J.; Xiong, R.; Shen, W.; Lu, J. State-of-charge estimation of LiFePO4 batteries in electric vehicles: A deep-learning enabled approach. *Appl. Energy* **2021**, *291*, 116812. [CrossRef]
12. Kröger, T.; Harte, P.; Klein, S.; Beuse, T.; Börner, M.; Winter, M.; Nowak, S.; Wiemers-Meyer, S. Direct investigation of the interparticle-based state-of-charge distribution of polycrystalline NMC532 in lithium ion batteries by classification-single-particle-ICP-OES. *J. Power Sources* **2022**, *527*, 231204. [CrossRef]
13. Zhang, Q.; Tong, Z.; Tong, S. Research on the influence of electrolytes on the low-temperature start-up performance of zinc-air battery for forklifts. *Int. J. Energy Res.* **2022**, *46*, 10169–10181. [CrossRef]
14. Chen, Z.; Zhou, J.; Zhou, F.; Xu, S. State-of-charge estimation of lithium-ion batteries based on improved H infinity filter algorithm and its novel equalization method. *J. Clean. Prod.* **2021**, *290*, 125180. [CrossRef]
15. Tian, J.; Xiong, R.; Shen, W.; Sun, F. Electrode ageing estimation and open circuit voltage reconstruction for lithium ion batteries. *Energy Storage Mater.* **2021**, *37*, 283–295. [CrossRef]
16. Wang, Y.; Zhang, C.; Chen, Z. A method for state-of-charge estimation of Li-ion batteries based on multi-model switching strategy. *Appl. Energy* **2015**, *137*, 427–434. [CrossRef]
17. Tong, Z.; Miao, J.; Mao, J.; Wang, Z.; Lu, Y. Prediction of Li-ion battery capacity degradation considering polarization recovery with a hybrid ensemble learning mode. *Energy Storage Mater.* **2022**, *50*, 533–542. [CrossRef]
18. Hannan, M.A.; Lipu, M.S.H.; Hussain, A.; Mohamed, A. A review of lithium-ion battery state of charge estimation and management system in electric vehicle applications: Challenges and recommendations. *Renew. Sust. Energ. Rev.* **2017**, *78*, 834. [CrossRef]

19. Wang, X.; Sun, Q.; Kou, X.; Ma, W.; Zhang, H.; Liu, R. Noise immune state of charge estimation of li-ion battery via the extreme learning machine with mixture generalized maximum correntropy criterion. *Energy* **2022**, *239*, 122406. [CrossRef]
20. Tong, Z.; Yang, Q.; Tong, S.; Chen, X. Two-stage thermal-hydraulic optimization for Pillow Plate Heat Exchanger with recirculation zone parameterization. *Appl. Therm. Eng.* **2022**, *215*, 119033. [CrossRef]
21. Wang, S.; Fernandez, C.; Yu, C.; Fan, Y.; Cao, W.; Stroe, D. A novel charged state prediction method of the lithium ion battery packs based on the composite equivalent modeling and improved splice Kalman filtering algorithm. *J. Power Sources* **2020**, *471*, 228450. [CrossRef]
22. Snihir, I.; Rey, W.; Verbitskiy, E.; Belfadhel-Ayeb, A.; Notten, P.H.L. Battery open-circuit voltage estimation by a method of statistical analysis. *J. Power Sources* **2006**, *159*, 1484–1487. [CrossRef]
23. Ng, K.S.; Moo, C.S.; Chen, Y.P.; Hsieh, Y.C. Enhanced coulomb counting method for estimating state-of-charge and state-of-health of lithium-ion batteries. *Appl. Energy* **2009**, *86*, 1506–1511. [CrossRef]
24. Li, M. Li-ion dynamics and state of charge estimation. *Renew. Energy* **2017**, *100*, 44–52. [CrossRef]
25. Xiong, R.; Li, L.; Yu, Q.; Jin, Q.; Yang, R. A set membership theory based parameter and state of charge co-estimation method for all-climate batteries. *J. Clean. Prod.* **2020**, *249*, 119380. [CrossRef]
26. Wang, S.; Fernandez, C.; Zou, C.; Yu, C.; Li, X.; Pei, S.; Xie, W. Open circuit voltage and state of charge relationship functional optimization for the working state monitoring of the aerial lithium-ion battery pack. *J. Clean. Prod.* **2018**, *198*, 1090–1104. [CrossRef]
27. Tian, Y.; Lai, R.; Li, X.; Xiang, X.; Tian, J. A combined method for state-of-charge estimation for lithium-ion batteries using a long short-term memory network and an adaptive cubature Kalman filter. *Appl. Energy* **2020**, *265*, 114789. [CrossRef]
28. Zheng, L.; Zhang, L.; Zhu, J.; Wang, G.; Jiang, J. Co-estimation of state-of-charge, capacity and resistance for lithium-ion batteries based on a high-fidelity electrochemical model. *Appl. Energy* **2016**, *180*, 424–434. [CrossRef]
29. Zheng, Y.; Ouyang, M.; Han, X.; Lu, L.; Li, J. Investigating the error sources of the online state of charge estimation methods for lithium-ion batteries in electric vehicles. *J. Power Sources* **2018**, *377*, 161–188. [CrossRef]
30. Chen, Z.; Sun, H.; Dong, G.; Wei, J.; Wu, J. Particle filter-based state-of-charge estimation and remaining-dischargeable-time prediction method for lithium-ion batteries. *J. Power Sources* **2019**, *414*, 158–166. [CrossRef]
31. Peng, J.; Luo, J.; He, H.; Lu, B. An improved state of charge estimation method based on cubature Kalman filter for lithium-ion batteries. *Appl. Energy* **2019**, *253*, 113520. [CrossRef]
32. Ren, H.; Zhang, H.; Gao, Z.; Zhao, Y. A robust approach to state of charge assessment based on moving horizon optimal estimation considering battery system uncertainty and aging condition. *J. Clean. Prod.* **2020**, *270*, 122508. [CrossRef]
33. Xiong, B.; Zhao, J.; Wei, Z.; Skyllas-Kazacos, M. Extended Kalman filter method for state of charge estimation of vanadium redox flow battery using thermal-dependent electrical model. *J. Power Sources* **2014**, *262*, 50–61. [CrossRef]
34. Wang, L.; Ma, J.; Zhao, X.; Li, X.; Zhang, K.; Jiao, Z. Adaptive robust unscented Kalman filter-based state-of-charge estimation for lithium-ion batteries with multi-parameter updating. *Electrochim. Acta* **2022**, *426*, 140760. [CrossRef]
35. Yang, X.; Wang, S.; Xu, W.; Qiao, J.; Yu, C.; Takyi-Aninakwa, P.; Jin, S. A novel fuzzy adaptive cubature Kalman filtering method for the state of charge and state of energy co-estimation of lithium-ion batteries. *Electrochim. Acta* **2022**, *415*, 140241. [CrossRef]
36. Wang, Y.; Zhang, C.; Chen, Z. A method for state-of-charge estimation of LiFePO$_4$ batteries at dynamic currents and temperatures using particle filter. *J. Power Sources* **2015**, *279*, 306–311. [CrossRef]
37. Liu, Y.; Guo, B.; Zou, X.; Li, Y.; Shi, S. Machine learning assisted materials design and discovery for rechargeable batteries. *Energy Storage Mater.* **2020**, *31*, 434–450. [CrossRef]
38. Zhao, J.; Xu, W.; Kuang, Z.; Long, R.; Liu, Z.; Liu, W. Segmental material design in thermoelectric devices to boost heat-to-electricity performance. *Energ. Conv. Manag.* **2021**, *247*, 114754. [CrossRef]
39. Tian, J.; Xiong, R.; Lu, J.; Chen, C.; Shen, W. Battery state-of-charge estimation amid dynamic usage with physics-informed deep learning. *Energy Storage Mater.* **2022**, *50*, 718–729. [CrossRef]
40. Xu, J.; Zhen, A.; Cai, Z.; Wang, P.; Gao, K.; Jiang, D. State of health diagnosis and remaining useful life prediction of lithium-ion batteries based on multi-feature data and mechanism fusion. *IEEE Access* **2021**, *9*, 85431–85441. [CrossRef]
41. Tong, Z.; Miao, J.; Tong, S.; Lu, Y. Early prediction of remaining useful life for Lithium-ion batteries based on a hybrid machine learning method. *J. Clean. Prod.* **2021**, *317*, 128265. [CrossRef]
42. Hu, J.N.; Hu, J.J.; Lin, H.B.; Li, X.P.; Jiang, C.L.; Qiu, X.H.; Li, W.S. State-of-charge estimation for battery management system using optimized support vector machine for regression. *J. Power Sources* **2014**, *269*, 682–693. [CrossRef]
43. Chemali, E.; Kollmeyer, P.J.; Preindl, M.; Ahmed, R.; Emadi, A. Long short-term memory networks for accurate state-of-charge estimation of Li-ion batteries. *IEEE Trans. Ind. Electron.* **2018**, *65*, 6730–6739. [CrossRef]
44. Zhu, J.; Chen, N.; Peng, W. Estimation of bearing remaining useful life based on multiscale convolutional neural network. *IEEE Trans. Ind. Electron.* **2018**, *66*, 3208–3216. [CrossRef]
45. Lu, L.; Jin, P.; Pang, G.; Zhang, Z.; Karniadakis, G.E. Learning nonlinear operators via DeepONet based on the universal approximation theorem of operators. *Nat. Mach. Intell.* **2021**, *3*, 218–229. [CrossRef]
46. Cao, Y.; Ding, Y.; Jia, M.; Tian, R. A novel temporal convolutional network with residual self-attention mechanism for remaining useful life prediction of rolling bearings. *Reliab. Eng. Syst. Saf.* **2021**, *215*, 107813. [CrossRef]
47. Liu, Y.; Li, J.; Zhang, G.; Hua, B.; Xiong, N. State of charge estimation of lithium-ion batteries based on temporal convolutional network and transfer learning. *IEEE Access* **2021**, *9*, 34177–34187. [CrossRef]

48. Jagtap, A.D.; Kawaguchi, K.; Karniadakis, G.E. Adaptive activation functions accelerate convergence in deep and physics-informed neural networks. *J. Comput. Phys.* **2020**, *404*, 109136. [CrossRef]
49. Jagtap, A.D.; Kawaguchi, K.; Karniadakis, G.E. Locally adaptive activation functions with slope recovery for deep and physics-informed neural networks. *Proc. R. Soc. A-Math. Phys. Eng. Sci.* **2020**, *476*, 20200334. [CrossRef]
50. Jagtap, A.D.; Shin, Y.; Kawaguchi, K.; Karniadakis, G.E. Deep Kronecker neural networks: A general framework for neural networks with adaptive activation functions. *Neurocomputing* **2022**, *468*, 165–180. [CrossRef]
51. Lipu, M.S.H.; Hannan, M.A.; Hussain, A.; Ayob, A.; Saad, M.H.M.; Karim, T.F.; How, D.N.T. Data-driven state of charge estimation of lithium-ion batteries: Algorithms, implementation factors, limitations and future trends. *J. Clean. Prod.* **2020**, *277*, 124110. [CrossRef]
52. United States Advanced Battery Consortium. *Electric Vehicle Battery Test Procedures Manual*; USABC: Southfield, MI, USA, 1996.
53. Peace, G.S. *Taguchi Methods: A Hands-On Approach*; Addison Wesley Publishing Company: Boston, MA, USA, 1993.
54. Zhao, W.; Gao, Y.; Ji, T.; Wan, X.; Ye, F.; Bai, G. Deep temporal convolutional networks for short-term traffic flow forecasting. *IEEE Access* **2019**, *7*, 114496–114507. [CrossRef]

Article

# State-of-Health Prediction of Lithium-Ion Batteries Based on CNN-BiLSTM-AM

Yukai Tian [1], Jie Wen [1,*], Yanru Yang [1], Yuanhao Shi [1] and Jianchao Zeng [2]

[1] School of Electrical and Control Engineering, North University of China, Taiyuan 030051, China
[2] School of Data Science and Technology, North University of China, Taiyuan 030051, China
* Correspondence: wenjie@nuc.edu.cn

**Abstract:** State-of-Health (SOH) prediction of lithium-ion batteries is crucial in battery management systems. In order to guarantee the safe operation of lithium-ion batteries, a hybrid model based on convolutional neural network (CNN)-bidirectional long short-term memory (BiLSTM) and attention mechanism (AM) is developed to predict the SOH of lithium-ion batteries. By analyzing the charging and discharging process of batteries, the indirect health indicator (HI), which is highly correlated with capacity, is extracted in this paper. HI is taken as the input of CNN, and the convolution and pooling operations of CNN layers are used to extract the features of battery time series data. On this basis, a BiLSTM depth model is built in this paper to collect the data coming from CNN forward and reverse dependencies and further emphasize the correlation between the serial data by AM to obtain an accurate SOH estimate. Experimental results based on NASA PCoE lithium-ion battery data demonstrate that the proposed hybrid model outperforms other single models, with the root mean square error (RMSE) of SOH prediction results all less than 0.01, and can accurately predict the SOH of lithium-ion batteries.

**Keywords:** lithium-ion battery; state of health; convolutional neural network; bidirectional long- and short-term memory; attention mechanism

**Citation:** Tian, Y.; Wen, J.; Yang, Y.; Shi, Y.; Zeng, J. State-of-Health Prediction of Lithium-Ion Batteries Based on CNN-BiLSTM-AM. *Batteries* **2022**, *8*, 155. https://doi.org/10.3390/batteries8100155

Academic Editor: Joeri Van Mierlo

Received: 15 August 2022
Accepted: 29 September 2022
Published: 3 October 2022

**Publisher's Note:** MDPI stays neutral with regard to jurisdictional claims in published maps and institutional affiliations.

**Copyright:** © 2022 by the authors. Licensee MDPI, Basel, Switzerland. This article is an open access article distributed under the terms and conditions of the Creative Commons Attribution (CC BY) license (https://creativecommons.org/licenses/by/4.0/).

## 1. Introduction

The lithium-ion battery industry is an essential precursor to the world's advanced technology development [1]. With the characteristics of higher energy density, higher power density, higher conversion rate, longer cycle time, and less pollution, lithium-ion batteries are extensively applied in electric vehicles and various energy storage systems [2]. Lithium-ion batteries are now being applied more widely in mobile communications, transportation, electrical energy storage, new energy resources in storage, and aerospace [3]. Considering the widespread application of lithium-ion batteries, the secure operation of lithium-ion batteries must be given paramount importance, and the state of health (SOH) is the most critical parameter for evaluating the current state and performance of lithium-ion batteries [4]. Consequently, optimizing the design and management of lithium-ion batteries and accurately predicting the SOH of lithium-ion batteries is essential to assessing degradation and aging mechanisms.

The methods for predicting lithium-ion batteries can be classified into three groups: model-based methods [5–10], data-driven methods [11–13], and hybrid methods [14,15]. Model-based methods require extensive knowledge in the field of physical chemistry, understanding the reaction mechanisms internal to the battery, accurately describing the mathematical equations for the internal reactions, and building efficient simulation models, which can be difficult in practical applications. Data-driven approaches have been demonstrated to be one of the most significant methods for modelling battery degradation and assessing battery SOH due to their flexibility and the lack of need to build models of physical mechanisms such as artificial neural networks [16–18], relevance vector machine [19,20],

Gaussian process regression [21,22], etc. Deep learning has been gaining more and more attention when it comes to SOH prediction for lithium-ion batteries [23–25]. For instance, Chaoui et al. proposed a simple Recurrent Neural Network (RNN)-based approach to estimate the SOH of lithium-ion batteries using a dynamically driven RNN [26]. Chen et al. used constant current discharge time, the charge/discharge cycle number, and charge capacity to build a long- and short-term memory network (LSTM) model to enable SOH prediction for lithium-ion batteries [27]. Hybrid methods are combinations of two or several models that use the same or different types of methods for SOH prediction. Bezha et al. combined Convolutional Neural Network (CNN) and LSTM for battery SOH prediction, taking the current–voltage profile as input and SOH as output, and it is demonstrated that the proposed hybrid method has the advantage of providing accurate estimates in terms of SOH [28]. Qu et al. developed an LSTM model and combined LSTM with particle swarm optimization (PSO) and Attention Mechanism (AM) to achieve monitoring and prediction of SOH for lithium-ion batteries. The results demonstrate a high level of accuracy of the estimates [29].

In summary, the single data-driven model cannot take into account feature extraction from lithium-ion batteries and accurate SOH prediction. Therefore, highly accurate models and methods are needed to achieve the SOH prediction of lithium-ion batteries. It is considered that the correlation between input data can be enhanced by AM and has an application to many forecasting tasks, for instance, stock forecasting and electricity forecasting [30–34]. The CNN-BiLSTM-AM model is presented in this paper to predict the SOH of lithium-ion batteries, and integrates the merits of CNN and BiLSTM. In the proposed CNN-BiLSTM-AM model, the convolution and pooling operations of the CNN layer are utilized to extract the features of the battery time series data, while the BiLSTM depth model is used to collect the forward and reverse dependencies of the CNN incoming data, which further emphasizes the correlation of the time series data and capturing long-term dependencies. In addition, the time series data associated with SOH are weighted by AM resulting in an accurate SOH prediction for lithium-ion batteries.

The remainder of this paper is organized as follows. The used basic theoretical knowledge is described in Section 2. Dataset description and data preprocessing for lithium-ion batteries SOH prediction are presented in Section 3. In Section 4, the proposed CNN-BiLSTM-AM model is described, while Section 5 presents the experimental settings. Based on a NASA dataset, the CNN-BiLSTM-AM model is used to predict the SOH of lithium-ion batteries in Section 6. Lastly, we summarize and discuss briefly possible future directions for this paper in Section 7.

## 2. Preliminaries

This section will briefly introduce the basic theoretical knowledge of CNN, BiLSTM, and AM used in the CNN-BiLSTM-AM model presented in this paper.

### 2.1. CNN

CNN has an exceptional ability to capture features of spatial data, and it has played an instrumental role in the recent development of deep learning. It comprises three main types of layers: the convolutional layer, the pooling layer, and the fully connected layer. Its output is as follows:

$$y_t = \tanh(W_t X_t + b_t) \qquad (1)$$

where $X_t$ and $y_t$ are the input and output, respectively; tanh is the activation function; $W_t$ and $b_t$ represent the weight and bias, respectively.

A prototypical CNN unit is shown in Figure 1. The convolutional layer extracts local features by the size of the filter, and the features extracted by the pooling layer selection, which reduces the sophistication of the network parameters and structure, while the fully connected layer is a neural layer with an activation function that maps the relationship between input and output in a non-linear way. Since the convolution operations use the same set of weights, this reduces the number of parameters in CNN and solves the problem of overfitting. As a consequence, CNN is extensively used to predict the time series. Nevertheless, the increased sensitivity to sparse data is a drawback of CNN, and CNN is likely to be restricted to cases where good data are easily available. Thus, this paper collects the forward and backward dependencies of the data coming from CNN by building a BiLSTM depth model.

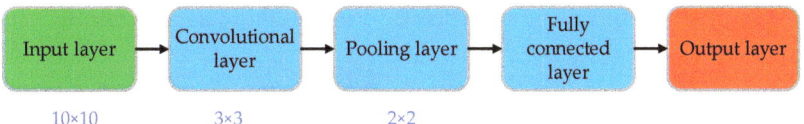

**Figure 1.** Prototypical CNN unit.

### 2.2. BiLSTM

Since lithium-ion battery data are collected during the charging and discharging cycles, they belong to the time series. RNN is applicable to processing time series, which help RNN process information in an orderly manner. To address the problem of exploding or disappearing gradients in RNN, the network structure of LSTM is proposed, substituting the state unit of classical RNN with the recurrent unit structure of LSTM.

A particular type of RNN model is LSTM. Compared to RNN, LSTM can better handle long-term continuous data. LSTM has three gate controls, i.e., the forget gate, the input gate, and the output gate, respectively. Figure 2 shows a prototype LSTM unit.

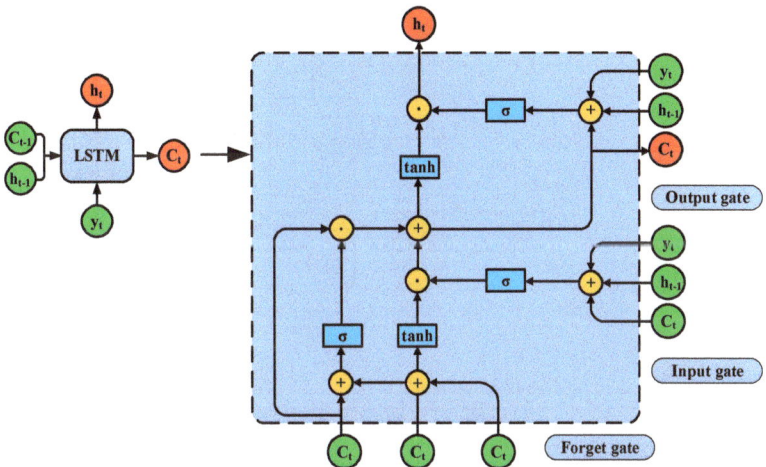

**Figure 2.** Recurrent unit structure of LSTM network.

The main role of the forgetting gate is to store information about when unit values ought to be forgotten, given by the following equation:

$$f_t = \sigma\left(W_f \cdot [y_t, h_{t-1}] + b_f\right) \tag{2}$$

where $y_t$ represents the input value; $h_{t-1}$ represents the output value; $W_f, b_f$ represents the weight and bias, respectively, and $\sigma$ is the activation function.

The input gate stores the values into a memory unit, which operates as follows:

$$i_t = \sigma(W_i \cdot [y_t, h_{t-1}] + b_i) \tag{3}$$

$$\widetilde{C}_t = \tanh(W_c \cdot [y_t, h_{t-1}] + b_c) \tag{4}$$

where $W_i$, $W_c$ are weights; tanh is the activation function, and $\tilde{C}_i$ is a one-dimensional matrix with values ranging from 0 to 1.

Combining the output of the forget and input gates, the information as $C_t$ is updated by:

$$C_t = f_t \cdot C_{t-1} + i_t \cdot \tilde{C}_t \tag{5}$$

The output gate controls the reading of the value of the memory unit.

$$o_t = \sigma(W_o \cdot [y_t, h_{t-1}] + b_o) \tag{6}$$

where $W_o$, $b_o$ are the weight and bias of the output gate, respectively, and $o_t$ is the output of the LSTM. The hidden state at time step $t$ is updated in the following manner:

$$h_t = o_t \cdot \tanh(C_t) \tag{7}$$

A BiLSTM consisting of a two-layer LSTM is shown in Figure 3, where the predicted system state is referred to a sequence of outputs from the bidirectional incoming LSTM layer. The predicted results are merged and assigned to the next LSTM layer; after the second LSTM layer, the ultimate prediction is determined by forward and backward propagation together. The BiLSTM depth model can better collect the bidirectional dependency of data from CNN than the LSTM model. Therefore, the model proposed in this paper is chosen as BiLSTM to enhance the accuracy of SOH prediction for lithium-ion batteries.

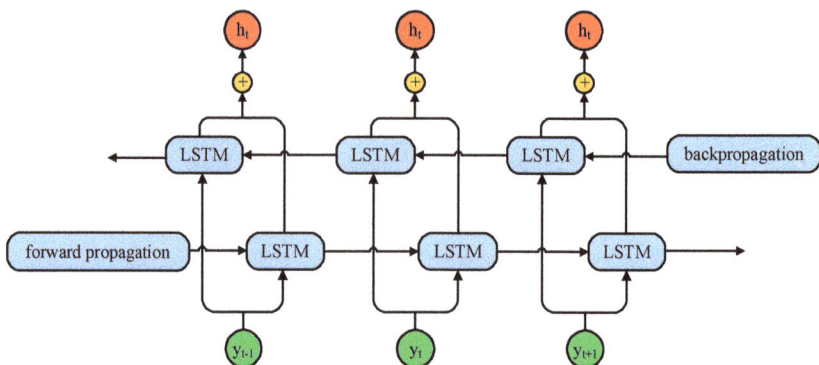

**Figure 3.** Recurrent unit structure of BiLSTM networks.

2.3. AM

Recently, scholars have applied AM to neural networks based on the concerns of the human brain AM and achieved excellent prediction results. In neural networks, each feature has a different impact on the outcome, but usually only a group of features determines the output. The main mechanism of AM is to follow the learning based on the attention level of the individual features in the series, and to integrate the features according to this attention level. To tackle the problem of attention distraction, this paper introduces AM, which sets weights for each feature according to its impact on the result. Figure 4 is a schematic diagram of the structure of the AM.

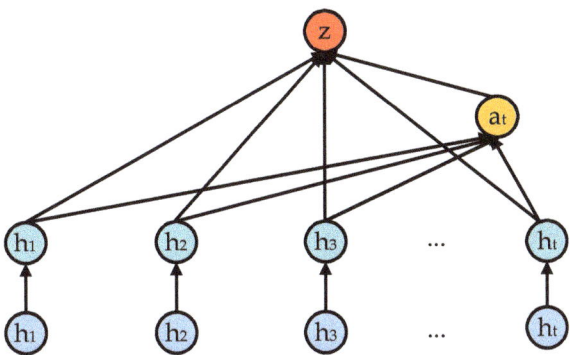

**Figure 4.** Structure diagram of AM.

The expressions for the calculation of AM are shown in (8)–(10).

$$u_t = \tanh(W_w h_t + b_w) \tag{8}$$

$$a_t = softmax\left(u_t^T, u_w\right) \tag{9}$$

$$z = \sum a_t h_t \tag{10}$$

where $W_w$ is first randomly initialized and then determined through the network training process; $u_w$ is weight; $b_w$ represents bias; $a_t$ represents the weight of each attribute; z represents the prediction result after weighted summation.

## 3. Dataset Description and Data Preprocessing

### 3.1. Dataset Description

To train and test the proposed model, B0005, B0006, and B0018 of the NASA PCoE battery dataset are selected in this paper [35]. Charge, discharge, and impedance operation of three lithium-ion batteries is carried out at room temperature (24 °C). To start with, there is a charging process in which each battery at a constant current of 1.5 A until it reaches a voltage of 4.2 V. This is followed by charging in a constant voltage mode until the charging current drops to 20 mA. Secondly, there is a discharging process in which each battery is discharged under a constant current of 2 A until each battery's voltage drops to 2.7 V, 2.5 V, and 2.5 V, respectively. Finally, there is an impedance process using electrochemical impedance spectra swept from 0.1 Hz to 5 kHz. The information of batteries B0005, B0006, and B0018 are displayed in Table 1.

**Table 1.** NASA dataset lithium-ion battery information.

| Battery | Number of Charges | Number of Discharges | Number of Impedances | Actual Life Expectancy |
|---|---|---|---|---|
| B0005 | 170 | 168 | 278 | 124 |
| B0006 | 170 | 168 | 278 | 108 |
| B0018 | 134 | 132 | 53 | 96 |

As can be observed in Figure 5, the three batteries gradually decrease in capacity through time and are accompanied by rebound in capacity during the degradation process. All three batteries are subjected to charge and discharge cycles, and once the batteries have dropped 30% of their nominal capacity, the end-of-life (EOL) point is reached, that is, from 2 Ah to 1.4 Ah.

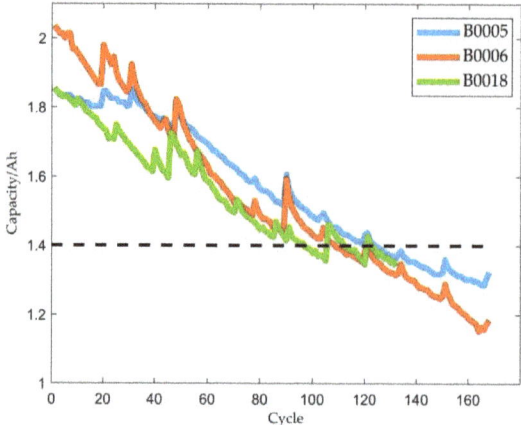

**Figure 5.** Capacity degradation curve of lithium-ion batteries.

*3.2. Data Preprocessing*

The most intuitive manifestation of battery degradation is the decay in capacity, which is predominantly related to the SOH of the battery. SOH is defined by capacity and given by the following equation [36]:

$$SOH = \frac{C_{actual}}{C_{nom}} \times 100\% \qquad (11)$$

where $C_{actual}$ and $C_{nom}$ represent the actual and nominal capacities, respectively.

Lithium-ion battery time series data for predicting SOH and the data preprocessing include data cleaning and normalization.

For better prediction accuracy and performance of deep learning models, experimental data need to be handled. To begin with, the data are cleaned by removing outliers and missing values, which are evaluated by moving averages or intermediate values, which cause the battery data to demonstrate periodic degradation characteristics. The battery specifications and data collection conditions have been summarized in Table 1. Data preprocessing assures that there are no erroneous values that could confound the model, as well as also being periodically averaged to avoid short-term fluctuations.

Data normalization is commonly applied in depth modelling algorithms where it is appropriate to improve the convergence of the model and the accuracy of the prediction. Normalization will be performed by the minimum–maximum method, where the data are scaled between 0 and 1. This is described by the following equation.

$$x_n = \frac{x - x_{min}}{x_{max} - x_{min}} \qquad (12)$$

where $x_n$ represents the processed data; $x$ represents the original data; $x_{max}, x_{min}$ represent the maximum and minimum values of the original data, respectively.

## 4. Methods

*4.1. CNN-BiLSTM-AM Model*

In this paper, the CNN-BiLSTM-AM model is proposed, which combines the characteristics and merits of CNN, BiLSTM, and AM to predict the SOH of lithium-ion batteries.

The model structure of CNN-BiLSTM-AM is presented in Figure 6, which primarily composed of the input layer, CNN layer, BiLSTM layer, AM layer, and output layer.

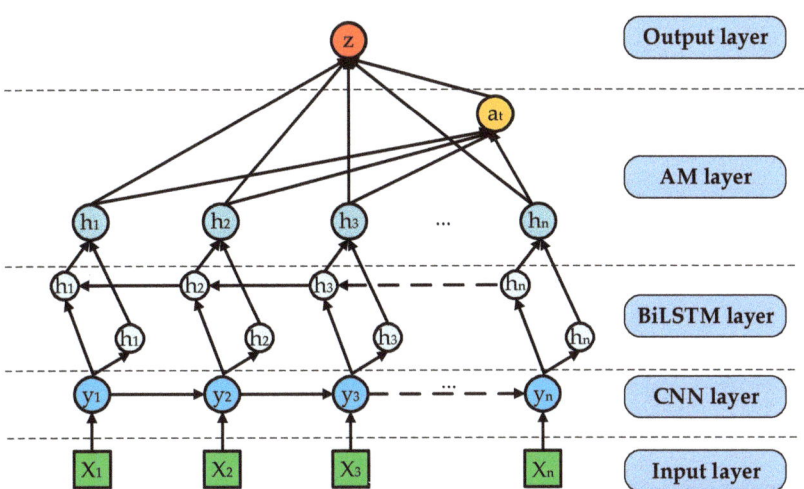

**Figure 6.** Structure diagram of CNN-BiLSTM-AM.

The model structure of CNN-BiLSTM-AM is described in detail as follows.

(1) Input layer: Firstly, HIs that can characterize the capacity during the charging and discharging process of the lithium-ion battery dataset are extracted, and those that are highly correlated with the capacity are selected as indirect HI; the indirect HI is preprocessed with the data, the processed dataset is divided, and the HI of the training set is used as the input of CNN.

(2) CNN layer: Including innovative concepts such as shared weight and local perceptual fields means CNN has unique benefits in processing battery datasets. In this paper, we use the convolution and pooling operations from the battery time series data to extract features.

(3) BiLSTM layer: The BiLSTM depth model is built, which is made up of forwarding and inverse LSTM. In comparison with the LSTM, the BiLSTM can extract time series in both directions and better collect the forward and reverse dependencies of the data coming from the CNN.

(4) AM layer: AM has been introduced into the hybrid model with the objective of enhancing the accuracy of the prediction model. AM assigns that each feature has a weight, further emphasizing the correlation between the data, which raises the accuracy of the prediction model.

(5) Output layer: The weighted summed prediction results from the AM layer are output and then the testing set is fed into the trained model for prediction to generate SOH prediction results.

*4.2. Prediction Procedure Based on CNN-BiLSTM-AM Model*

Flow chart of SOH prediction based on CNN-BiLSTM-AM model is in Figure 7, from which one can see that the prediction procedure of SOH is comprised of the following five steps.

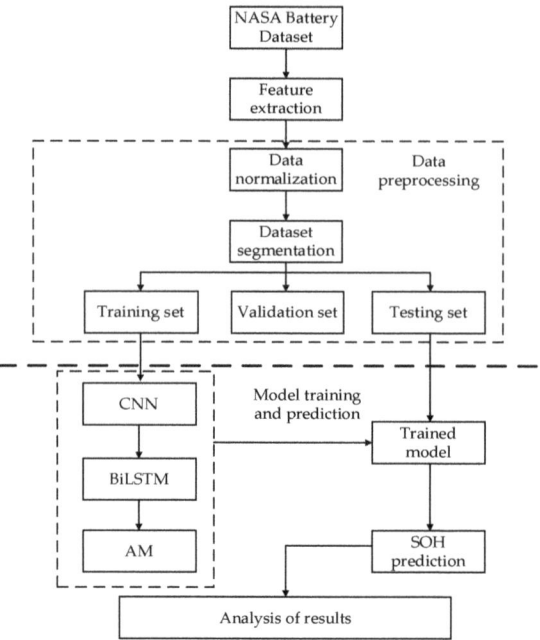

**Figure 7.** Flow chart of SOH prediction based on CNN-BiLSTM-AM model.

- Step 1. Feature Extraction: Initially, indirect HIs that reflect battery capacity degeneration are extracted by considering the charge and discharge voltage, current, and temperature curves of lithium-ion batteries, and those that are highly correlated with capacity are selected as indirect HIs.
- Step 2. Data Preprocessing: The extracted HI is normalized as well as followed by segmentation of the dataset.
- Step 3. Build the model and Train: The convolution and pooling operations of the CNN layer are used to extract features, and the bidirectional dependencies of the data coming from CNN are collected by the BiLSTM depth model, followed by the time series data related to SOH weighted by AM, so as to build the CNN-BiLSTM-AM model, and the hyperparameters of the model are determined using the grid search method during the model training process to derive the optimal model.
- Step 4. Model Prediction: The testing set is fed into the trained model for prediction, resulting in SOH prediction results.
- Step 5. Analysis of Results: Lastly, to further verify the validity of the proposed model, three comparison models of CNN, BiLSTM, and CNN-BiLSTM are designed, and the SOH prediction model is quantitatively evaluated using error evaluation metrics.

## 5. Experimental Settings

### 5.1. Experimental Equipment and Model Parameter Settings

The experimental environment utilized in this paper has been described as follows. Hardware environment: Intel (R) Core (TM) i5-6300HQ CPU @ 2.30 GHz 2.30 GHz, 8 GB RAM, 64-bit operating system. The model is implemented in Python 3.7 using Keras.

In this paper, three datasets are used for the SOH prediction of lithium-ion batteries. The datasets are divided into training set, validation set, and testing set. The training set serves to train the model, validation set to adjust the parameters, and testing set for assessing the performance of the model. The details of the division are presented in Table 2.

Table 2. Dataset segmentation.

| Dataset | Training Set | Validation Set | Testing Set |
|---|---|---|---|
| B0005 | (70%) 50<br>(80%) 74<br>(90%) 99 | 37<br>25<br>12 | (30%) 37<br>(20%) 25<br>(10%) 13 |
| B0006 | (70%) 43<br>(80%) 65<br>(90%) 86 | 32<br>22<br>11 | (30%) 33<br>(20%) 21<br>(10%) 11 |
| B0018 | (70%) 38<br>(80%) 58<br>(90%) 77 | 29<br>19<br>10 | (30%) 29<br>(20%) 19<br>(10%) 9 |

A number of appropriate hyperparameters need to be selected in the model to guarantee the accuracy of the model predictions. Model performance is frequently validated using grid search and cross-validation to obtain optimum parameters. The K values for cross-validation may impact how sensitive they are to changes in the training set, hence affecting the hyperparameter results. The grid search method will be applied to determine the hyperparameters. The parameters are set as in Table 3.

Table 3. Parameter setting.

| Parameter | Setting Value |
|---|---|
| Optimizer | Adam |
| Loss function | MSE |
| Activation function | RELU |
| Filter size | 10 |
| Batch size | 16 |
| Epochs | 1000 |
| Dropout rate | 0.200 |
| Learning rate | 0.001 |
| Number of neurons | 160 |

In CNN convolutional model, too few convolutional and pooling layers will contribute to the inadequate extraction of critical local information, while too many will result in a longer run time and extraction of too much invalid information. When the batch size is too large, the optimization of the loss function and gradient descent is detrimental and can cause large errors. If the batch size is too small, the time consumption of the neural network may be greatly extended, and eventually, the dropout layer is inserted to prevent overfitting and boost the training speed.

5.2. Performance Evaluation Indicators

Three error evaluation metrics are used in this paper, i.e., Root Mean Square Error (RMSE), Mean Absolute Error (MAE), and Mean Absolute Percentage Error (MAPE), which are presented to provide a quantitative assessment of the accuracy of the proposed SOH prediction model and defined as:

$$RMSE = \sqrt{\frac{1}{N}\sum_{i=1}^{N}(y_i - \hat{y}_i)^2} \qquad (13)$$

$$MAE = \frac{1}{N}\sum_{i=1}^{N}|y_i - \hat{y}_i| \qquad (14)$$

$$MAPE = \frac{1}{N}\sum_{i=1}^{N}\left|\frac{y_i - \hat{y}_i}{y_i}\right| \times 100\% \qquad (15)$$

where $y_i$ is the real SOH value and $\hat{y}_i$ denotes the SOH predicted value. Specifically, for indicators such as RMSE, MAE, and MAPE, the closer they approached zero, the more accurate the prediction.

## 6. Experiment and Result Analysis of SOH Prediction
### 6.1. HI Extraction

We have successfully extracted the capacity degradation data of a group of three lithium-ion batteries of the same type from the NASA PCoE public dataset. By analyzing charge and discharge characteristics of the B0005 battery in the NASA dataset, as an illustration, it is demonstrated in Figure 8.

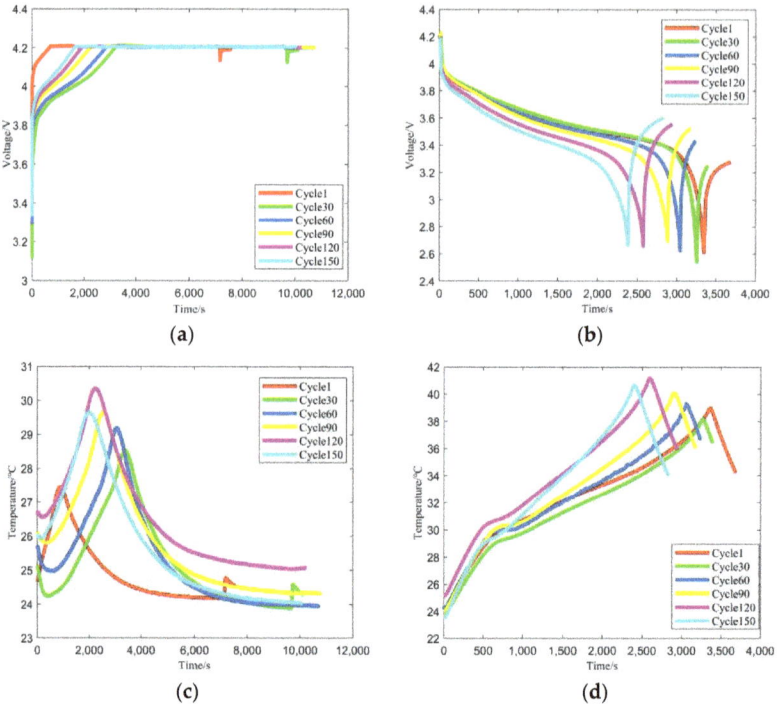

**Figure 8.** B0005 charge and discharge curves at different cycles. (**a**) Charge voltage curve; (**b**) Discharge voltage curve; (**c**) Charge temperature curve; (**d**) Discharge temperature curve.

In Figure 8a, the charging process is constant voltage charging, so we have analyzed the discharging process. From Figure 8b, the voltage variation of the B0005 battery at different cycles in the discharging process, it is clear that there is rich degradation information in the voltage between 3.4 V and 3.8 V. In order to avoid information redundancy, the Time Interval of an Equal Discharging Voltage Difference (TIEDVD) is chosen as the time difference corresponding to 3.4–3.8 V, and the 150th cycle is taken as an illustration.

On this basis, we have extracted four HIs: the time at which the discharge voltage reaches its minimum point, the maximum gradient of the voltage curve in the initial stage of the discharge process, the discharge power, and the time it takes for the temperature to reach the peak value during the discharge process. Table 4 presents the results of the Pearson correlation analysis of HI and capacity of B0005, from which it can be noticed that TIEDVD has the highest correlation with the capacity of lithium-ion battery.

Table 4. Correlation analysis of HI and capacity of B0005.

| HI | Pearson |
|---|---|
| TIEDVD | 0.9972 |
| Time at which the discharge voltage reaches its minimum point | 0.9928 |
| Maximum gradient of the voltage curve in the initial stage of the discharge process | 0.8050 |
| Discharge power | 0.9132 |
| Time it takes for the temperature to reach the peak value during the discharge process | 0.9886 |

Figure 9 presents a qualitative analysis of SOH and TIEDVD, from which one can see that TIEDVD follows the same trend as that of SOH, and the rebound part can be followed better. As a result, TIEDVD works as HI to predict the SOH of lithium-ion batteries.

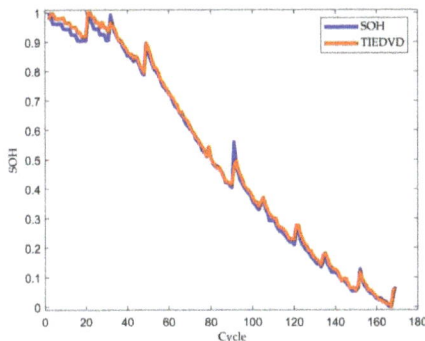

Figure 9. Qualitative analysis of SOH and TIEDVD.

### 6.2. Results and Analysis of SOH Prediction

To demonstrate the effectiveness of the proposed CNN-BiLSTM-AM model, three models, i.e., CNN model, BiLSTM model, CNN-BiLSTM model, are designed to make a comparison with the CNN-BiLSTM-AM model for different prediction Starting Points (SPs) in this subsection, and the CNN model, BiLSTM model, CNN-BiLSTM model, and CNN-BiLSTM-AM model are represented as M1, M2, M3, and M4, respectively.

Firstly, the B0005 battery is selected for prediction at SP = 70%, and the prediction results of the B0005 battery are presented in Figure 10. It is evident from Figure 10 that the M1 model for the B0005 battery shows the largest error, of which the M2 and M3 models are the next largest. With the best prediction performance of the M4 model for trend degradation and capacity rebound, and the predicted values being closest to the real SOH value, the validity of the CNN-BiLSTM-AM model introduced in this paper is verified.

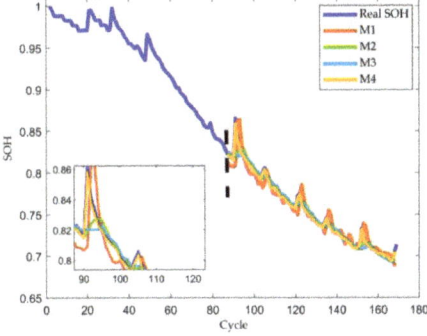

Figure 10. Different models of SOH prediction results for B0005.

A setup of SOH predictions for the same battery at different SPs has been carried out to demonstrate further the accuracy of the presented model (M4). The prediction results of SOH for the B0005 battery at different SPs are shown in Figure 11. The BiLSTM model in the M4 model collects the bidirectional dependence of the incoming CNN data, AM further emphasizes the correlation between the serial data by weighting, and the SP = 90% follows the same pattern of variation. In addition, the prediction results of M4 are similar to the real SOH values at different SPs, and more accurate predictions are obtained, especially for the capacity reversion. As the SPs increase, the prediction results become more and more accurate. Furthermore, Figure 11 presents that the error of the prediction results for SP = 70%, SP = 80%, and SP = 90% are smaller, which further proves the high accuracy of the CNN-BiLSTM-AM model.

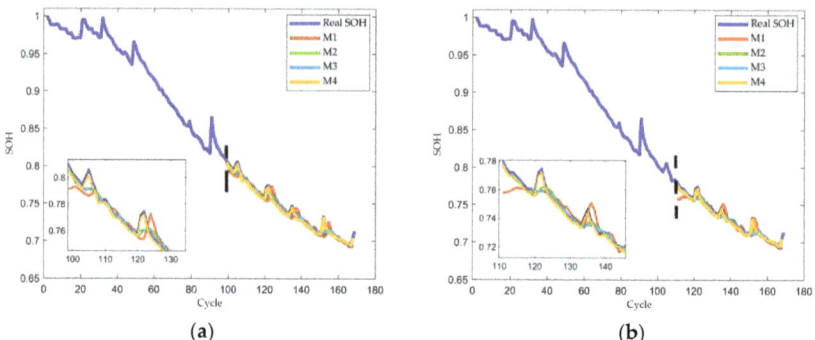

**Figure 11.** SOH prediction results for B0005 at different SPs. (**a**) SP = 80%; (**b**) SP = 90%.

RMSE, MAE, and MAPE have been adopted as error evaluation metrics to quantitatively assess the accuracy of the CNN-BiLSTM-AM prediction model. Table 5 illustrates the prediction results of SOH based on different models for batteries B0005, B0006, and B0018 with different SPs. In Table 5, the smallest RMSE, MAE, and MAPE are 0.00487, 0.00307, and 0.420%, respectively, corresponding to the M4 model for the B0005 battery at SP = 90%; the largest RMSE, MAE, and MAPE are 0.0153, 0.0121, and 1.60%, respectively, corresponding to the M1 model for the B0018 battery at SP = 90%. While the magnitude of the capacity rebound portion of the B0018 battery is larger, the M4 model is also accurately validated. The error of the M4 model for all three batteries are lower than the comparison models, with RMSE lower than 0.0120, MAE lower than 0.007, and MAPE lower than 0.9%. According to the results, the M4 model has the highest accuracy and the lowest prediction error.

**Table 5.** Prediction results of SOH based on different models for batteries B0005, B0006, and B0018 with different SPs.

| Battery | Prediction SP | Model | RMSE | MAE | MAPE (%) |
|---|---|---|---|---|---|
| B0005 | 87 (70%) | M1 | 0.00893 | 0.00565 | 0.738 |
| | | M2 | 0.00867 | 0.00540 | 0.702 |
| | | M3 | 0.00794 | 0.00531 | 0.692 |
| | | M4 | **0.00737** | **0.00382** | **0.496** |
| | 99 (80%) | M1 | 0.00863 | 0.00661 | 0.886 |
| | | M2 | 0.00620 | 0.00432 | 0.582 |
| | | M3 | 0.00595 | 0.00432 | 0.581 |
| | | M4 | **0.00481** | **0.00316** | **0.425** |
| | 111 (90%) | M1 | 0.00813 | 0.00633 | 0.825 |
| | | M2 | 0.00600 | 0.00409 | 0.558 |
| | | M3 | 0.00598 | 0.00408 | 0.558 |
| | | M4 | **0.00487** | **0.00307** | **0.420** |

**Table 5.** *Cont.*

| Battery | Prediction SP | Model | RMSE | MAE | MAPE (%) |
|---|---|---|---|---|---|
| B0006 | 75 (70%) | M1 | 0.0151 | 0.0101 | 1.49 |
| | | M2 | 0.0133 | 0.00838 | 1.24 |
| | | M3 | 0.0133 | 0.00846 | 1.25 |
| | | M4 | **0.0114** | **0.00485** | **0.709** |
| | 87 (80%) | M1 | 0.0134 | 0.00864 | 1.28 |
| | | M2 | 0.0129 | 0.00812 | 1.21 |
| | | M3 | **0.0117** | **0.00779** | **1.15** |
| | | M4 | 0.0105 | 0.00491 | 0.727 |
| | 97 (90%) | M1 | 0.00914 | 0.00708 | 1.11 |
| | | M2 | 0.00845 | 0.00617 | 0.963 |
| | | M3 | 0.00759 | 0.00567 | 0.883 |
| | | M4 | **0.00572** | **0.00361** | **0.561** |
| B0018 | 67 (70%) | M1 | 0.0145 | 0.0110 | 1.44 |
| | | M2 | 0.0139 | 0.00997 | 1.30 |
| | | M3 | 0.0138 | 0.0118 | 1.421 |
| | | M4 | **0.0108** | **0.00612** | **0.797** |
| | 77 (80%) | M1 | 0.0151 | 0.0119 | 1.56 |
| | | M2 | 0.0133 | 0.00981 | 1.29 |
| | | M3 | 0.0141 | 0.0109 | 1.44 |
| | | M4 | **0.0109** | **0.00623** | **0.619** |
| | 87 (90%) | M1 | 0.0153 | 0.0121 | 1.60 |
| | | M2 | 0.0141 | 0.0109 | 1.45 |
| | | M3 | 0.0150 | 0.0121 | 1.60 |
| | | M4 | **0.0112** | **0.00627** | **0.828** |

To investigate the effect of EOL on prediction error, we take B0005 as an example, where the EOLs are selected 70%, 75%, and 80% respectively, and the dataset has been repartitioned. The prediction results are presented in Figure 12 and Table 6, from which one can see that the largest RMSE, MAE, and MAPE are 0.00737, 0.00382, and 0.496%, respectively, corresponding to the M4 model for B0005 batteries with EOL = 70%; the smallest RMSE, MAE, and MAPE are 0.00621, 0.00432, and 0.343%, respectively, corresponding to the M4 model for EOL = 80%. Thus, the prediction error becomes higher when the EOL is set as the smaller percent of the health indicator, which is particularly evident in MAPE.

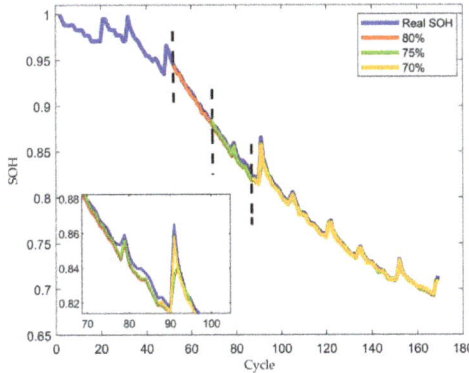

**Figure 12.** Different EOLs of SOH prediction results for B0005.

Table 6. Prediction results of SOH based on the M4 model for the B0005 battery with different EOLs.

| EOL | Prediction SP | RMSE | MAE | MAPE (%) |
|---|---|---|---|---|
| 70% | 87 (70%) | 0.00737 | 0.00382 | 0.496 |
|  | 99 (80%) | 0.00481 | 0.00316 | 0.425 |
|  | 111 (90%) | 0.00487 | 0.00307 | 0.420 |
| 75% | 69 (70%) | 0.00660 | 0.00465 | 0.364 |
|  | 79 (80%) | 0.00665 | 0.00454 | 0.355 |
|  | 89 (90%) | 0.00740 | 0.00383 | 0.499 |
| 80% | 52 (70%) | 0.00621 | 0.00432 | 0.343 |
|  | 60 (80%) | 0.00627 | 0.00435 | 0.343 |
|  | 67 (90%) | 0.00654 | 0.00457 | 0.357 |

## 7. Conclusions

Considering the security and dependability of lithium-ion batteries in real-world applications, a hybrid model based on CNN, BiLSTM, and AM is advanced to predict the SOH of lithium-ion batteries in this paper. In the CNN-BiLSTM-AM model, CNN is utilized to extract the features of the battery time series, BiLSTM to collect the bidirectional relationships, and AM to assign weights to achieve accurate SOH estimation of lithium-ion batteries. The prediction results of SOH by investigating different batteries with different SPs demonstrate that the proposed CNN-BiLSTM-AM model outperforms the CNN model, BiLSTM model, and CNN-BiLSTM model, with RMSE lower than 0.0120, MAE lower than 0.007, and MAPE lower than 0.9%, which can more accurately predict SOH for lithium-ion batteries.

The further works can be considered as follows: (1) SOH and RUL prediction for lithium-ion batteries considering practical applications; (2) Application of the latest machine learning techniques in the prediction of SOH and RUL for lithium-ion battery, such as various variants and improvements of the transformer; (3) Ways and ideas to combine the physical–chemical laws of lithium batteries and artificial intelligence technologies.

**Author Contributions:** Conceptualization, methodology, software, validation, formal analysis, investigation, writing—original draft preparation, visualization, Y.T.; conceptualization, methodology, validation, investigation, writing—review and editing, supervision, project administration, funding acquisition, J.W.; methodology, validation, investigation, Y.Y.; conceptualization, formal analysis, Y.S.; supervision, project administration, funding acquisition, J.Z. All authors have read and agreed to the published version of the manuscript.

**Funding:** This work was supported by the National Natural Science Foundation of China under Grant 72071183, and the Research Project Supported by Shanxi Scholarship Council of China under Grant 2020-114.

**Data Availability Statement:** Not applicable.

**Acknowledgments:** In particular, thanks to the NASA Ames Center of Excellence Diagnostic Center for providing the experimental data.

**Conflicts of Interest:** The authors declare no conflict of interest.

## References

1. Rajaeifar, M.A.; Ghadimi, P.; Raugei, M.; Wu, Y.; Heidrich, O. Challenges and recent developments in supply and value chains of electric vehicle batteries: A sustainability perspective. *Resour. Conserv. Recycl.* **2022**, *180*, 106144. [CrossRef]
2. Xiong, R.; Li, L.; Tian, J. Towards a smarter battery management system: A critical review on battery state of health monitoring methods. *J. Power Sources* **2018**, *405*, 18–29. [CrossRef]
3. Shen, M.; Gae, Q. A review on battery management system from the modeling efforts to its multiapplication and integration. *Int. J. Energy Res.* **2019**, *43*, 5042–5075. [CrossRef]
4. Shen, P.; Ouyang, M.; Lu, L.; Li, J.; Feng, X. The co-estimation of state of charge, state of health, and state of function for lithium-ion batteries in electric vehicles. *IEEE Trans. Veh. Technol.* **2017**, *67*, 92–103. [CrossRef]
5. Lai, X.; Gao, W. A comparative study of global optimization methods for parameter identification of different equivalent circuit models for Li-ion batteries. *Electrochim. Acta* **2019**, *295*, 1057–1066. [CrossRef]

6. Wang, Y.; Gao, G.; Li, X.; Chen, Z. A fractional-order model-based state estimation approach for lithium-ion battery and ultra-capacitor hybrid power source system considering load trajectory. *J. Power Sources* **2020**, *449*, 227543. [CrossRef]
7. Cheng, G.; Wang, X.; He, Y. Remaining useful life and state of health prediction for lithium batteries based on empirical mode decomposition and a long and short memory neural network. *Energy* **2021**, *232*, 121022. [CrossRef]
8. Rechkemmer, S.; Zang, X. Empirical Li-ion aging model derived from single particle model. *J. Energy Storage* **2019**, *21*, 773–786. [CrossRef]
9. Li, K.; Wang, Y.; Chen, Z. A comparative study of battery state-of-health estimation based on empirical mode decomposition and neural network. *J. Energy Storage* **2022**, *54*, 105333. [CrossRef]
10. Geng, Z.; Wang, S.; Lacey, M.J.; Brandell, D.; Thiringer, T. Bridging physics-based and equivalent circuit models for lithium-ion batteries. *Electrochim. Acta* **2021**, *372*, 137829. [CrossRef]
11. Xu, N.; Xie, Y.; Liu, Q.; Yue, F.; Zhao, D. A Data-Driven Approach to State of Health Estimation and Prediction for a Lithium-Ion Battery Pack of Electric Buses Based on Real-World Data. *Sensors* **2022**, *22*, 5762. [CrossRef] [PubMed]
12. Ng, M.F.; Zhao, J.; Yan, Q.; Conduit, G.J.; Seh, Z.W. Predicting the state of charge and health of batteries using data-driven machine learning. *Nat. Mach. Intell.* **2020**, *2*, 161–170. [CrossRef]
13. Alipour, M.; Tavallaey, S. Improved Battery Cycle Life Prediction Using a Hybrid Data-Driven Model Incorporating Linear Support Vector Regression and Gaussian. *ChemPhysChem* **2022**, *23*, e202100829. [CrossRef] [PubMed]
14. Li, X.; Wang, Z. Prognostic health condition for lithium battery using the partial incremental capacity and Gaussian process regression. *J. Power Sources* **2019**, *421*, 56–67. [CrossRef]
15. Li, Y.; Abdel-Monem, M. A quick on-line state of health estimation method for Li-ion battery with incremental capacity curves processed by Gaussian filter. *J. Power Sources* **2018**, *373*, 40–53. [CrossRef]
16. Ströbel, M.; Pross-Brakhage, J.; Kopp, M.; Birke, K.P. Impedance Based Temperature Estimation of Lithium Ion Cells Using Artificial Neural Networks. *Batteries* **2021**, *7*, 85. [CrossRef]
17. Wu, B.; Han, S.; Shin, K.G.; Lu, W. Application of artificial neural networks in design of lithium-ion batteries. *J. Power Sources* **2018**, *395*, 128–136. [CrossRef]
18. Dai, H.; Zhao, G.; Lin, M.; Wu, J.; Zheng, G. A novel estimation method for the state of health of lithium-ion battery using prior knowledge-based neural network and Markov chain. *IEEE Trans. Ind. Electron.* **2019**, *66*, 7706–7716. [CrossRef]
19. Yang, Y.; Wen, J.; Shi, Y.; Zeng, J. State of Health Prediction of Lithium-Ion Batteries Based on the Discharge Voltage and Temperature. *Electronics* **2021**, *10*, 1497. [CrossRef]
20. Feng, X.; Weng, C.; He, X.; Han, X.; Lu, L.; Ren, D.; Ouyang, M. Online state-of-health estimation for li-ion battery using partial charging segment based on support vector machine. *IEEE Trans. Veh. Technol.* **2019**, *68*, 8583–8592. [CrossRef]
21. Jia, J.; Liang, J.; Shi, Y.; Wen, J.; Pang, X.; Zeng, J. SOH and RUL Prediction of Lithium-Ion Batteries Based on Gaussian Process Regression with Indirect Health Indicators. *Energies* **2020**, *13*, 375. [CrossRef]
22. Yang, D.; Zhang, X.; Pan, R.; Wang, Y.; Chen, Z. A novel Gaussian process regression model for state-of-health estimation of lithium-ion battery using charging curve. *J. Power Sources* **2018**, *384*, 387–395. [CrossRef]
23. Kheirkhah-Rad, E.; Moeini-Aghtaie, M. A novel data-driven SOH prediction model for lithium-ion batteries. In Proceedings of the 2021 31st Australasian Universities Power Engineering Conference (AUPEC), Perth, Australia, 26–30 September 2021; pp. 1–6. [CrossRef]
24. Tan, Y.; Zhao, G. Transfer Learning With Long Short-Term Memory Network for State-of-Health Prediction of Lithium-Ion Batteries. *IEEE Trans. Ind. Electron.* **2020**, *67*, 8723–8731. [CrossRef]
25. Khumprom, P.; Yodo, N. Data-driven Prognostic Model of Li-ion Battery with Deep Learning Algorithm. In Proceedings of the 2019 Annual Reliability and Maintainability Symposium (RAMS), Orlando, FL, USA, 28–31 January 2019; pp. 1–6. [CrossRef]
26. Chaoui, H.; Ibe-Ekeocha, C.C. State of Charge and State of Health Estimation for Lithium Batteries Using Recurrent Neural Networks. *IEEE Trans. Veh. Technol.* **2017**, *66*, 8773–8783. [CrossRef]
27. Chen, Z.; Song, X.; Xiao, R.; Shen, J.; Xia, X. State of Health Estimation for Lithium-Ion Battery Based on Long Short Term Memory Networks. In Proceedings of the 2018 Joint International Conference on Energy, Ecology and Environment (ICEEE) and International Conference on Electric and Intelligent Vehicles (ICEIV), Melbourne, Australia, 21–25 November 2018; pp. 1–6. [CrossRef]
28. Bezha, M.; Nanahara, T.; Nagaoka, N. Development of Fast SoH Estimation of Li-Ion Battery Pack/Modules Using Multi Series-Parallel based ANN Structure. In Proceedings of the 2021 IEEE 12th Energy Conversion Congress & Exposition-Asia (ECCE-Asia), Singapore, 24–27 May 2021; pp. 1719–1724. [CrossRef]
29. Qu, J.T.; Liu, F.; Ma, Y.X.; Fan, J.M. A Neural-Network-Based Method for RUL Prediction and SOH Monitoring of Lithium-Ion Battery. *IEEE Access* **2019**, *7*, 87178–87191. [CrossRef]
30. Cai, W.; Wang, Y.; Ma, J.; Jin, Q. CAN: Effective cross features by global attention mechanism and neural network for ad click prediction. *Tsinghua Sci. Technol.* **2022**, *27*, 186–195. [CrossRef]
31. He, M.; Xue, X.; Zhang, X.; Zhou, C. A Bike-sharing Demand Predicting Model with Integrating Temporal Convolutional Network and Self-Attention. In Proceedings of the International Conference on Electronic Information Engineering and Computer Science (EIECS), Changchun, China, 23–26 September 2021; pp. 278–281. [CrossRef]
32. Feng, A.; Zhang, X.; Song, X. Unrestricted Attention May Not Be All You Need–Masked Attention Mechanism Focuses Better on Relevant Parts in Aspect-Based Sentiment Analysis. *IEEE Access* **2022**, *10*, 8518–8528. [CrossRef]

33. Liu, D.; Liu, J.; Luo, Y.; He, Q.; Lei, D. MGATMDA: Predicting microbe-disease associations via multi component graph attention network. In *IEEE/ACM Transactions on Computational Biology and Bioinformatics*; IEEE: Piscataway, NJ, USA, 2021; p. 29. [CrossRef]
34. Zhu, Y.; Zhao, C.; Guo, H.; Wang, J.; Zhao, X.; Lu, H. Attention CoupleNet: Fully Convolutional Attention Coupling Network for Object Detection. *IEEE Trans. Image Process.* **2018**, *28*, 113–126. [CrossRef]
35. Saha, B.; Goebel, K. Battery Data Set. Available online: http://ti.arc.nasa.gov/tech/dash/pcoe/prognostic-data-repository/ (accessed on 18 February 2020).
36. El-Dalahmeh, M.; Al-Greer, M.; El-Dalahmeh, M.; Short, M. Time-Frequency Image Analysis and Transfer Learning for Capacity Prediction of Lithium-Ion Batteries. *Energies* **2020**, *13*, 5447. [CrossRef]

*Perspective*

# Smart Battery Technology for Lifetime Improvement

Remus Teodorescu [1], Xin Sui [1,*], Søren B. Vilsen [2], Pallavi Bharadwaj [1], Abhijit Kulkarni [1] and Daniel-Ioan Stroe [1]

[1] AAU Energy, Aalborg University, 9220 Aalborg, Denmark
[2] Department of Mathematical Sciences, Aalborg University, 9220 Aalborg, Denmark
* Correspondence: xin@energy.aau.dk

**Abstract:** Applications of lithium-ion batteries are widespread, ranging from electric vehicles to energy storage systems. In spite of nearly meeting the target in terms of energy density and cost, enhanced safety, lifetime, and second-life applications, there remain challenges. As a result of the difference between the electric characteristics of the cells, the degradation process is accelerated for battery packs containing many cells. The development of new generation battery solutions for transportation and grid storage with improved performance is the goal of this paper, which introduces the novel concept of Smart Battery that brings together batteries with advanced power electronics and artificial intelligence (AI). The key feature is a bypass device attached to each cell that can insert relaxation time to individual cell operation with minimal effect on the load. An advanced AI-based performance optimizer is trained to recognize early signs of accelerated degradation modes and to decide upon the optimal insertion of relaxation time. The resulting pulsed current operation has been proven to extend lifetime by up to 80% in laboratory aging conditions. The Smart Battery unique architecture uses a digital twin to accelerate the training of performance optimizers and predict failures. The Smart Battery technology is a new technology currently at the proof-of-concept stage.

**Keywords:** Smart Battery; artificial intelligence; pulse current; lifetime extension; second-life applications

## 1. Introduction

Due to their high power density (≈1500 W/kg) and energy density (≈250 Wh/kg), high energy efficiency (>95%), and also relatively long cycle life measured in thousands of cycles, Li-ion batteries are the accepted solution for electronics, transportation, and grid storage. Battery packs are composed of a string of series and parallel connected cells to meet the power requirements of the applications. Cells cannot be manufactured with identical electrical characteristics, and these differences get amplified during operation, leading to a large unbalance in state of health (SOH) and premature lifetime termination. Therefore, it is essential to find a strategy that is able to operate with cells having unequal characteristics without limitation in performance. For achieving this goal, the concept of Smart Battery technology is proposed in this paper, using power electronics for the bypass device and artificial intelligence for performance optimization.

In the first stage, we explored several pulsed current charging strategies and their effect on battery lifetime. As shown in Figure 1, an up to 80% lifetime extension can be achieved by charging with a 2C-rate, 50% duty cycle, and 0.05 Hz current in comparison with a 1C-rate constant current with equivalent average charging power. It is generally believed that the pulsed current favorable effect is due to the slowing down of the degradation modes associated with high C rate operation over long periods, such as loss of active material. In addition, a Smart Battery lifetime prediction framework is proposed, as described in Section 4. Under the framework, the short-term (daily or weekly) state of health (SOH) can be accurately estimated based on a partial charging curve. With the help of the short-term SOH estimates, the established model is updated by transfer learning to track the long-term degradation behavior of batteries under varied working conditions.

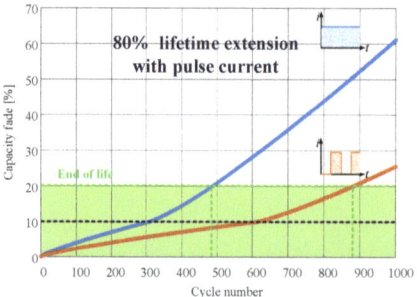

**Figure 1.** Lifetime extension using pulsed current charging compared with constant current [1].

The structure of the paper includes the hardware architecture and realization of the bypass device and cell controller with wireless communication, State of Temperature (SOT), SOH estimation and prediction, and Digital Twin as tools to improve safety and performance followed by a conceptual approach of the complex performance optimization problem definition. In the end, applications of the Smart Battery are identified.

## 2. The Hardware Architecture

The Smart Battery system aims to develop an integrated battery solution with increased safety, fault-tolerant operation, improved lifetime, and software reconfiguration for second life applications. The high-level architecture of a Smart Battery system is shown in Figure 2 and consists of a cell connected to a half-bridge circuit, which is controlled by a digital controller termed a slave CPU. The cell is connected to the battery string via the output ports of the half-bridge, as shown in Figure 2. The switching state of the half-bridge determines if the cell is inserted into the string or bypassed. Figure 3a,b show the state of the output terminals of the half-bridge when the cell is inserted or bypassed, respectively. By turning on the top device the cell will be inserted, and by turning on the bottom device, the cell will be bypassed. Note that the two devices are switched in complimentary PWM and can be switched at any frequency and duty ratio to realize pulsed charging or discharging of the cell. The slave controller provides the switching commands to the half-bridge, and monitors the cell voltage, current, and temperature using the appropriate sensors. Using the measurements, the slave estimates the state of charge (SOC) and communicates the measurements to the master controller shown in Figure 2. Note that the master controller (shown as master CPU/GPU in Figure 2) uses AI-based algorithms to estimate the state of health (SOH) and remaining useful life (RUL) for the cells and communicates the same information to respective cells. The master controller performs the functions of SOC and SOH balancing and lifetime control. The balancing process is done by bypassing one cell at a time and thus not affecting the load current. In contrast to other active balancing methods, this balancing method does not use bidirectional DC–DC converters. It has better efficiency due to the absence of additional inductors/capacitors used in active balancing methods. The proposed bypass device only needs to order the cells according to their SOC and SOH states and then decides which cell should be bypassed. The method for balancing control is simple and effective. This provides a fault-tolerant operation mode, which can improve the safety and reliability at the system level.

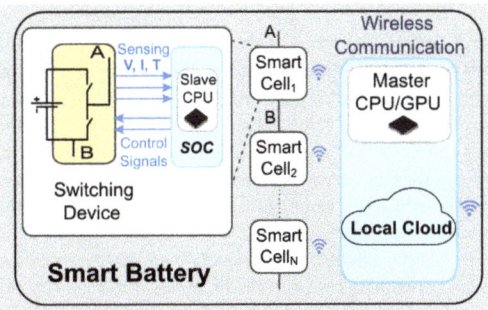

**Figure 2.** High-level architecture showing Smart Battery with slave controllers and a master controller.

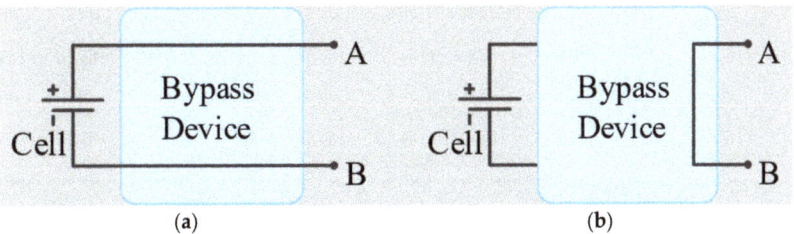

**Figure 3.** The operation mode of the switching device: (**a**) inserted; (**b**) bypassed.

*2.1. Smart Battery Cell—Hardware Implementation Approach*

Individual cells are integrated with a half-bridge circuit to provide the bypass capability as described above. The overall control electronics consist of the following subgroups:

1. MOSFETs: Low on-resistance MOSFETs are used in the half-bridge across the cell. It is important to have low-on resistance to limit the power loss in the MOSFETs, which acts as an undesirable load on the batteries. Note that MOSFETs with sub-milliohm on-resistances ($r_{ds,on}$) are available and they introduce negligible losses. It is also possible to parallel additional MOSFETs to reduce the resistance further and to provide redundancy for improving the reliability. Table 1 shows some of the commercially available MOSFETs with sub-milliohm $r_{ds,on}$ and conduction loss at 50 A. It would be good to use automotive-certified MOSFETs (e.g., AUIRF8739L2TR [2] in Table 1) such as electric vehicles (EVs) is one of the major applications for the Smart Battery.
2. Sensors: To monitor the cell voltage, current, and temperature, appropriate sensors are used. Additional electronic circuits are essential to interface the sensors with the slave controller. The sensors can be interfaced with the analog-to-digital converter (ADC) channels of the controller or to the appropriate digital communication channels depending on the output format.
3. Voltage regulators: Switching voltage regulators are necessary to convert the battery voltage to the required regulated DC voltage to supply the control electronics and to the gate drivers of the half-bridge circuit. Note that linear regulators cannot be used because typically they can only step down the cell voltage and they have poor efficiency.
4. Gate driver: A smart gate driver is necessary to implement the insert/bypass functionality of the Smart Battery. This gate driver receives the commands from the slave controller, which in turn obtains the commands from the master controller wirelessly. The gate driver will also prevent any shoot through of the DC voltage, hence avoiding any short circuit.

5. Slave controller: The slave controller performs the computation of SOC, provides commands to the gate driver, implements protection algorithms, and communicates with the master wirelessly. For wireless communication, protocols such as wifi, Bluetooth Low Energy (BLE), and Zigbee are possible. The range required for the wireless communication for the Smart Battery is in the order of a few meters considering the application area of electric vehicles, wherein the Smart Battery will be tightly packed and the master controller will be at close proximity within the vehicle. Considering these points, BLE communication can be one of the options. However, since the Smart Battery architecture for EV involves a large number of slaves (>100), custom wireless protocols such as IEEE TSCH may be a good compromise between performance and power consumption. Texas Instruments offers a number of wireless controllers suitable for BMS applications. One popular series is the Simplelink controller CC26 × 2 [3,4].

Table 1. Commercially available MOSFETs with very low on-resistance for minimizing the power loss in a Smart Battery.

| MOSFET | $R_{ds,on}$ (mΩ) | Rated Current (A) | Power Loss at 50 A (W) |
| --- | --- | --- | --- |
| IPT004N03L | 0.4 | 300 | 1 |
| IST006N04NM6 | 0.6 | 475 | 1.5 |
| IRL40SC209 | 0.8 | 478 | 2 |
| AUIRF8739L2TR | 0.35 | 545 | 0.875 |

*2.2. Layout Design for Low Electromagnetic Interference (EMI)*

Figure 4 shows the detailed components of the single cell in the Smart Battery architecture and its hardware components described above. The electronic circuits in Figure 4 contain EMI sources as well as sensitive electronics whose performance may be impacted by the EMI. For example, the switching regulators to supply regulated voltage produce both conducted and radiated noise. As these converters are switched at a high frequency in the order of a few MHz, the radiated noise may interfere with the BLE communication. Thus, it is important to protect the communication and the sensing circuits from the noise generated by the switching regulators. Note that the switching by the MOSFETs of the half-bridge may not result in any appreciable EMI because these MOSFETs are switched at a very low frequency in the order of a few hertz or less. As a result, this switching does not radiate any significant energy at the BLE frequencies of interest and can be ignored as the source of conducted or radiated EMI.

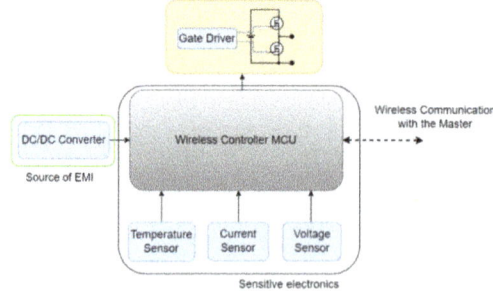

**Figure 4.** Hardware components in a Smart Battery cell.

In order to minimize the impact of EMI and ensure high fidelity communication, a multi-layer routing is followed for the printed circuit board (PCB) to provide good ground planes and to minimize the loop areas that can cause unwanted radiation [5]. The components of the switching regulators are physically placed in a predefined area and they are covered by an EMI shield. This ensures that the BLE communication is not impacted by the radiations from the switching regulators. A green border shown around the DC–DC converter in Figure 4 illustrates the EMI shield across it. As the Smart Battery cells are in close physical proximity in applications such as EV battery packs, a proper layout design prevents noise from one PCB from impacting the communication with the neighboring PCBs and with the master controller.

*2.3. Hardware Challenges and Design for High Current*

For cells with high Ah capacity such as 50 Ah or more, the design of the slave board is to be done to minimize the losses and impact of parasitics such as stray inductances. The current in such cells can be hundreds of amperes for any operation beyond 2C. Thus, the hardware design needs to provide a low resistance path for the current and low conduction loss in the MOSFETs. A conceptual diagram illustrating the slave board design for a prismatic cell is shown in Figure 5. A combination of copper bus bars and copper in-lays is used, as shown in Figure 5 (shown in orange) to provide a low resistance path for the high current. The half-bridge is shown with two MOSFETs, Q1 and Q2. Note that a parallel combination of multiple MOSFETs may be necessary to minimize the losses. The bus bar design should also ensure that parasitic inductances are very low as they can cause large voltage spikes on the MOSFETs during insert/bypass operations.

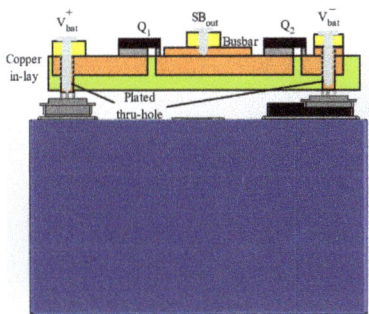

**Figure 5.** Smart Battery hardware design concept for high Ah prismatic cells.

## 3. SOT Estimation

Li-ion batteries, due to their high energy/power density, long cycle life, and high efficiency, have been widely used in electric vehicles, portable electronics, and smart grid systems. However, thermal-related technological bottlenecks, including thermal runaway [6], extreme fast charging (XFC) [7,8], reduced performance in cold climates [9,10], and accelerated aging at high temperatures [11,12], still hinder the large-scale application of Li-ion batteries. Such bottlenecks stem from the complex effect of temperature on the safety, performance, and lifespan of Li-ion batteries. For instance, when battery temperature exceeds the threshold under extreme situations, thermal runaway might be triggered and accompanied by safety problems such as smoke, fire, and explosion [6]. In cold climates, the performance of Li-ion batteries is severely reduced due to slow electrochemical reactions inside the cell [10,13], and thus the available energy and power of Li-ion batteries decline dramatically [9,10]. Additionally, XFC at relatively lower temperatures is likely to trigger lithium plating, which leads to accelerated battery degradation [11,12]. At elevated temperatures, side reactions such as the growth of a solid electrolyte interface become significant, giving rise to the consumption of cyclable lithium and accelerated battery capacity fade [8,12].

Battery management systems (BMS) are indispensable for managing the charging/discharging patterns and regulating battery temperature in a smart way, where temperature monitoring serves as the basis of the BMS. Typically, battery temperature can be monitored by temperature sensors placed on the battery surface. However, in real-life battery packs, it is impractical to place temperature sensors on the surface of each battery cell due to cost and complexity considerations. Furthermore, the surface-mounted sensors cannot track the rapid variation of internal temperatures because of heat transfer delay from the battery core to the surface caused by the thermal mass of the battery, especially at high charging/discharging rates. Hence, it is of great significance to estimate the battery temperature in a battery pack, and accurate SOT estimation benefits battery management in several ways.

From the perspective of battery safety, accurate monitoring of internal temperature helps keep the battery within the safety threshold and gives an early warning of potential hazards that could trigger thermal runaway. In particular, nowadays, Li-ion batteries are designed to have large capacities and high power/energy densities, which could inevitably enhance the risk of thermal hazards. From the perspective of fast charging, SOT estimation helps regulate battery temperature actively to a charging favorable temperature range so that XFC can be achieved and the lifetime of the battery can also be extended [14]. From the perspective of battery health management, knowing battery SOT makes it possible to develop a temperature-independent SOH estimation by decoupling the temperature effect during the extraction of health indicators, which gives rise to a more accurate and robust SOH estimation. This will further allow for accurate lifetime prediction and improvement of the operation of the Smart Battery. All of these features of the Smart Battery enabled by accurate SOT estimation are illustrated in Figure 6.

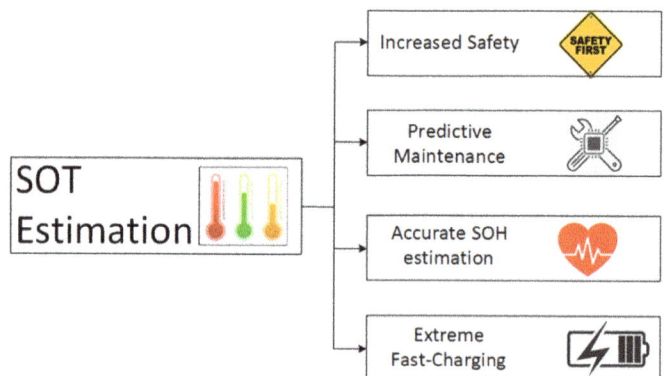

**Figure 6.** Smart Battery functionalities enabled by SOT estimation.

Existing methods for SOT estimation can be classified into three categories: impedance-based estimation, thermal model-based estimation, and data-driven estimation. Impedance-based estimation exploits the relationship between battery temperature and impedance parameters such as phase, real part, and imaginary part, to estimate SOT according to the measured impedance [15]. By modeling the heat generation and heat transfer models inside the cell, thermal model-based methods realize the internal temperature estimation based on battery current, voltage, and possibly a surface-mounted sensor [16]. Data-driven approaches ignore the thermal dynamics of the cell and explore the data patterns of battery temperature evolution to realize highly accurate estimation [17]. However, these three methods have limitations. Impedance-based estimation can only provide information about the average temperature of the cell but neglect the temperature distribution inside the cell. Therefore, the maximum internal temperature is likely to be underestimated, especially for large-format cells with high energy/power density or cells operating at high rates (e.g., XFC). As for thermal model-based estimation, it is a great challenge to balance the model complexity and accuracy. In addition, parameterization is sometimes complex due to many required model parameters. For data-driven approaches, obtaining a considerable training dataset is sometimes technically challenging and unattainable. For instance, the temperature value at random points inside the cell cannot be measured. Generalization is another problem for many data-driven approaches since the training dataset cannot cover all of the operation scenarios.

To address the challenges existing SOT estimation methods face, there is a growing trend to combine model-based approaches with data-driven methods to realize accurate and robust estimation [18,19]. There are many ways to combine physics-based models and machine learning models, as discussed in [20]. A competitive candidate, which will be used in the framework of the Smart Battery, is the physics-informed neural network (PINN). PINN can rapidly solve the underlying nonlinear heat transfer partial differential equation (PDE) with small amounts of data and provide insights into battery internal temperature distribution. Typically, the temperature data used for neural network training is limited to the surface temperature and possibly the core temperature measured through sensor intrusion [21,22], making it difficult for conventional neural networks to estimate the temperature distribution between the core and the surface. PINN can overcome the limitations traditional neural networks face and mimic the data patterns governed by the heat transfer PDE so that the temperature distribution inside the cell can be estimated. The framework of a PINN for estimating the battery temperature distribution is shown in Figure 7, where a cylindrical cell is used as an example. When collecting the training data, the current, voltage, surface, and core temperature can be measured (by inserting a sensor into the battery core). The measured data are treated as training data for the deep neural network, and the loss is calculated based on the predicted temperature and the measured temperature. Additionally, the temperature at the core, the surface, and any point inside the domain should follow the heat transfer PDE and its initial and boundary conditions. The differentials of temperature with respect to time and location can be calculated through automatic differentiation so that the physics loss can be obtained accordingly based on heat transfer PDE. In PINN, the loss function consists of the loss of training data and the loss of physics. By minimizing the total loss, the weights and biases of the neural networks can be adjusted, and the unknown coefficient in the PDE can be identified. For a trained PINN, the predictions can have high accuracy while also following the heat transfer law so that it can be used to estimate internal temperature distribution under other operating conditions.

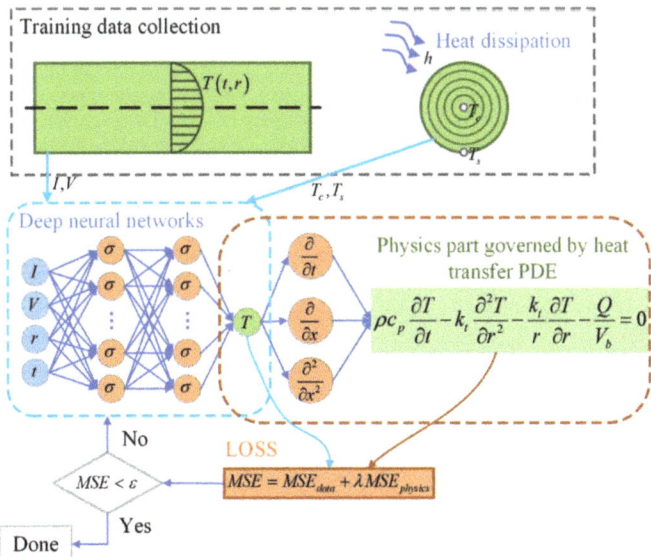

**Figure 7.** Framework of temperature distribution estimation using PINN: Measured temperature data at the battery surface and core are used for training the data-driven model (e.g., based on deep

neural networks); then a physics part is added as a regularization term to further train the deep neural network so that the estimated temperature not only follows the patterns of data in the training set but also obeys the spatiotemporal physical law in the physics part. The network is trained by minimizing both the loss of the measured data and the loss of the physics part after 10,000 epochs or more, where an appropriate $\lambda$ should be selected to adjust the relative importance between data loss and physics loss. The training process can adjust the weights and bias in neural networks, as well as identify unknown parameters in PDE simultaneously (where $T$ represents the temperature, $t$ represents the time, $I$ represents the current, $V$ represents the voltage, $r$ represents the space distribution, $Q$ represents the heat generation, $c_p$ represents the specific heat capacity, $V_b$ represents the volume of the battery, MSE represents the mean squared error, $\text{MSE}_{\text{data}}$ represents the mean squared error of the temperature estimation based on the measured data, $\text{MSE}_{\text{physics}}$ represents the mean squared error of the temperature estimation in physics part, $k_t$ represents the thermal conductivity of the cell, and $\lambda$ represents coefficient between data loss and physics loss).

Accurate SOT estimation using PINN represents the first step toward the goal of realizing long-term (e.g., 10 min ahead) temperature prediction, which will lead to optimal operation and reduce safety concerns of the Smart Battery.

## 4. SOH Estimation and Lifetime Prediction

The degradation of the battery is unavoidable, and it is caused by complex aging mechanisms that are happening in parallel inside the battery. At the macroscopic scale, the degradation of the battery is manifested as capacity fade and power fade [23] that are therefore often used as indicators of a battery's SOH. Using these two measures of degradation, a battery is considered at the end-of-life (EOL) when its capacity reaches 70–80% of the initial capacity. Given an EOL criterion, the RUL of the battery can be defined as the time (or the number of cycles) until the battery reaches its EOL [24]. It follows that to lower the cost of Li-ion batteries, both environmentally and economically, it is imperative to control their RUL [25]. Accurately predicting the RUL of the battery will also help reduce the cost through predictive maintenance, reduce the risk of failure guaranteeing safer operation, and improve the reliability of the system [25]. However, the degradation of batteries begins the moment they exit the production line, resulting in reduced lifetime. Additionally, Li-ion batteries undergo a wide range of aging conditions during real-world operations, from calendar aging (idling) to cycling aging (charging or discharging), which is non-deterministic and difficult to predict. These uncertainties create a bottleneck in the large-scale acceptance and deployment of Li-ion batteries in critical applications, such as transportation.

*4.1. SOH Estimation*

After a decade of research on battery SOH estimation, SOH estimation methods are slowly becoming mature [26–28]. SOH estimation methods typically fall into one of three categories: (1) Empirical methods, (2) physics-based models, and (3) AI data-driven methods. While empirical methods such as directly measuring the charge throughput or indirectly analyzing the incremental capacity have been used to quantify the SOH mechanisms, their stability severely limits their use in real-life applications. The physics-based models are designed to estimate the SOH through state-space models typically built using electrochemical models, or equivalent electrical circuit models [29]. The physics-based models use filters to effectively update the dynamic characteristics of the system, but are entirely dependent on the accuracy of the underlying physics-based model, which introduces an unavoidable and cumbersome parameter identification process requiring extensive laboratory testing. Lastly, recent years have seen the rise of more data-driven methods through statistics, machine learning, and artificial intelligence methods. These methods have the ability to effectively learn any non-linear regression problem, given enough of the right data. Among the most popular are methods such as support vector machine, artificial neural network, deep learning (DL), and random forest [30].

The battery degradation process is accompanied by a series of side reactions involving various parts of the battery, such as the anode, cathode, electrolyte, and electrode–electrolyte interface. As a result, the battery will exhibit different aging behaviors as the operating conditions change. This leads to the biggest challenge of data-driven methods for SOH estimation: features extracted under laboratory conditions might be invalid in real-life applications. There are three ways to account for this discrepancy: (1) To define and extract robust features, (2) to adapt models from the laboratory to the field by transfer learning, and (3) to use automatic feature extraction through DL. A viable method for creating and extracting robust features is the fuzzy entropy method as proposed in [31]. It has been shown that fuzzy entropy-based features are effective in both SOH estimation and SOH prediction. Additionally, it has been proven to have strong robustness against parameter selection, data size, working conditions, and noise [31]. Moreover, noise suppression methods were used to pre-process the SOH data, improving not just the accuracy but also the speed of the fuzzy entropy-based feature extraction [32,33]. An alternative to creating robust features is to account for the change in domain by transferring the model [34]. There are two approaches to adapting models from one domain to another—during the training of the model the discrepancy between the features in the two domains is accounted for, or the model is trained in the original domain and then re-trained in the new domain. The second approach will give better results but requires knowledge of the SOH in both domains unlike the first approach [35]. Lastly, the feature failure problem may be almost entirely avoided using DL. Deep neural networks have the ability to extract global features from raw multi-dimensional data. However, due to the latent nature of SOH, obtaining the amount of SOH information required to train such a neural network is usually impossible in real-life applications. In order to improve the estimation accuracy on small data sizes, a bagging-based ensemble method was proposed in [36]. Bagging creates augmented samples by resampling from the original dataset, and a series of ELMs are trained based on these samples. The bagging ELM method has many of the upsides of DL, such as the automatic feature extraction, while requiring much less data to train and perform well when estimating SOH. In addition to using fuzzy entropy-based features and the ensemble ELM method, the Smart Battery framework aims to increase the amount of useful information extracted from a single partial charge of the battery by data augmentation. The augmentation will allow for the extraction of not only the charging voltage, but also every partial charging voltage sequence found within any charge (no matter how large). The general framework for data cleaning, augmentation of partial charges, feature extraction, and SOH estimation is outlined in the top panel of Figure 8.

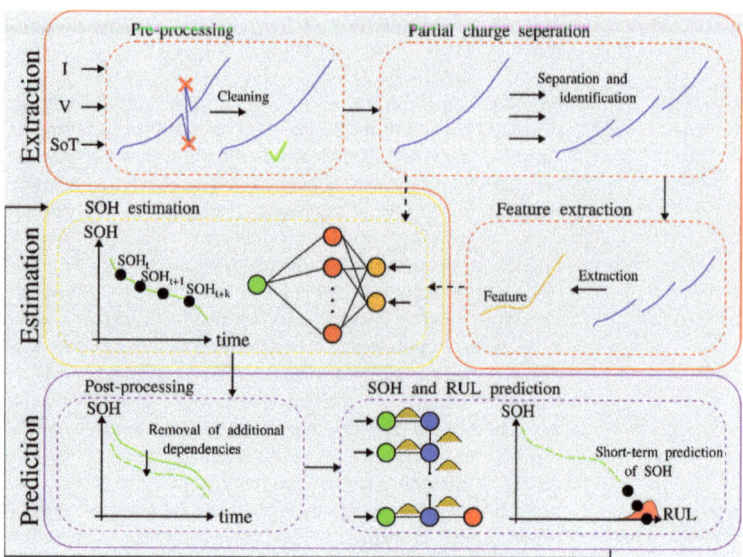

**Figure 8.** A flowchart of the Smart Battery SOH and RUL prediction framework. In order to stabilize the predictions of the SOH, the time dependence of the system is moved from the SOH to the features. To predict the SOH, the features are predicted forward in time, and a SOH estimation model is then used to predict the SOH.

## 4.2. SOH and Lifetime Prediction

The aim of any RUL algorithm is to predict the time to EOL of a battery. However, before the EOL can be predicted, it is necessary to predict the SOH; given a mission profile and a short-term SOH prediction method, the long-term behavior of the SOH can be predicted to the EOL. SOH and RUL prediction methods are usually divided into physics-based and data-driven AI-based methods. In the physics-based lifetime models, while the non-linear and time-varying characteristics of the electrochemical system can be explained, the parameters of these models are very difficult to identify since they rely on destructive testing methodologies. Consequently, the development of physics-based models is time and resource-demanding, and thus not necessarily a viable option for use in real-time prediction. These methods are more suitable to study the aging mechanisms of the battery, provide a theoretical basis for data-driven methods, and make suggestions on battery design [36]. The data-driven AI-based methods used for lifetime modeling and prediction are unlike SOH estimation, and are usually more probabilistic in nature. Among the most common methods are Gaussian process regression and dynamic Bayesian networks. Their main advantage is that they do not need access to the mechanical and electrochemical behavior of the battery [37]. The disadvantage is the need to specify the structure of the probabilistic structure of these models. Therefore, recent years have seen an increase in the use of DL methods. A DL lifetime model can be established based on the collected data and continually updated using gradient optimization [38]. That is, the relationship between the features (i.e., the health indicators such as voltage, current, and temperature) and the cycle and calendar life of the battery cell can be established. The main disadvantages of DL methods are their computational cost, and that they are not probabilistic by nature, making RUL uncertainty prediction difficult.

However, as cloud computation becomes cheaper and more readily available, many DNN algorithms have shown promise, such as deep neural networks [39], convolutional neural networks [40], and recurrent neural networks (RNNs) [41]. RNN will be a suitable algorithm for RUL prediction because of its intrinsic modeling of time-dependent parameters. Furthermore, to accommodate the need for probabilistic predictions, the Smart Battery framework will attempt to combine approximate Bayesian methods, such as approximate Bayesian computation [42,43], Bayesian synthetic likelihood [44], or variational inference with RNNs.

The biggest challenge with the SOH and RUL prediction methods mentioned above is that they need SOH measurements to function. However, in real-life applications, obtaining SOH measurements means stopping the operation of the battery and running an entire cycle, i.e., fully charge and discharge the battery. Furthermore, for these prediction methods to be effective, this needs to be performed on a regular schedule, and as often as possible. As this is not a possibility in most applications, the predictions created in most applications would be extremely unreliable (i.e., the uncertainty intervals of their predictions would be large). Therefore, in the Smart Battery framework, the SOH and RUL prediction will not operate directly on the SOH, but can instead operate on the SOH estimation model, which can provide an estimate of the SOH for every partial charge of the battery. Given the estimated SOH, a post-processing may need to be applied to remove effects of dependencies such as temperature and C-rate (if this effect cannot be removed through the construction of invariant features). The use of the estimated SOH when predicting future SOH should stabilize the uncertainty predictions of SOH and RUL Furthermore, as new measurements of SOH are made, the differences between the predicted and measured SOH will be used to update the SOH estimation model, ultimately leading to better SOH prediction. The general framework of post-processing as well as SOH and RUL prediction can be seen in the bottom panel of Figure 8.

The methodology described above is both a data- and computationally-intensive process, which would be very difficult to implement for most battery architectures. However, as outlined in Section 2, the Smart Battery technology will have the ability to collect raw signals of current, voltage, and temperature directly. Furthermore, the computational cost of the Smart Battery SOH prediction methodology will be offset through local cloud computation.

## 5. Digital Twin

Digital twins are virtual models of physical objects that reflect them accurately and can be used to verify if a planned operational change will produce the desired effect. The concept was first used in the 1960s by NASA, which used an analog twin of the Apollo spacecraft to test in almost real-time certain changes or reactions to certain faults in a realistic environment before testing it for real with human safety at stake. The technology eventually went digital and became very popular in manufacturing, where a virtual product can be designed and presented to customers virtually, for example in the construction, mobility, and even wind turbine industries.

*5.1. Digital Twin as an Optimization Tool in Smart Battery*

In the case of a Smart Battery concept, the Battery Digital Twin (BDT) is defined as an online digital platform based on an AI core (GPU/TPU) capable of replicating the sensed signals (voltage, state of temperature) of a real cell in all possible operating conditions in terms of loading (current), ambient temperature, and aging. A full cell aging model (CAM) is developed using a sparse laboratory testing dataset (full charging/discharging curves at relevant temperature) that is further expanded by using AI techniques of domain adaptation (part of transfer learning) to cover the whole working/aging domain, as depicted in Figure 9.

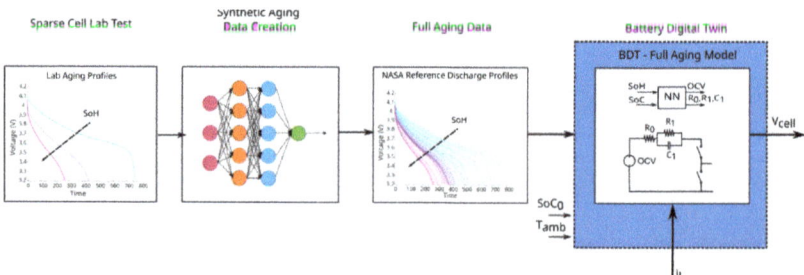

**Figure 9.** Concept of CAM BDT.

The CAM BDT is first developed in Python and then implemented in an AI-core platform (Google Coral Edge), and then it can be used as a development tool for the Smart Battery for:

1. Providing a training dataset for SOH estimation/prediction;
2. Validation of battery performance optimization (BPO);
3. Predictive maintenance.

Training of SOH estimation/prediction using AI using CAM BDT is shown in Figure 10. The CAM BDT can not only synthetically generate a full aging data set, but as also runs in a virtual space in which time can be accelerated, and thus the required lab testing time for the conventional approach can be reduced by several orders of magnitude.

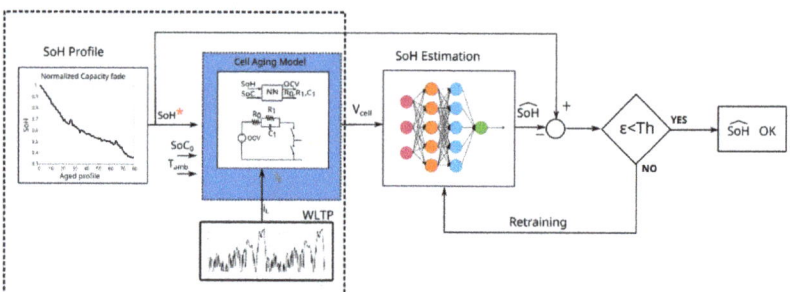

**Figure 10.** Training of SOH estimation using CAM BDT. SOH* represents the reference value.

*5.2. Validation of Battery Performance Optimization*

The Smart Battery allows performance optimization due to the unique feature of cell-level load management enabled by the bypass device. The action of bypassing a cell in the pack during charging or discharging mode can improve balancing in SOC, SOH, and SOT and maximize the SOH, both actions leading to lifetime maximization. As the processes are very complex, AI techniques are used for both training and operational optimization. The BDT is used to validate the performance optimization in an HIL environment including a battery cell simulator (BCS), as illustrated in Figure 11.

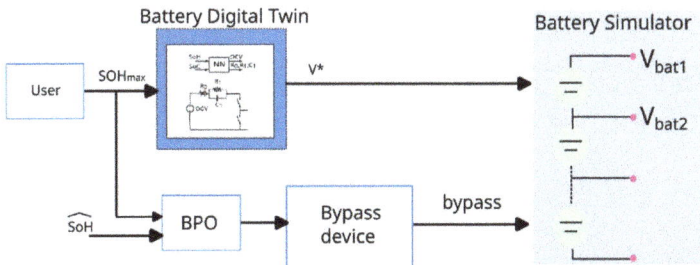

**Figure 11.** SOH maximization as an example of BPO using BDT. $SOH_{max}$ represents the maximum value of the optimization objective SOH, and $V^*$ is the voltage reference.

*5.3. Predictive Diagnostic*

The BDT is implemented online in each cell processor as shown in Figure 12. The idea is that the BDT is fed with the real current measurement and temperature estimation, and calculates the output voltage, which is compared with the real voltage measurement. Any large deviation will be interpreted as a potential condition for failure and will be processed accordingly. With this approach, dangerous events such as thermal runaway events can be avoided.

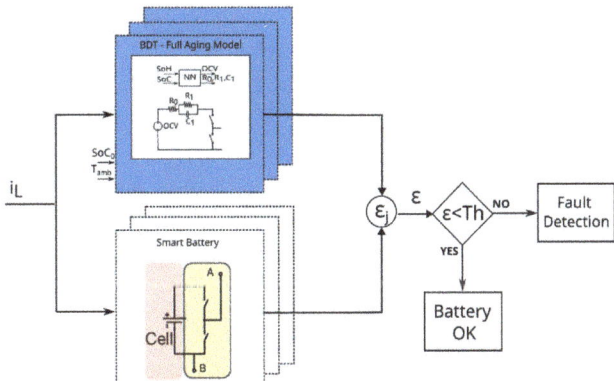

**Figure 12.** BDT used in predictive diagnostics.

## 6. Performance Optimization of the Smart Battery

The electrochemical performance of a battery is defined in terms of three parameters, namely the battery capacity, which measures the total charge stored in a battery, the open circuit voltage or the maximum terminal voltage with no current flow, and the internal resistance, which represents the degree to which the component materials impede the flow of ions during battery operation [45]. Battery performance degrades as the battery ages due to repetitive cycling of lithium ions, which leads to degradation modes such as loss of lithium ions and loss of lithium inventory to set in. This battery aging phenomenon leads to increased internal resistance along with capacity and power fade during a battery's lifetime. To optimize this threefold battery performance, it is crucial to understand the degradation phenomenon and correlate it with measurable battery states. Based on the pre-trained AI-based battery aging models as discussed in previous sections, from measured data, namely terminal current, voltage, and surface temperature, the battery internal states can be evaluated, namely SOH, SOP, SOE, and SOT.

Using these states of the battery, which we define as health indicators, a numerical optimization can be developed that considers operational and power constraints of the battery with the potential to maximize SOH for example. Here, we use reward-based learning to adaptively learn from the battery environment or the balance of the system such as the EV power train and desired user performance to ensure the states of the battery are maximized. It must be noted that since the discharging profile is not in our control, we focus on the charging profile and use the bypass action of Smart Battery slave boards to charge or bypass a battery cell at any given point in time. A complete representation of the aforementioned methodology of battery performance optimization is presented in Figure 13. The goal of this complete optimization as shown in Figure 11 is to extend the lifetime of the batteries, keep the operational cost minimal and maximize the system reliability by embedding fault diagnosis within the system architecture.

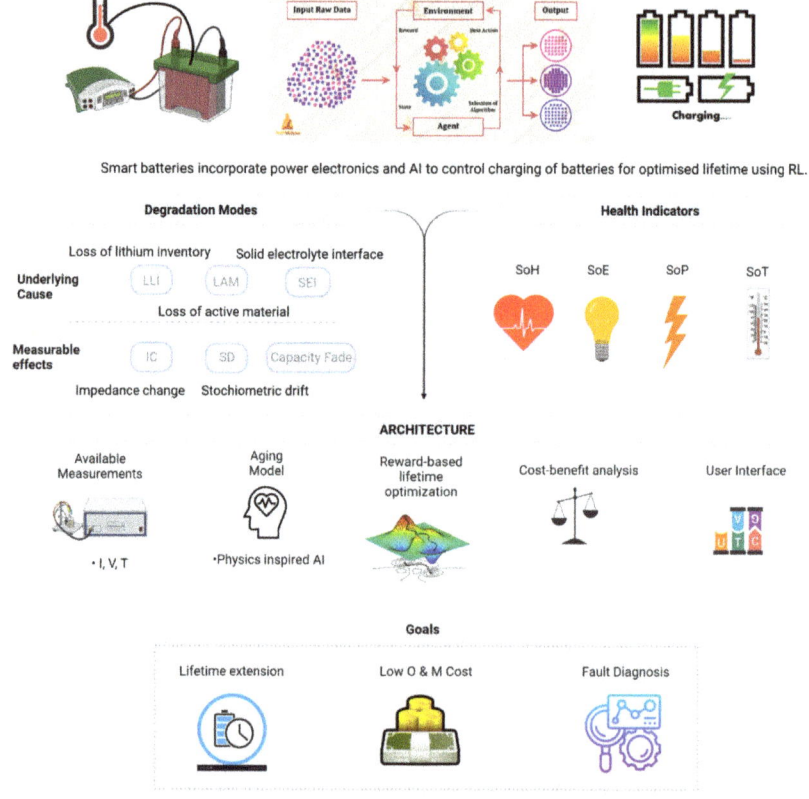

**Figure 13.** A flowchart of the BPO framework, showing the use of reward-based learning for optimizing the charging profile of Li-ion batteries by using measured V, I, and T. The brain of the system is physics-informed and maps degradation modesto quantifiable health indicators to achieve an extended lifetime of batteries along with added features of fault diagnosis and cost minimization. A user interface makes the entire approach more pragmatic for an EV scenario with varying load profiles and desired performance as per the user's needs and constraints.

A more detailed overview of how this battery performance optimization works is shown below with a simplified flow diagram in Figure 14, wherein the performance optimization, preventive diagnostics, and BDT containing the CAM act together in the local cloud to achieve the user-desired and constrained available performance metric, which is an output of the optimization algorithm. The comparison of these two parameters determines whether the bypass switches should connect or disconnect a battery cell while charging. This is in line with the previously discussed results of pulse charging leading to an extended lifetime of the battery, and bypass switches make it possible.

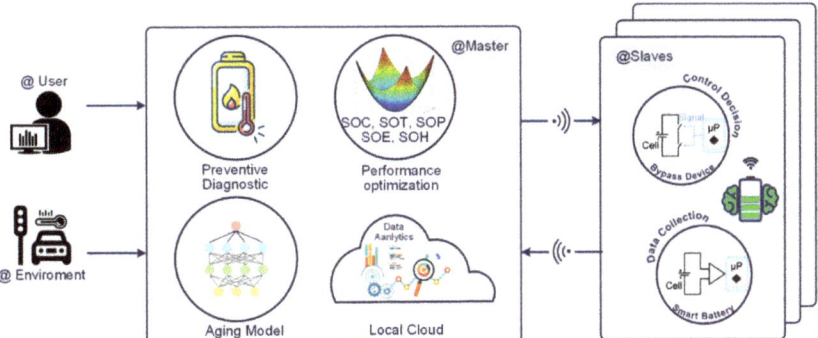

**Figure 14.** A detailed overview of the Smart Battery system with various attributes including optimization of SOX (SOP, SOE, SOT, and SOH). Based on whether the performance is optimized or not, the bypass switches act to activate or provide rest to a cell. The cell aging model is built on AI trained neural network blocks that fully emulate a real battery aging mechanism for given user load profiles and desired performance with a given aging condition of the battery as the initial state of operation.

## 7. Applications of the Smart Battery

A natural question that arises regards the contribution of Smart Batteries to the field of power and energy systems, which spans many directions as shown in Figure 15. Starting from energy storage in power grids to maximum power point tracking in solar photovoltaics, the Smart Battery widely covers the generation, transmission, and distribution sectors of electrical energy. Vehicle-to-grid is an upcoming industry and is anticipated to be limited by battery cycling and aging constraints, but with the Smart Battery, this can be effectively overcome. With green transition as the target of many developing and developed nations, transportation electrification for rail, road, and airways is being investigated. This requires electric vehicles to have reliable, high-performing, and long-lasting batteries, requirements that are the core fundamentals on which the Smart Battery is designed. One major challenge in the EV industry is the rising fast charging industry, which is known to degrade batteries and accelerate their aging. With lifetime extension as a key objective in battery performance optimization, this challenge can be positively overcome using the Smart Battery. We also argue that with Smart Battery technologies, Li-ion batteries can be easily reconfigured for residential energy storage due to lower power and capacity fade in Smart Batteries. Overall, the Smart Battery technology can revolutionize the green energy transition by making disruptive ideas such as ultra-fast charging, second lifetime, and V2G a reality.

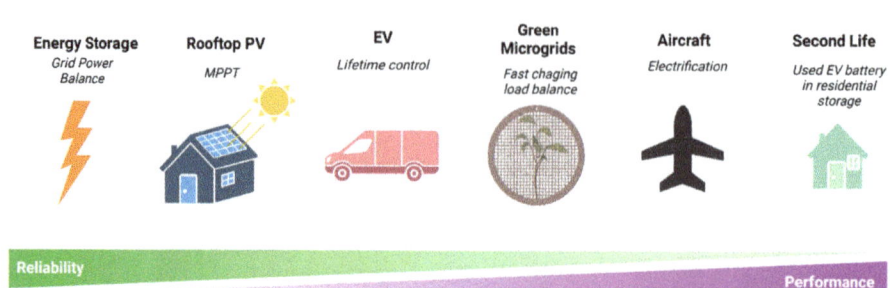

**Figure 15.** The performance-optimized Smart Batteries find applications in energy storage for modern power grids and green microgrids. They can also be readily applied in maximum power point tracking in photovoltaic applications by acting as a controlled voltage source. With the fast-growing EV industry, the role of a high-performing Smart Battery is inevitable for fast charging and lifetime extension, which is also applicable to the electric aircraft industry. The second lifetime of batteries is the sustainable way of reusing EV batteries in residential energy storage with reduced capacity fade using the Smart Battery system.

**Author Contributions:** Conceptualization, R.T.; methodology, R.T., X.S., S.B.V. and D.-I.S.; software, X.S., S.B.V. and D.-I.S.; validation, X.S., S.B.V. and A.K.; formal analysis, R.T., X.S., A.K., P.B. and D.-I.S.; investigation, R.T., X.S., S.B.V., A.K., P.B. and D.-I.S.; writing—original draft preparation, R.T, X.S., S.B.V., P.B., A.K. and D.-I.S.; writing—review and editing, X.S., D.-I.S. and R.T.; supervision, R.T.; funding acquisition, R.T. All authors have read and agreed to the published version of the manuscript.

**Funding:** This research was funded by the "SMART BATTERY" project, granted by Villum Foundation in 2021 (project number 222860).

**Conflicts of Interest:** The authors declare no conflict of interest.

## References

1. Teodorescu, R.; Sui, X.; Acharya, A.B.; Stroe, D.I.; Huang, X. Smart battery concept: A battery that can breathe. In Proceedings of the 5th E-Mobility Power System Integration Symposium (EMOB 2021), Hybrid Conference, Berlin, Germany, 27 September 2021; pp. 214–220.
2. Datasheet, Automotive grade MOSFET AUIRF8739L2TR, International Rectifier. Available online: https://www.infineon.com/cms/en/product/power/mosfet/automotive-mosfet/auirf8739l2/ (accessed on 10 August 2022).
3. Application Note, SimpleLink Microcontroller Platform, Texas Instruments. Available online: https://www.ti.com/lit/pdf/swat002 (accessed on 10 August 2022).
4. Datasheet, CC2642R SimpleLink™ Bluetooth®5.2 Low Energy Wireless MCU, Texas Instruments. Available online: https://www.ti.com/document-viewer/cc2642r/datasheet (accessed on 10 August 2022).
5. Archambeault, B.R.; Drewniak, J. *PCB Design for Real-World EMI Control*; Springer Science & Business Media: Berlin, Germany, 2013; Volume 696.
6. Feng, X.; Ouyang, M.; Liu, X.; Lu, L.; Xia, Y.; He, X. Thermal runaway mechanism of lithium ion battery for electric vehicles: A review. *Energy Storage Mater.* **2018**, *10*, 246–267. [CrossRef]
7. Liu, Y.; Zhu, Y.; Cui, Y. Challenges and opportunities towards fast-charging battery materials. *Nat. Energy* **2019**, *4*, 540–550. [CrossRef]
8. Tomaszewska, A.; Chu, Z.; Feng, X.; O'Kane, S.; Liu, X.; Chen, J.; Ji, C.; Endler, E.; Li, R.; Liu, L.; et al. Lithium-ion battery fast charging: A review. *ETransportation* **2019**, *1*, 100011. [CrossRef]
9. Hu, X.; Zheng, Y.; Howey, D.A.; Perez, H.; Foley, A.; Pecht, M. Battery warm-up methodologies at subzero temperatures for automotive applications: Recent advances and perspectives. *Prog. Energy Combust. Sci.* **2020**, *77*, 100806. [CrossRef]
10. Rodrigues, M.T.F.; Babu, G.; Gullapalli, H.; Kalaga, K.; Sayed, F.N.; Kato, K.; Joyner, J.; Ajayan, P.M. A materials perspective on Li-ion batteries at extreme temperatures. *Nat. Energy* **2017**, *2*, 17108. [CrossRef]
11. Lin, X.; Khosravinia, K.; Hu, X.; Li, J.; Lu, W. Lithium plating mechanism, detection, and mitigation in lithium-ion batteries. *Prog. Energy Combust. Sci.* **2021**, *87*, 100953. [CrossRef]

12. Edge, J.S.; O'Kane, S.; Prosser, R.; Kirkaldy, N.D.; Patel, A.N.; Hales, A.; Ghosh, A.; Ai, W.; Chen, J.; Yang, J.; et al. Lithium ion battery degradation: What you need to know. *Phys. Chem. Chem. Phys.* **2021**, *23*, 8200–8221. [CrossRef] [PubMed]
13. Zhu, G.; Wen, K.; Lv, W.; Zhou, X.; Liang, Y.; Yang, F.; Chen, Z.; Zou, M.; Li, J.; Zhang, Y.; et al. Materials insights into low-temperature performances of lithium-ion batteries. *J. Power Sources* **2015**, *300*, 29–40. [CrossRef]
14. Yang, X.G.; Liu, T.; Gao, Y.; Ge, S.; Leng, Y.; Wang, D.; Wang, C.Y. Asymmetric Temperature Modulation for Extreme Fast Charging of Lithium-Ion Batteries. *Joule* **2019**, *3*, 3002–3019. [CrossRef]
15. Zhu, J.G.; Sun, Z.C.; Wei, X.Z.; Dai, H.F. A new lithium-ion battery internal temperature on-line estimate method based on electrochemical impedance spectroscopy measurement. *J. Power Sources* **2015**, *274*, 990–1004. [CrossRef]
16. Kim, Y.; Mohan, S.; Siegel, J.B.; Stefanopoulou, A.G.; Ding, Y. The estimation of temperature distribution in cylindrical battery cells under unknown cooling conditions. *IEEE Trans. Control Syst. Technol.* **2014**, *22*, 2277–2286.
17. Ojo, O.; Lang, H.; Kim, Y.; Hu, X.; Mu, B.; Lin, X. A neural network based method for thermal fault detection in lithium-ion batteries. *IEEE Trans. Ind. Electron.* **2021**, *68*, 4068–4078. [CrossRef]
18. Li, M.; Dong, C.; Mu, Y.; Yu, X.; Xiao, Q.; Jia, H. Data-model alliance network for the online multi-step thermal warning of energy storage system based on surface temperature diffusion. *Patterns* **2022**, *3*, 100432. [CrossRef]
19. Surya, S.; Samanta, A.; Marcis, V.; Williamson, S. Hybrid electrical circuit model and deep learning-based core temperature estimation of lithium-ion battery cell. *IEEE Trans. Transp. Electrif.* **2022**, *8*, 3816–3824. [CrossRef]
20. Aykol, M.; Gopal, C.B.; Anapolsky, A.; Herring, P.K.; van Vlijmen, B.; Berliner, M.D.; Bazant, M.Z.; Braatz, R.D.; Chueh, W.C.; Storey, B.D. Perspective—Combining physics and machine learning to predict battery lifetime. *J. Electrochem. Soc.* **2021**, *168*, 030525. [CrossRef]
21. Richardson, R.R.; Ireland, P.T.; Howey, D.A. Battery internal temperature estimation by combined impedance and surface temperature measurement. *J. Power Sources* **2014**, *265*, 254–261. [CrossRef]
22. Richardson, R.R.; Howey, D.A. Sensorless battery internal temperature estimation using a Kalman filter with impedance measurement. *IEEE Trans. Sustain. Energy* **2015**, *6*, 1190–1199. [CrossRef]
23. Broussely, M.; Biensan, P.; Bonhomme, F.; Blanchard, P.; Herreyre, S.; Nechev, K.; Staniewicz, R.J. Main aging mechanisms in Li ion batteries. *J. Power Sources* **2005**, *146*, 90–96. [CrossRef]
24. Escobar, L.A.; Meeker, W.Q. A review of accelerated test models. *Stat. Sci.* **2006**, *21*, 552–577. [CrossRef]
25. Ren, L.; Zhao, L.; Hong, S.; Zhao, S.; Wang, H.; Zhang, L. Remaining useful life prediction for lithium-ion battery: A deep learning approach. *IEEE Access* **2018**, *6*, 50587–50598. [CrossRef]
26. Farmann, A.; Waag, W.; Marongiu, A.; Sauer, D.U. Critical review of on-board capacity estimation techniques for lithium-ion batteries in electric and hybrid electric vehicles. *J. Power Sources* **2015**, *281*, 114–130. [CrossRef]
27. Ng, M.F.; Zhao, J.; Yan, Q.; Conduit, G.J.; Seh, Z.W. Predicting the state of charge and health of batteries using data-driven machine learning. *Nat. Mach. Intell.* **2020**, *2*, 161–170. [CrossRef]
28. Sui, X.; He, S.; Vilsen, S.B.; Meng, J.; Teodorescu, R.; Stroe, D.I. A review of non-probabilistic machine learning-based state of health estimation techniques for lithium-ion battery. *Appl. Energy* **2021**, *300*, 117346. [CrossRef]
29. Zou, C.; Manzie, C.; Nešić, D.; Kallapur, A.G. Multi-time-scale observer design for state-of-charge and state-of-health of a lithium-ion battery. *J. Power Sources* **2016**, *335*, 121–130. [CrossRef]
30. Murphy, K.P. *Machine Learning: A Probabilistic Perspective*; MIT Press: Cambridge, MA, USA, 2012.
31. Sui, X.; He, S.; Meng, J.; Teodorescu, R.; Stroe, D.I. Fuzzy entropy-based state of health estimation for Li-ion batteries. *IEEE Trans. Emerg. Sel. Topics Power Electron* **2021**, *9*, 5125–5137. [CrossRef]
32. Sui, X.; Stroe, D.I.; He, S.; Huang, X.; Meng, J.; Teodorescu, R. The effect of voltage dataset selection on the accuracy of entropy-based capacity estimation methods for lithium-ion batteries. *Appl. Sci.* **2019**, *9*, 4170.
33. Sui, X.; He, S.; Huang, X.; Teodorescu, R.; Stroe, D.I. Data smoothing in fuzzy entropy-based battery state of health estimation. In Proceedings of the IECON 2020 The 46th Annual Conference of the IEEE Industrial Electronics Society, Singapore, 18–21 October 2020; pp. 1779–1784.
34. Che, Y.; Zheng, Y.; Wu, Y.; Sui, X.; Bharadwaj, P.; Stroe, D.I.; Yang, Y.; Hu, X.; Teodorescu, R. Data efficient health prognostic for batteries based on sequential information-driven probabilistic neural network. *Appl. Energy* **2022**, *323*, 119663. [CrossRef]
35. Pan, S.J.; Yang, Q. A survey on transfer learning. *IEEE Trans. Knowl. Data Eng.* **2009**, *22*, 1345–1359. [CrossRef]
36. Sui, X.; He, S.; Teodorescu, R.; Stroe, D.I. Fast and robust estimation of Lithium-ion batteries state of health using ensemble learning. In Proceedings of the 2021 IEEE Energy Conversion Congress and Exposition (ECCE), Vancouver, BC, Canada, 10–14 October 2021; pp. 1393–1399.
37. Lin, C.; Tang, A.; Wang, W. A review of SOH estimation methods in Lithium-ion batteries for electric vehicle applications. *Energy Procedia* **2015**, *75*, 1920–1925. [CrossRef]
38. Wu, Y.; Xue, Q.; Shen, J.; Lei, Z.; Chen, Z.; Liu, Y. State of health estimation for lithium-ion batteries based on healthy features and long short-term memory. *IEEE Access* **2020**, *8*, 28533–28547. [CrossRef]
39. You, G.W.; Park, S.; Oh, D. Real-time state-of-health estimation for electric vehicle batteries: A data-driven approach. *Appl. Energy* **2016**, *176*, 92–103. [CrossRef]
40. Ma, G.; Zhang, Y.; Cheng, C.; Zhou, B.; Hu, P.; Yuan, Y. Remaining useful life prediction of lithium-ion batteries based on false nearest neighbors and a hybrid neural network. *Appl. Energy* **2019**, *253*, 113626. [CrossRef]

41. Chaoui, H.; Ibe-Ekeocha, C.C. State of charge and state of health estimation for lithium batteries using recurrent neural networks. *IEEE Trans. Veh. Technol.* **2017**, *66*, 8773–8783. [CrossRef]
42. Rubin, D.B. Bayesianly justifiable and relevant frequency calculations for the applied statistician. *Ann. Stat.* **1984**, *12*, 1151–1172. [CrossRef]
43. Diggle, P.J.; Gratton, R.J. Monte Carlo methods of inference for implicit statistical models. *J. R. Stat. Soc. Ser. B (Methodol.)* **1984**, *46*, 193–212. [CrossRef]
44. Price, L.F.; Drovandi, C.C.; Lee, A.; Nott, D.J. Bayesian synthetic likelihood. *J. Comput. Graph. Stat.* **2018**, *27*, 1–11. [CrossRef]
45. Harmon, J.E. *Assessing Battery Performance: Compared to What?* Argon National Lab.: Lemont, IL, USA, 2019. Available online: https://www.anl.gov/article/assessing-battery-performance-compared-to-what (accessed on 10 August 2022).

Article

# Development of a Data-Driven Method for Online Battery Remaining-Useful-Life Prediction

Sebastian Matthias Hell and Chong Dae Kim *

Technische Hochschule Köln, 50679 Köln, Germany
* Correspondence: chong.kim@th-koeln.de; Tel.: +49-221-8275-2947

**Abstract:** Remaining-useful-life (RUL) prediction of Li-ion batteries is used to provide an early indication of the expected lifetime of the battery, thereby reducing the risk of failure and increasing safety. In this paper, a detailed method is presented to make long-term predictions for the RUL based on a combination of gated recurrent unit neural network (GRU NN) and soft-sensing method. Firstly, an indirect health indicator (HI) was extracted from the charging processes using a soft-sensing method that can accurately describe power degradation instead of capacity. Then, a GRU NN with a sliding window was applied to learn the long-term performance development. The method also uses a dropout and early stopping method to prevent overfitting. To build the models and validate the effectiveness of the proposed method, a real-world NASA battery data set with various battery measurements was used. The results show that the method can produce a long-term and accurate RUL prediction at each position of the degradation progression based on several historical battery data sets.

**Keywords:** lithium-ion batteries; remaining-useful-life (RUL); gated recurrent unit neural network (GRU NN); real-world data

## 1. Introduction

Li-ion batteries have become an essential part of our everyday lives. Due to their low cost, high energy density and long service life, they are already an essential component of cell phones, laptops and electric cars [1]. In particular, the current progressive political developments away from combustion engines in the direction of electric mobility increasingly support this spread of batteries [2], so that energy-efficient and at the same time safe use of these energy storage devices is essential for an environmentally friendly, resource-saving and economic future. The lifetime of these batteries is not unlimited, because conductivity decreases with repeated charging and discharging processes. As soon as the battery falls below its end-of-life (EOL) threshold, the risk of battery failure or even battery fire increases [3]. By monitoring the condition and predicting the expected EOL of the battery, the risk of battery failure can be reduced, thereby increasing safety [4]. On the other hand, replacing the battery too early leads to a waste of valuable resources, which contradicts the claim of efficient use. Therefore, a precise prognosis is of essential importance.

To predict the remaining-useful-life (RUL) of Li-ion batteries, a differentiation is usually made between model-based and data-driven methods. For the application of model-based methods, detailed prior knowledge of the respective battery is required [5]. Mainly, an electrochemical model is used, which is represented by differentiated mathematical models in order to be able to represent the internal chemical process reactions [6]. Model-internal variables can be represented precisely with this method, whereby a high accuracy in the prediction can be achieved [6]. However, these models are highly complex and the battery has to be disassembled for the parameterization of the electrochemical model [7], which makes them difficult to integrate into real applications [8]. In contrast, data-driven methods do not focus on the complex internal electrochemical reactions and failure mechanisms of the battery [9]. Rather, the internal behavior of a battery is considered as a black box to model and simplify electrochemical dynamics. For this purpose, a model is generally created first and then it is refined and optimized using plenty of historical data [10] so that the model can learn battery performance degradation behavior directly from the monitoring data.

In recent years, the data-driven method and in particular the AI-based methods have attracted much attention in the research area of RUL prediction for Li-ion batteries. These methods can be divided into the meta-areas of Neural Networks [11–13], Support Vector Machines [14–16] and Deep Learning. One widely used method for predicting time-series data is the Recurrent Neural Network (RNN). However, this method tends to explode and vanish gradients due to its structure. Therefore, the RNN-based and improved variant, the Long-Short-Term-Memory Neural Network (LSTM NN) is often used. For example, Zhang et al. presented an LSTM NN that predicts the RUL based on historical capacity data [17]. Park et al. introduced an LSTM model using multi-channel charging profiles. However, the prediction interval is set to a fixed value [18]. LSTMs are well suited to store and transfer information from long data sequences, but LSTMs require a large number of parameters for training.

To overcome this issue, the Gated Recurrent Unit Neural Network (GRU NN) was developed. This method is similar to the LSTM but has a simplified structure and fewer parameters, making it especially suitable for online RUL prediction. Previous works with this approach are mostly based only on using classical performance indicators such as the capacity to predict the RUL. However, these direct health indicators are difficult to measure in real applications, because the particular battery must be separated from the original application [19]. To overcome this challenge, indirect health indicators (HI) were used, which can be obtained from the monitoring sensor data to represent the direct HIs. The authors of [20] use the voltage-measured data of the discharge process for this approach. In real applications, discharges mostly correspond to dynamic behavior that can lead to large prediction errors. In contrast, the use of charging data is more static and thus more controllable, which can lead to more reliable results. An open aspect in many of these works is the detailed design and algorithm for the longer-term RUL prediction.

Therefore, this paper proposes a detailed described RUL method from the combination of soft sensing and deep learning. To avoid difficulties in measuring HI directly, an indirect health indicator is extracted from the monitoring data. In addition, a GRU NN is presented using the sliding window method and detailed procedure. The major contributions of this paper are listed as follows:

- A specific indirect HI is extracted from the charge monitoring data. A correlation analysis is used to show that these indirect HIs accurately reflect the capacity. Therefore, complicated measurements or elaborate calculations are no longer needed.
- The combination of soft-sensing and GRU NN with sliding window produces a model capable of both accurate state-of-health estimation and reliable long-term RUL prediction using historical data sets.
- Dropout and early stopping methods were also used to prevent overfitting.
- The effectiveness of the method is validated and verified by the real-world NASA data set.

The structure of this paper is as follows: Section 2 shows the general structure of GRU NN. Section 3 presents data preparation and the construction of indirect HI. The algorithm and the approach of the GRU NN model are proposed in Section 4. Section 5 includes the results and discussion and Section 6 represents the conclusion.

## 2. Gated Recurrent Unit Neural Network

A GRU NN is a neural network based on Gated Recurrent Unit and is a further development of RNN to overcome the exploding and vanishing gradient problem in long-term dependencies. The structure of the GRU NN is also a simplified version of the

LSTM since no cell state is needed anymore. The hidden cell state takes over the data transfer tasks.

The output of the GRU NN depends on the parameters update gate and reset gate. The update gate decides which new information should be added and which information should be dropped. The reset gate decides which and how much information from the past should be forgotten. The general architecture of a GRU NN is shown in Figure 1. Moreover, it is described by the following equations [21]:

$$z_t = \sigma(W_z \cdot [h_{t-1}; x_t] + b_z) \tag{1}$$

$$r_t = \sigma(W_r \cdot [h_{t-1}; x_t] + b_r) \tag{2}$$

$$\widetilde{h}_t = tanh(W_h \cdot [(r_t \odot h_{t-1}); x_t] + b_h) \tag{3}$$

$$h_t = (1 - z_t) \odot h_{t-1} + z_t \odot \widetilde{h}_t \tag{4}$$

where $z_t$ represents the update gate and $r_t$ the reset gate. Both gates depend on the current state $x_t$ and hidden state $h_{t-1}$ at the previous time. For the output of the hidden state $h_t$ at time t, the candidate state $\widetilde{h}_t$ is also required. $W_z$, $W_r$, $W_h$ are weight matrices and $b_z$, $b_r$, $b_h$ indicates the biases for the update gate, candidate state and reset gate. The symbol $\odot$ shows an element-wise multiplication, $\sigma$ is the sigmoid function and ; indicates a vector-concatenation operation.

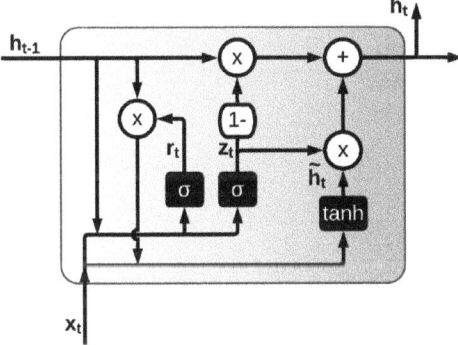

**Figure 1.** The architecture of the proposed GRU NN.

## 3. Data Preparation

### 3.1. Test Data

The data used here were derived from a battery data set from NASA Ames Research Center [22]. Four Li-ion batteries (Bat. 5, Bat. 6, Bat. 7, and Bat. 18) were fully charged cyclically at room temperature (24 °C) and then fully discharged. The batteries were charged with the constant current-constant voltage (CC-CV) mode with a constant current of 1.5 A until the battery voltage reached 4.2 V. The battery was then charged in a constant current mode. Then, charging was continued in a constant voltage mode until the charge current dropped to 20 mA. Discharging was performed at a constant current mode of 2 A until the discharge voltage dropped to 2.7 V for Bat. 5, 2.5 V for Bat. 6, 2.2 V for Bat. 7, and 2.5 V for Bat. 18. The nominal capacity of the batteries is equal to 2 Ah. EOL is reached when the capacity value of the respective battery falls below 70% of the nominal capacity (from 2 Ah to 1.4 Ah).

Another step of data preparation is data cleaning. Since the tests for Bat. 5, Bat. 6, and Bat. 7 were recorded at the same time, these batteries have the same documented irregular behavior. To maintain a regular and cyclically ordered sequence, the first measurement series for both charging and discharging was removed because it was an outlier. Measurement series 12 and 33 were dropped for charging, as the batteries were charged twice here without any documented discharge in the meantime. Measurement series 90 was dropped for discharging, due to a double discharge without documented charging. The charging process of cycle 170 was an incomplete charge. For Bat. 18, the first charge and discharge measurement series was also removed because it was also an outlier. In addition, charge cycles 47 and 58 were dropped because here the batteries were charged twice, without documented discharge. After data cleaning, the records for Bat. 5, Bat. 6, and Bat. 7 had 166 complete cycles and Bat. 18 had 131 complete charge and discharge cycles.

*3.2. Health Indicator Extraction*

Capacity and internal resistance are direct health indicators of power degradation, but they are difficult to measure in real time. Therefore, an effective indirect health indicator is needed to reflect the performance degradation of the battery. To achieve this, a soft-sensing method was used. In this method, a directly measured variable that is difficult to measure is represented by an easily measurable variable or several variables of the existing monitoring data [23].

The charging process was used for this since it is more stable than the discharging process. As an example, the voltage and current values for cycles 10, 60, 100, and 160 are shown for Bat. 5 in Figure 2. The time range of the constant current charge time (CCCT) decreases as the number of cycles increases. The documented voltages started at different voltage values. To create the same conditions, the start point of CCCT is the time value when the voltage value exceeded 3.8 V for the first time. The end point is the time value when the voltage value of 4.2 V was exceeded for the first time. The respective value for CCCT was calculated by the difference between the end time value and the start time value. Figure 3 shows an example of the CC-CV procedure for the 10th cycle of Bat. 5, and CCCT is also shown. All extracted CCCT values of the 4 batteries are shown in Figure 4.

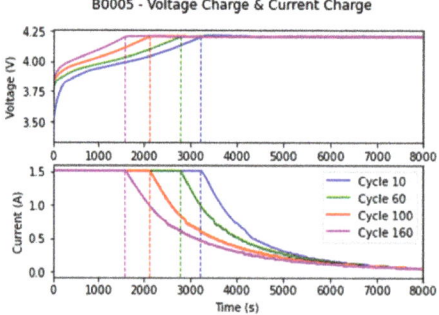

**Figure 2.** The charge voltage and charge current curves for different cycles.

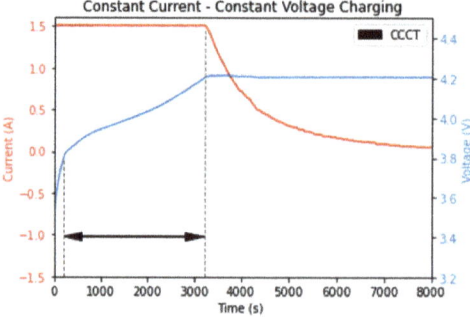

**Figure 3.** The charge voltage and charge current curves for Cycle 10.

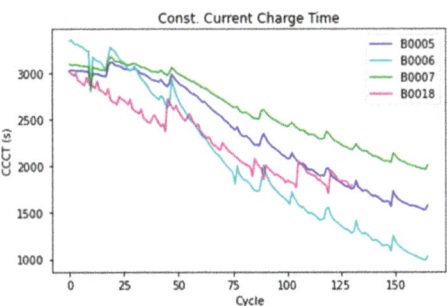

**Figure 4.** The constant current charge time for all cycles for the four batteries.

*3.3. Correlation Analysis*

The correlation analysis between CCCT and capacity can be used to show the degree to which the two variables are related. The correlation coefficient indicates the strength of the correlation in a value range between 0 and 1. A value of 1 indicates a strong correlation and a value of 0 indicates a low correlation.

Table 1 shows correlation coefficients close to 1 for both the Spearman and the Pearson analyses for all batteries. This proves that there is a significant linear correlation between CCCT and capacity. Accordingly, the indirect HI is able to represent the battery performance degradation instead of the capacity.

**Table 1.** The Spearman and Pearson correlation analysis.

| Correlation between CCCT and Capacity | Bat. 5 | Bat. 6 | Bat. 7 | Bat. 18 |
|---|---|---|---|---|
| Spearman: | 0.993 | 0.996 | 0.992 | 0.975 |
| Pearson: | 0.997 | 0.993 | 0.990 | 0.986 |

## 4. Algorithm and Approach

*4.1. General Algorithm*

Figure 5 shows the schematic structure of the RUL prediction model. This process has three different phases. In the preparation phase, data are imported, data are cleaned, the indirect HI is extracted, and the linear relationship between HI and capacity is tested using correlation analysis. In the state-of-health (SOH) estimation phase, the GRU NN is built and trained using the extracted HI. In this work, the SOH describes the expected performance capability of the battery to a next cycle. The last phase describes the RUL prediction process. Here, it is checked whether the initially determined threshold value is reached. If the last prediction value is higher than this, then the predicted value is fed back into the neural network. A prediction is then made based on this value. This process is then iteratively repeated until the system falls below the EOL threshold. The RUL describes the expected cycles during which the battery is still capable of performing under the current conditions.

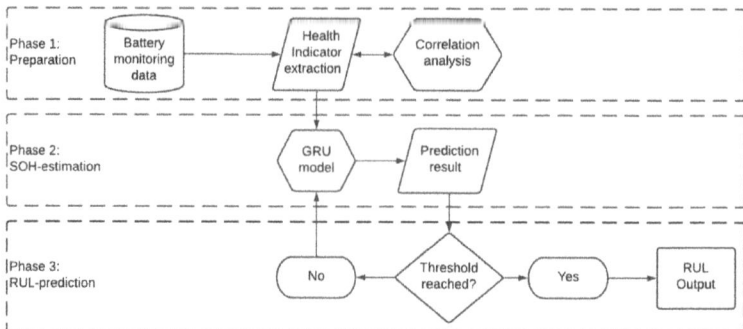

Figure 5. The RUL prediction model.

### 4.2. SOH Estimation Framework

The SOH estimation can be shown in more detail. Figure 6 shows a flowchart of the SOH estimation process. After HI extraction, data sets were selected that were later used for training, validation or testing (cf. Section 4.3.1). These data were normalized between 0 and 1 using the min-max scaler to obtain the same scaling for the different data sets. Subsequently, the selected data sets were split into the train, test, and validation sets and prepared for the recurrent neural network. The next step is to transform the different data sets into the appropriate format for Recurrent Neural Networks. For this purpose, the sliding window method with a constant window size is used in order to create a temporal reference. Hyperparameters and network architecture were determined to create the GRU NN. The detailed approach is explained in Section 4.3.2. Then, the model was trained and evaluated for SOH prediction. Based on this trained model, the SOH and RUL prediction was subsequently created.

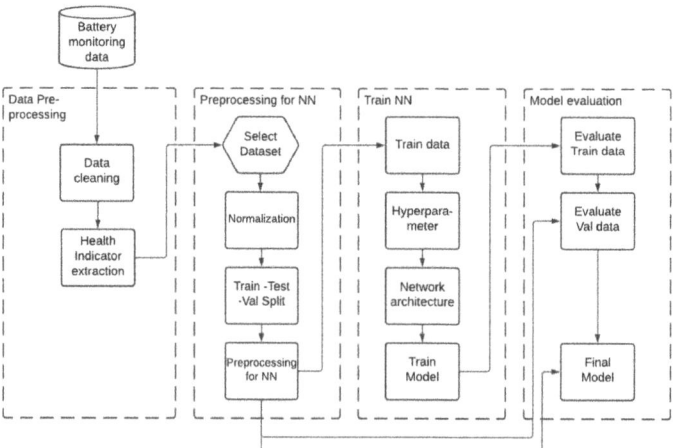

Figure 6. Flowchart for SOH estimation.

### 4.3. Approach

#### 4.3.1. Data Set Selection

The training of two data sets is a challenging task. Since we were dealing here with cyclically ordered data sets and the sequence was accordingly decisive, two data sets were not simply combined into one large data set because this would not correspond to the actual sequence of this data structure. Therefore, the training was split. The first training data set was used to build a base model. Since recursive neural networks have the possibility to store and pass weights and settings, this opportunity was used to train the second data set based on the pretrained model and to update the model with additional data.

The data sets were divided according to the following principle: One battery data set was used for the test in each case. Two batteries were used for training and the fourth battery was used as a validation set. This procedure was repeated three times so that each battery had been used twice as a training set and once as a validation set. The combination with the best result was used for the final model. Table 2 shows an overview of the final combinations.

**Table 2.** Overview of the test, training and validation set combination.

| Test | Train | Val |
|---|---|---|
| Bat. 5 | Bat. 7 and 18 | Bat. 6 |
| Bat. 6 | Bat. 5 and 18 | Bat. 7 |
| Bat. 7 | Bat. 6 and 18 | Bat. 5 |
| Bat. 18 | Bat. 6 and 7 | Bat. 5 |

4.3.2. Hyperparameter Optimization

The hyperparameters were determined and optimized using the following procedure. First, the sequence length and batch size were selected. The number of epochs was also defined, but the training of the epochs was stopped early by the early stopping method. For the optimizer, the superior Adam optimizer was used. The common Mean Square Error (MSE) was used to determine the loss function. In addition, the architecture of the neural network was defined using various tests and experiments. Subsequently, with the help of the learning rate, the training data set was adjusted so that the evaluation metrics produced values as good as possible. In addition, a visual comparison of the generated regression with the real curves of the training and validation set was made. In this context, this means that the difference between predicted graphs and expected graphs was examined. In the process of different experiments, it was shown that especially here a small difference is an essential factor for the quality of the later test results. As soon as suitable values were shown and the trend was correct, the trained model was used to make predictions for the SOH and RUL of the test set. Table 3 shows the overview of the hyperparameters used.

**Table 3.** Hyperparameters for RUL prediction model.

| Description | Parameter |
|---|---|
| Sequence length | 10 |
| Learning rate | $9 \times 10^{-4}$ |
| Number of Epochs | 100 |
| Batch size | 16 |
| Optimizer | Adam |
| Loss | Mean Square Error |

The network architecture has 2 GRU layers with 50 neurons. The tanh function was used for the activation function. Each GRU layer is followed by a dropout layer with a dropout rate of 0.2 to prevent overfitting. The output layer has a dense layer with a single output neuron.

## 5. Results and Discussion

*5.1. Evaluation Parameters*

In this paper, the quality measures for evaluating the predictions are root mean square error ($RMSE$), mean absolute error ($MAE$), coefficient of determination ($R2$), and Actual Error ($AE$).

$$RMSE = \sqrt{\frac{1}{n} \sum_{i=1}^{n} (y_i - \hat{y}_i)^2} \qquad (5)$$

$$MAE = \frac{1}{n}\sum_{i=1}^{n}|y_i - \hat{y}_i| \quad (6)$$

$$R2 = 1 - \frac{\sum_{i=1}^{n}(y-\hat{y})^2}{\sum_{i=1}^{n}(y-\overline{y})^2} \quad (7)$$

$$AE = R - \hat{R} \quad (8)$$

where $\hat{y}$ describes the predicted value and $\overline{y}$ represents the mean value of $y$. $R$ and $\hat{R}$ denote the real and predicted number of cycles until the EOL threshold is reached.

### 5.2. SOH Results Analysis

Figure 7 shows the predictions of the four batteries created by the GRU NN. The blue graph indicates the real values of the extracted health indicator. The prediction is shown in magenta. In addition, the EOL threshold is shown by a horizontal red line and the starting point is shown by a black vertical line, from which the training is completed and the test area begins. The prediction is shown as an example for the starting position from 30% of the test data set.

For Bat. 5, Bat. 6, and Bat. 7, the results show that the SOH estimations approximate the real trends. The power regeneration peaks can be partially mapped. The recorded performance curve of Bat. 18 was significantly different relative to the other three batteries. The curve also shows several local variations. Therefore, multiple and larger differences can be seen for the SOH estimation, since the model takes longer to represent the real values. This can be illustrated using the evaluation parameters. This is shown in Table 4 for start position 0.3. The denormalized values of RMSE, MAE and R2 refer to the SOH scale.

**Table 4.** The evaluation metrics for the SOH estimations at starting points 0.3.

| Battery | RMSE | MAE | R2 |
|---|---|---|---|
| Bat. 5 | 0.0060 | 0.0041 | 0.993 |
| Bat. 6 | 0.0103 | 0.0065 | 0.984 |
| Bat. 7 | 0.0056 | 0.0037 | 0.991 |
| Bat. 18 | 0.0121 | 0.0089 | 0.925 |

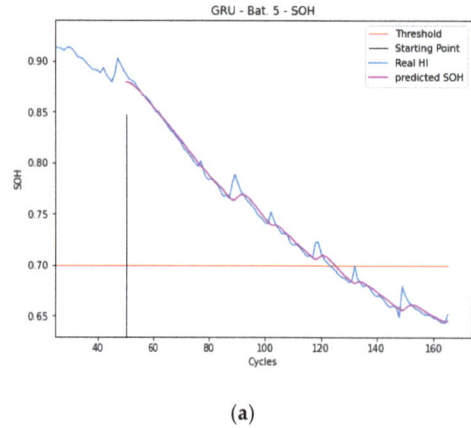

(a)

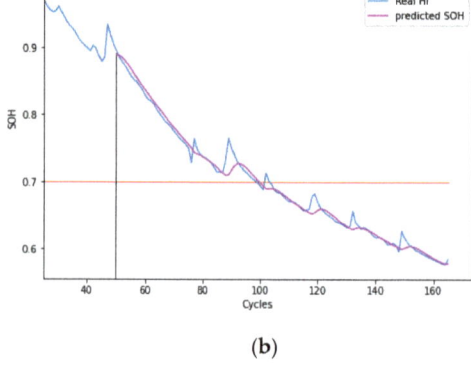

(b)

**Figure 7.** *Cont.*

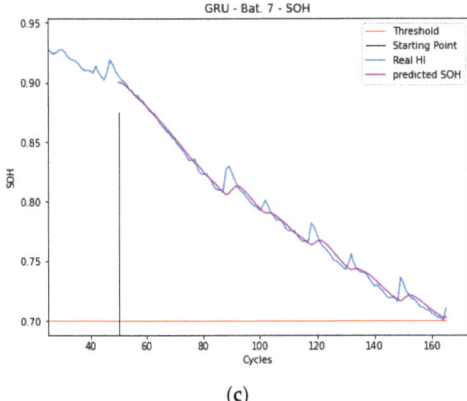
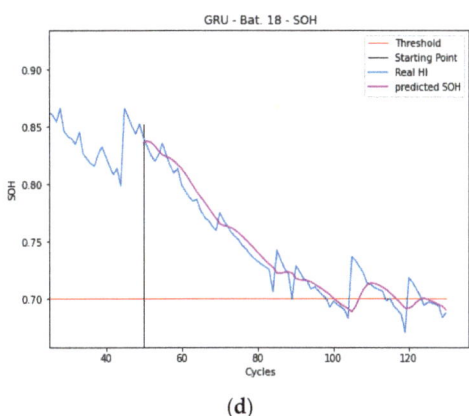

(c)                                    (d)

**Figure 7.** The SOH estimation result at starting point 0.3 for batteries (**a**) Bat. 5; (**b**) Bat. 6; (**c**) Bat. 7; (**d**) Bat. 18.

Bat. 5, Bat. 6, and Bat. 7 show comparable results for an R2 score close to 1, which means that the estimated values are close to the actual values. Bat. 6 shows higher deviations for RMSE and MAE in relation to the other two batteries. This is due to the stronger characteristics of the power regeneration peaks. The parameters also show the outliers of Bat. 18. Overall, the results demonstrate that training for the test data set for a step forward is successful and the GRU NN produces an accurate state-of-health estimation. This is significant because the subsequent RUL prediction is based on the trained model.

### 5.3. RUL Results Analysis

Figure 8 shows the RUL prediction based on two pre-trained battery data sets. The start position of 0.3 of the total data set is shown in green, in magenta, the start position is 0.5 and in orange, the start position is 0.7. In addition, the EOL threshold is also shown in red.

The predictions show the descending trend of the individual batteries. Bat. 5 to 7 show a decent curve progression, which becomes flatter as the number of cycles increases. The predictions of Bat. 18 each follow an almost linear course. The individual predictions in each battery data set start at the designated position and each shows a similar curve thereafter. From this can be derived that the focus of the model is on the wide and longer-term evolution of performance degradation. Table 5 shows an overview of the RUL prediction for the four batteries at the three different starting positions.

The table shows precise prediction values for Bat. 5–7. For example, at 75 steps into the future, Bat. 5 is exactly the real value. With 42 steps ahead, the prediction is off by only 2 cycles. For Bat. 6, the AE RUL is 4 cycles for start point 0.3. For start point 0.5, the prediction is short by 5 cycles. The prediction value for battery 7 is also accurate. For example, at 115 steps into the future, the prediction is 5 cycles off the real value and at 82 steps, it is only 3 steps off. Due to the nature of the data, the experiments of Bat. 18 show ambivalent results. Since the predicted values are close to the actual values, the proposed model can produce an accurate and long-term RUL prediction.

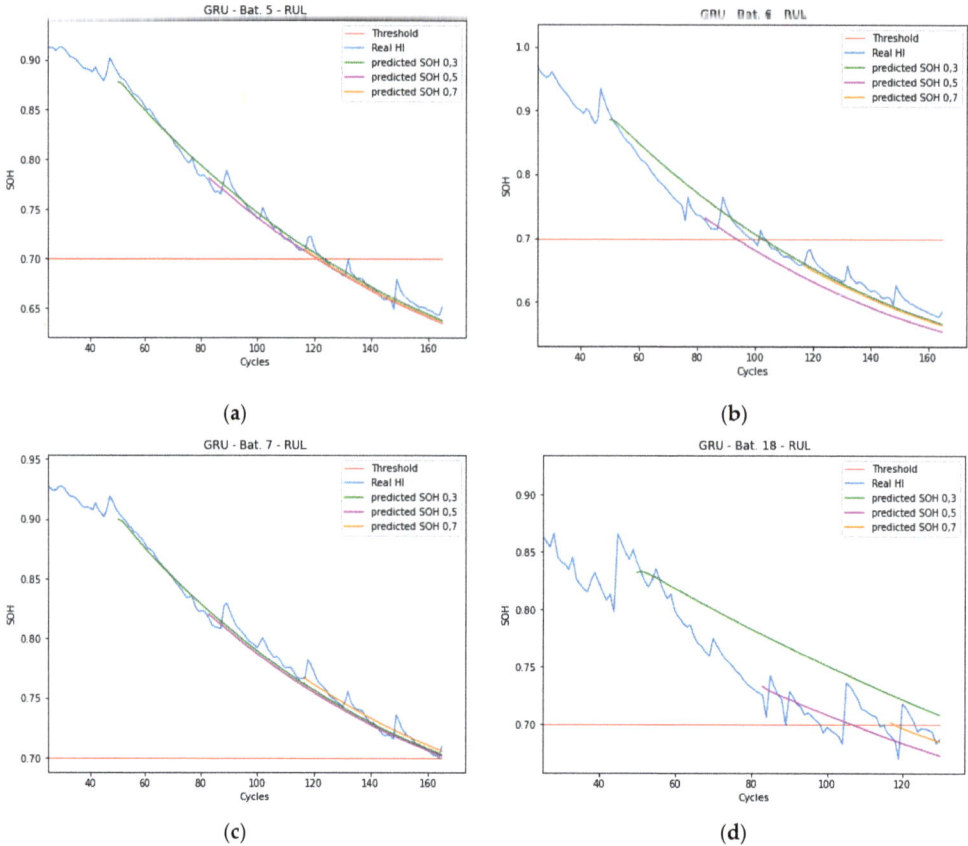

**Figure 8.** The RUL prediction results at different starting points for batteries (**a**) Bat. 5; (**b**) Bat. 6; (**c**) Bat. 7; (**d**) Bat. 18.

**Table 5.** The RUL prediction results with different starting points.

| Battery | Starting Point | Real RUL | Pred. RUL | AE RUL |
|---|---|---|---|---|
| Bat. 5 | 0.3 | 75 | 75 | 0 |
|  | 0.5 | 42 | 40 | −2 |
|  | 0.7 | 8 | 6 | −2 |
| Bat. 6 | 0.3 | 50 | 54 | 4 |
|  | 0.5 | 17 | 12 | −5 |
|  | 0.7 | - | - | - |
| Bat. 7 | 0.3 | 115 | 120 | 5 |
|  | 0.5 | 82 | 85 | 3 |
|  | 0.7 | 48 | 56 | 8 |
| Bat.18 | 0.3 | 47 | - | - |
|  | 0.5 | 14 | 25 | 11 |
|  | 0.7 | - | - | - |

## 6. Conclusions

In this paper, an RUL prediction method for Li-ion batteries based on the combination of deep learning and soft-sensing is presented and described in detail. For this purpose, an indirect HI is extracted from the monitoring data of the charging area, which can reflect the performance degradation instead of capacity. At the same time, the GRU NN was trained on the basis of various historical data sets to learn the long-term dependencies. For verification and validation, several experiments were created and presented using the real NASA Li-ion battery data set. This leads to precise results for SOH estimation and accurate results for long-term RUL prediction trends. In reality, the variable loads of actual user charging on the battery are one of the main challenges. In the future, the presented method will be validated with more practical measured data and compared with further time-series methods. In addition, the prediction accuracy will be improved by incorporating more of the existing historical data from the current battery.

**Author Contributions:** Conceptualization, S.M.H. and C.D.K.; Formal analysis, S.M.H.; Methodology, S.M.H.; Software, S.M.H.; Supervision, C.D.K.; Validation, S.M.H. and C.D.K.; Visualization, S.M.H.; Writing—original draft, S.M.H.; Writing—review and editing, C.D.K. All authors have read and agreed to the published version of the manuscript.

**Funding:** This research received no external funding.

**Institutional Review Board Statement:** Not applicable.

**Informed Consent Statement:** Not applicable.

**Acknowledgments:** The authors would like to thank the reviewers and the editor for their comments and suggestions that helped to improve the paper significantly. The authors thank Hans Willi Langenbahn, the Dean of the Faculty of Process Engineering, Energy and Mechanical Systems of Technische Hochschule Köln, for financial assistance in publishing our research paper.

**Conflicts of Interest:** The authors declare no conflict of interest.

# References

1. Korthauer, R. *Handbuch Lithium-Ionen-Batterien*; Springer: Berlin/Heidelberg, Germany, 2013.
2. Bundes-Ministerium für Wirtschaft und Klimaschutz, "bmwk.de". Available online: https://www.bmwk.de/Redaktion/DE/Dossier/elektromobilitaet.html (accessed on 15 July 2022).
3. Wang, D.; Miao, Q.; Pecht, M. Prognostics of lithium-ion batteries based on relevance vectors and a conditional three-parameter capacity degradation model. *J. Power Sources* **2013**, *239*, 253–264. [CrossRef]
4. Hao, J.; Jing, L.; Ke, H.-L.; Wang, Y.; Gao, Q.; Wang, X.-X.; Sun, Q.; Xu, Z.-J. Determination of cut-off time of accelerated aging test under temperature stress for LED lamps. *Front. Inf. Technol. Electron. Eng.* **2017**, *18*, 1197–1204. [CrossRef]
5. Liu, K.; Shang, Y.; Ouyang, Q.; Widanage, W.D. A Data-Driven Approach With Uncertainty Quantification for Predicting Future Capacities and Remaining Useful Life of Lithium-ion Battery. *IEEE Trans. Ind. Electron.* **2021**, *68*, 3170–3180. [CrossRef]
6. Wang, Y.; Tian, J.; Sun, Z.; Wang, L.; Xu, R.; Li, M.; Chen, Z. A comprehensive review of battery modeling and state estimation approaches for advanced battery management systems. *Renew. Sustain. Energy Rev.* **2020**, *131*, 110015. [CrossRef]
7. Qu, J.; Liu, F.; Ma, Y.; Fan, J. A Neural-Network-Based Method for RUL Prediction and SOH Monitoring of Lithium-Ion Battery. *IEEE Access* **2019**, *7*, 87178–87191. [CrossRef]
8. Jin, S.; Sui, X.; Huang, X.; Wang, S.; Teodorescu, R.; Stroe, D.-I. Overview of Machine Learning Methods for Lithium-Ion Battery Remaining Useful Lifetime Prediction. *Electronics* **2021**, *10*, 3126. [CrossRef]
9. Lui, Y.H.; Li, M.; Downey, A.; Shen, S.; Nemani, V.P.; Ye, H.; VanElzen, C.; Jain, G.; Hu, S.; Laflamme, S.; et al. Physics-based prognostics of implantable-grade lithium-ion battery for remaining useful life prediction. *J. Power Sources* **2021**, *485*, 229327. [CrossRef]
10. Wang, S.; Jin, S.; Deng, D.; Fernandez, C. A Critical Review of Online Battery Remaining Useful Lifetime Prediction Methods. *Front. Mech. Eng.* **2021**, *7*, 719718. [CrossRef]
11. Wu, Y.; Li, W.; Wang, Y.; Zhang, K. Remaining Useful Life Prediction of Lithium-Ion Batteries Using Neural Network and Bat-Based Particle Filter. *IEEE Access* **2019**, *7*, 54843–54854. [CrossRef]
12. She, C.; Wang, Z.; Sun, F.; Liu, P.; Zhang, L. Battery Aging Assessment for Real-World Electric Buses Based on Incremental Capacity Analysis and Radial Basis Function Neural Network. *IEEE Trans. Ind. Inform.* **2019**, *16*, 3345–3354. [CrossRef]
13. Zhou, D.; Li, Z.; Zhu, J.; Zhang, H.; Hou, L. State of Health Monitoring and Remaining Useful Life Prediction of Lithium-Ion Batteries Based on Temporal Convolutional Network. *IEEE Access* **2020**, *8*, 53307–53320. [CrossRef]
14. Zhao, Q.; Qin, X.; Zhao, H.; Feng, W. A novel prediction method based on the support vector regression for the remaining useful life of lithium-ion batteries. *Microelectron. Reliab.* **2018**, *85*, 99–108. [CrossRef]
15. Wang, F.-K.; Mamo, T. A hybrid model based on support vector regression and differential evolution for remaining useful lifetime prediction of lithium-ion batteries. *J. Power Sources* **2018**, *401*, 49–54. [CrossRef]

16. Wei, J.; Dong, G.; Chen, Z. Remaining Useful Life Prediction and State of Health Diagnosis for Lithium-Ion Batteries Using Particle Filter and Support Vector Regression. *IEEE Trans. Ind. Electron.* **2018**, *65*, 5634–5643. [CrossRef]
17. Zhang, Y.; Xiong, R.; He, H.; Pecht, M.G. Long Short-Term Memory Recurrent Neural Network for Remaining Useful Life Prediction of Lithium-Ion Batteries. *IEEE Trans. Veh. Technol.* **2018**, *67*, 5695–5705. [CrossRef]
18. Park, K.; Choi, Y.; Choi, W.J.; Ryu, H.-Y.; Kim, H. LSTM-Based Battery Remaining Useful Life Prediction With Multi-Channel Charging Profiles. *IEEE Access* **2020**, *8*, 20786–20798. [CrossRef]
19. Lee, C.-J.; Kim, B.-K.; Kwon, M.-K.; Nam, K.; Kang, S.-W. Real-Time Prediction of Capacity Fade and Remaining Useful Life of Lithium-Ion Batte-ries Based on Charge/Discharge Characteristics. *Electronics* **2021**, *10*, 846. [CrossRef]
20. Chen, L.; Chen, J.; Wang, H.; Wang, Y.; An, J.; Yang, R.; Pan, H. Remaining Useful Life Prediction of Battery Using a Novel Indicator and Framework With Fractional Grey Model and Unscented Particle Filter. *IEEE Trans. Power Electron.* **2020**, *35*, 5850–5859. [CrossRef]
21. Abdulwahab, S. Deep Learning Models for Paraphrases Identification. Master's Thesis, Universitat Rovira I Virgili, Tarragona, Spain, 2017.
22. Saha, B.; Goebel, K. NASA Ames Prognostics Data Repository. 2007. Available online: https://ti.arc.nasa.gov/tech/dash/groups/pcoe/prognostic-data-repository/ (accessed on 28 April 2022).
23. Kadlec, P.; Gabrys, B.; Strandt, S. Data-driven Soft Sensors in the process industry. *Comput. Chem. Eng.* **2009**, *33*, 795–814. [CrossRef]

Article

# Reducing the Computational Cost for Artificial Intelligence-Based Battery State-of-Health Estimation in Charging Events

Alessandro Falai [1,2,*], Tiziano Alberto Giuliacci [1,3,*], Daniela Anna Misul [1,2,*] and Pier Giuseppe Anselma [2,4]

1 Department of Energy, Politecnico di Torino, Corso Duca Degli Abruzzi 24, 10129 Turin, Italy
2 Interdepartmental Center for Automotive Research and Sustainable Mobility (CARS@PoliTO), Politecnico di Torino, Corso Duca Degli Abruzzi 24, 10129 Turin, Italy
3 Addfor S.P.A., Piazza Solferino 7, 10121 Turin, Italy
4 Department of Mechanical and Aerospace Engineering, Politecnico di Torino, Corso Duca Degli Abruzzi 24, 10129 Turin, Italy
* Correspondence: alessandro.falai@polito.it (A.F.); tiziano.giuliacci@polito.it (T.A.G.); daniela.misul@polito.it (D.A.M.)

**Abstract:** Powertrain electrification is bound to pave the way for the decarbonization process and pollutant emission reduction of the automotive sector, and strong attention should hence be devoted to the electrical energy storage system. Within such a framework, the lithium-ion battery plays a key role in the energy scenario, and the reduction of lifetime due to the cell degradation during its usage is bound to be a topical challenge. The aim of this work is to estimate the state of health (SOH) of lithium-ion battery cells with satisfactory accuracy and low computational cost. This would allow the battery management system (BMS) to guarantee optimal operation and extended cell lifetime. Artificial intelligence (AI) algorithms proved to be a promising data-driven modelling technique for the cell SOH prediction due to their great suitability and low computational demand. An accurate on-board SOH estimation is achieved through the identification of an optimal SOC window within the cell charging process. Several Bi-LSTM networks have been trained through a random-search algorithm exploiting constant current constant voltage (CCCV) test protocol data. Different analyses have been performed and evaluated as a trade-off between prediction performance (in terms of RMSE and customized accuracy) and computational burden (in terms of memory usage and elapsing time). Results reveal that the battery state of health can be predicted by a single-layer Bi-LSTM network with an error of 0.4% while just monitoring 40% of the entire charging process related to 60–100% SOC window, corresponding to the constant-voltage (CV) phase. Finally, results show that the amount of memory used for data logging and processing time has been cut by a factor of approximately 2.3.

**Keywords:** lithium-ion battery; SOH estimation; artificial intelligence; lifetime prediction; neural networks; supervised learning; LSTM; data mining; battery aging

**Citation:** Falai, A.; Giuliacci, T.A.; Misul, D.A.; Anselma, P.G. Reducing the Computational Cost for Artificial Intelligence-Based Battery State-of-Health Estimation in Charging Events. *Batteries* **2022**, *8*, 209. https://doi.org/10.3390/batteries8110209

Academic Editors: Remus Teodorescu and Xin Sui

Received: 20 September 2022
Accepted: 26 October 2022
Published: 2 November 2022

**Publisher's Note:** MDPI stays neutral with regard to jurisdictional claims in published maps and institutional affiliations.

**Copyright:** © 2022 by the authors. Licensee MDPI, Basel, Switzerland. This article is an open access article distributed under the terms and conditions of the Creative Commons Attribution (CC BY) license (https:// creativecommons.org/licenses/by/ 4.0/).

## 1. Introduction

The necessity of reducing pollutant emissions caused by internal combustion engines of road vehicles and to increase the efficiency of the energy use in vehicles has led researchers to find new propulsion solutions. Electric motors have been used for road vehicle-propulsion systems for a long time ('La Jamais Contente' in 1899 was the first car in history that went beyond 100 km/h, and it was electric [1]; however, it was never used in production because of the difficulties in storing a large quantity of electric energy on vehicles). Recent technological advancements in Li-ion batteries partially fixed this problem and allowed electric motors to be employed for automotive traction. As a matter of fact, in contrast to many other electrical storage systems such as lead-acid batteries, Li-ion batteries have quite high energy and power density, a low level of self-discharge, a low need for maintenance, and good load characteristics, and they can be partially charged and discharged without being damaged [2–4]. If appropriately managed by a battery management system (BMS), Li-ion batteries can ensure an acceptable level of safety and valid lifespan, as essential requirements for automotive applications.

On the other hand, the Li-ion battery package is definitely the most critical and fragile component of the electric vehicle. In order to preserve battery health, it is extremely important to monitor and oversee its status while in operation. This is done by the BMS, which ensures that the battery pack works within its safe range and optimal conditions [5]. Cells must always operate within a specific range of temperature and voltage, and they cannot deliver excessively high currents. These conditions change from cell to cell depending on many factors, such as chemistry type. For instance, when batteries operate at excessively high temperatures, they may bloat with gas, causing leakage or explosion, or a thermal runaway may even occur [6,7]. As a result, the BMS must guarantee vehicle safety. Thermal management of cells is another key issue: at high temperatures, the battery degrades faster, leading to degradation of performance over time [8–10], while at low temperatures, the efficiency is lower due to the higher internal resistance of the cell [11]. Overvoltage and undervoltage conditions can also damage the battery chemistry [12,13]. In general, the more the batteries work far from their optimal temperature range which is commonly between $-20\ °C$ and $60\ °C$, the faster they degrade. According to the literature, although safe conditions are respected, batteries degrade at varying rates depending on the stress cycles. This is referred to as cyclic aging [14].

The BMS is critical to safeguard as much of the health and efficiency of the battery as possible, but it is also very important to know the battery health condition at any given time. When the battery degrades, its capacity reduces, producing a decrease in vehicle range, and its internal resistance increases. Specifically, the decrease in capacity impacts the amount of energy a battery can store, although the rise in internal resistance restricts the amount of power that can be generated [15]. For this reason, when battery capacity reaches 80% of its initial value or internal resistance reached 200% of the initial value, they are ordinarily not used any longer for automotive applications, and this is considered the conventional battery's end of life (EOL). They can then be used for a variety of stationary applications, such as grid energy distribution, thus giving them a second life before recycling [16]. The health condition can be described by the state-of-health (SOH) parameter. In some applications where the power capacity is more significant than the energy amount, the internal resistance is generally regarded a SOH metric, and the SOH is therefore defined by the ratio between EOL and real internal resistance and EOL and fresh state internal resistance. In contrast, the SOH is defined as the ratio between the actual battery capacity and the capacity at the beginning of its life for applications wherein the available energy plays a significant role [17].

Therefore, depending on the application, capacity or internal resistance should be measured. Several techniques are proposed in the literature concerning the battery SOH estimation on board electric vehicles [18–20]. The battery aging state can be theoretically evaluated by knowing the history of the battery. A semi-empirical formula for the SOH identification has been exploited in [21,22], taking inspiration from the Arrhenius equation for ideal gases' behaviour and considering as the main aging agent the lithium-ion loss. This describes the dependency of the battery capacity loss on the number of cycles, temperature, charge, and discharge rate and depth of discharge. This formula may be useful both for estimating battery lifetime and for on-board applications. The equivalent circuit models (ECMs) are well-known model-based strategies that exhibit simplicity and good accuracy [23]. This method parameterizes the model variables in relation to the battery SOH by using experimental data [24]. For the battery aging status analysis, these models take into account the internal resistance increase. This can be accurately measured by using the electrochemical impedance spectroscopy (EIS) technology [25–27]. The EIS is a precise and reliable technique; however, nowadays it is rarely exploited for online applications. The high cost of the instruments required does not allow a large-scale use. Therefore, the internal resistence needs to be estimated on board with a different method. Many empirical data-driven models were developed to this end [28,29]. Among these, autoregressive models [30,31] and state observer models coupled with the extended Kalman filter [32,33] provided good results. A large number of different algorithms can be found, such as the particle swarm support vector machine algorithm (PSO-SVM) [34], a particle filter method [35], or even statistical methods [36]. Moreover, neural networks (NNs) seem to be a promising solution in giving accurate results [37–41]. In these studies, it is demonstrated that NNs continue to be research hotspots, exhibiting great potential in estimating SOH under complex aging conditions, particularly when the data are sufficiently abundant, owing to the advantages of approximation and learning speed. Briefly, ML-based SOH estimation approaches are research focuses and will have a significant impact on the future of transportation electrification. In particular, the feed-forward neural networks (FNNs), the convolutional neural networks (CNNs), and the recurrent long short-term memory (LSTM) are the best-performing NNs according to the literature. A comparison between them is given by Sungwoo Jo et al. (2021) [42] which shows the best performance belonging to LSTM compared with the others two types. However, LSTMs may require high computational cost and memory use due to the dimension of the memory cell and its complex structure.

The BMS handles the battery SOH identification task, as well as numerous other functions, including safety control, failure avoidance, and energy consumption optimization. Typically, automotive boards are supplied with ARM processors embedding a 32-bit architecture (multi-core), which can provide adequate processing power [43,44]. However, it is expected that more and more data and tasks will need to be stored and fulfilled as a consequence of technological advancements [43]. Performing some tasks by using external cloud devices could be a solution for this issue, yet it requires effective and reliable internet communication [45]. In [46], an LSTM for remaning useful life (RUL) estimation by using multichannel full charge profiles is presented, with considerable improvements over the baseline LSTM and a significant reduction in the amount of the parameters considered for the model. However, entire charge cycle data is employed, resulting in a large amount of memory and processing space. As a result, [47] develops an RNN-LSTM to estimate the RUL based on partial charge data in the voltage domain range, setting boundary limits. However, the complete SOC domain is not explored, and it is unclear how much memory and computational cost may be saved by varying the different SOC window lengths during charge for SOH estimation. Hence, the computational and memory use reduction for SOH estimation through a data-driven model is a current research gap.

To help fill the highlighted research gap, the main contribution of this study relates to the estimation of the battery SOH from partial charging data and varying the SOC window length through a bidirectional LSTM (Bi-LSTM) in order to reduce the on-board computational cost and memory use while maintaining a high degree of precision, consistent with other research studies in the literature that use the full charge data [48]. In particular, sensitivity analyses are performed to determine the minimum amount of data required in a battery charge process to ensure a good SOH estimation. Furthermore, the charge phase (in terms of SOC range) which is most reliable for the SOH estimation is assessed. This is done by training several Bi-LSTM NNs with data of charging made up of different lengths and considering different SOC windows. The Bi-LSTM neural network (NN) is a wide temporal prediction technique used for SOH estimation, and its predictive powers are derived from learning the forward and backward temporal correlation information in the input data [49,50]. As part of the learning process, many model parameters are automatically tuned based on the user-defined hyperparameters selected from a large pool of solutions. The best hyperparameter training combination has been determined separately for each experiment with a random-search algorithm. The final aim is to provide a light methodology from the computational and memory use points of view for the on-board estimation of the battery SOH, exploiting a data-logged time series that is as short as possible. Final results presented in the last part of this activity highlight that the method of partial charging data can be taken into account for the SOH on-board estimation to have a reduction of the computational demand at the control unit level with a saving of memory usage thanks to shorter data logging.

## 2. Materials and Methods

The current investigation involved estimating the remaining battery life while performing charge–discharge cycles. Several different cycle aging experimental tests have been performed in the literature [51] considering different cell chemistries. For our purpose, the selected aging dataset was that of Sandia National Laboratories [52]. This study has focused on the influence of cell operating conditions on long-term degradation of 18,650 nickel–manganese–cobalt (NMC) cells. Several bidirectional LSTM networks were created to investigate the accuracy in predicting the SOH prediction during partial charging phases with variable time lengths. The best SOC range for the SOH estimation during a single partial charging was finally evaluated. As a result, by knowing the optimal battery SOC window for SOH prediction, the battery's health management system may be improved.

In this section, the proposed method composed by sequential steps is discussed and shown in Figure 1.

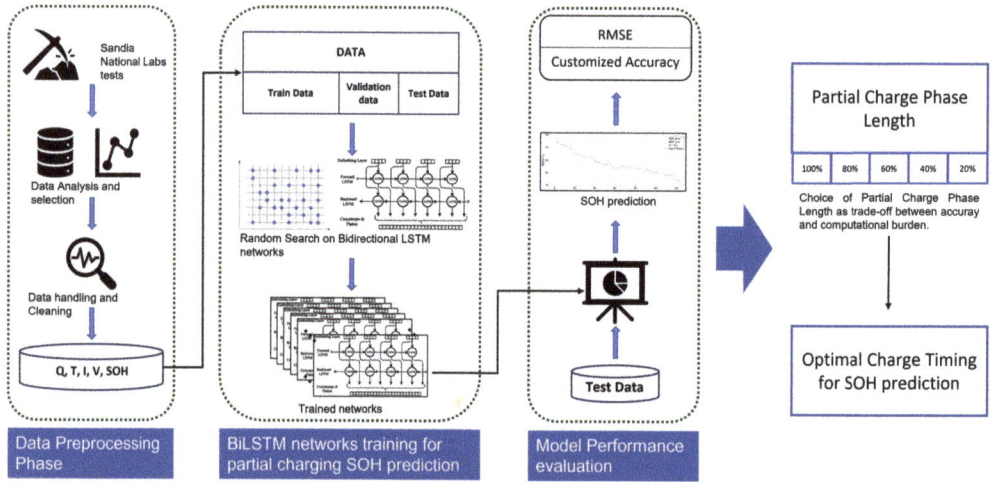

**Figure 1.** The proposed methodology for battery SOH prediction during partial charging processes.

In the data processing phase, the cell signals acquired from cycle aging tests were analysed, handled, and cleaned. In the second step related to Bi-LSTM networks training phase, the data were split in training and validation datasets, and then exploited to perform the learning process of several Bi-LSTM architectures. The random search algorithm was used as a powerful hyperparameter tuning technique to find the most accurate network layout. Grid search and random search are often the most prevalent hyperparameter optimization approaches utilized for this purpose. From a computational cost standpoint, the latter enables the analysis of a larger number of neural networks to choose the best, hence lowering the time required to find the optimal hyperparameters [53]. During the learning process, the created dataset is randomly divided in training, validation, and test sets in order to train and validate the selected AI logics. Bi-LSTM NNs were used in this work due to their excellent capability and performance in time-series forecasting and learning the key paths in cell cycle aging events [54]. The Bi-LSTM is an extendend form of the baseline LSTM NN, and it is composed of two LSTM networks which process data in both forward and backward directions. An LSTM-based model contains a "gate" block that enables storing longer time sequences of data in the memory. Because Bi-LSTM models enable additional training by forward and backward data processing, Bi-LSTM-based modeling gives better performance and prediction with respect to regular LSTM-based models [55]. As with any AI model, the Bi-LSTM is defined by a set of hyperparameters that must be specified in order to tailor the model for the specific application. The random search optimization technique was used here to tune the hyperparameters [56]. In the final step, the performance of the identified best Bi-LSTM network was evaluated by considering a test dataset according to two different metrics, i.e., RMSE and a customized accuracy parameter. The described method was then exploited to find the best partial charging time length for the accurate and computationally lightweight estimation of the battery SOH. Finally, the most suitable SOC window for estimating the battery SOH during a vehicle charging event was assessed.

### 2.1. Data Preprocessing Phase

In the present work, the cell under examination was of 18,650 type with NMC chemistry on the cathode and graphite on the anode. The cycle aging tests have been performed by using a multi-channel battery testing system. Moreover, the cycle aging protocol is reported from Sandia National Laboratories [52]. A summary of test equipment and test operating conditions are, respectively, reported in Tables 1 and 2.

**Table 1.** Sandia National Laboratories equipment for cycle aging experiments [52].

| Cell Type | Cathode | Anode | Capacity (Ah) | Test Equipment |
|---|---|---|---|---|
| 18,650 NMC | NMC | graphite | 3.00 | High-precision Arbin |

**Table 2.** Test operating conditions of cycle aging experiments. N° Cycles is the number of charge and discharge cycles that a cell can process before it reaches its end-of-life condition (20% of capacity loss).

| Charge C Rate | Discharge Crate | SOC Range | Environmental Temperature (°C) | N° Cycles |
|---|---|---|---|---|
| 0.50 | 2.00 | 0–100 | 25 | 661 |

The cell has been charged through a constant current constant voltage (CCCV) protocol, with 0.5 C current during CC phase and current taper to 0.05 A on CV. The NMC cell has been cycled from 2 to 4.2 V during all cycling tests for the whole SOC domain. A portion of the experimental test acquisition and the exploited CCCV protocol are reported in Figure 2.

The data acquisition system collected the following signals over time:
- Cycle index, number of charge–discharge cycle;
- Cell current [A];
- Cell voltage [V];
- Charge and discharge capacity [Ah];
- Charge and discharge energy [Wh];
- Cell temperature [°C];
- Environmental temperature [°C].

The SOH parameter was computed after the cell residual capacity has been determined at the end of each $i$th cycle by using Equation (1),

$$SOH_i = \frac{Q_{actual,i}}{Q_{rated}}, \qquad (1)$$

where $Q_{actual,i}$ is the capacity computed at the $i_{th}$ charge–discharge cycle and $Q_{rated}$ is the cell nominal capacity.

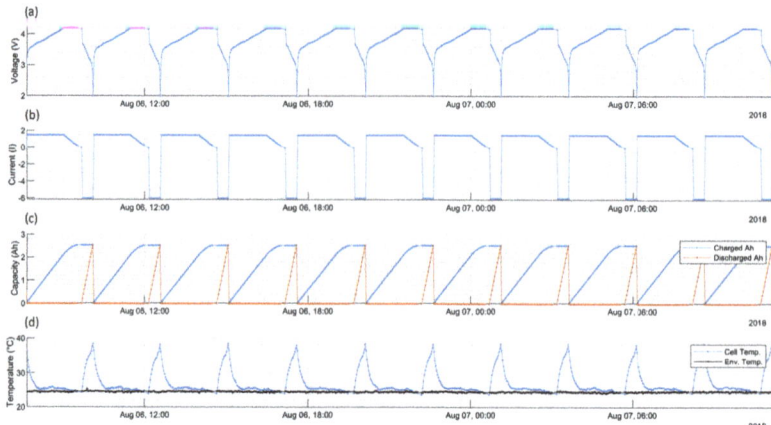

**Figure 2.** Example of constant current constant voltage charge profiles measured by Sandia National Laboratories during the performed experimental cell aging campaign. The cell operating parameters measured during the tests are (**a**) cell voltge (V), (**b**) current (A), (**c**) charged and discharged capacity (Ah), and (**d**) cell and environmental temperature (°C).

Before being used for training the AI algorithms, the acquired data were preprocessed by checking their robustness and quality. Examples include the identification and removal of anomalous traces, identified nans, and outliers of output signals. Additionally, the signals were cut so that the various case studies under consideration could take into account only the specific data of interest over time in order to accelerate the neural network training process. Finally, the acquired data were resempled from varying to constant frequency over time. Specifically, a sample time of 5 s is used to interpolate the considered data. Finally, the obtained dataset included a number of charging cycles from new cell conditions up to their EOL. Each cycle comprised of signals over time for cell temperature, voltage, current, charged capacity, and the corresponding SOH value.

### 2.2. AI Neural Networks Learning Process

The developed AI model, Bi-LSTM architectures, and the hyperparameters involved are shown in Figure 3.

The Bi-LSTM model had an input layer with the dimension of that input data, a batch normalization layer, a Bi-LSTM layer for learning long-term dependencies between cell parameters, and the SOH value to be predicted, a dropout layer to prevent overfitting [58], a fully connected layer for the SOH output forecasting, and the output layer, i.e., a regression layer that computed the loss function. As far as the training process is concerned, the training method reported in the box of Figure 3 was the algorithm used to perform optimization and is by far the most common way to optimize neural networks. An overview of all optimization technique can be seen in [59]. The learning phase lasts a certain number of epochs, which specifies how many times the whole dataset has been thoroughly processed.

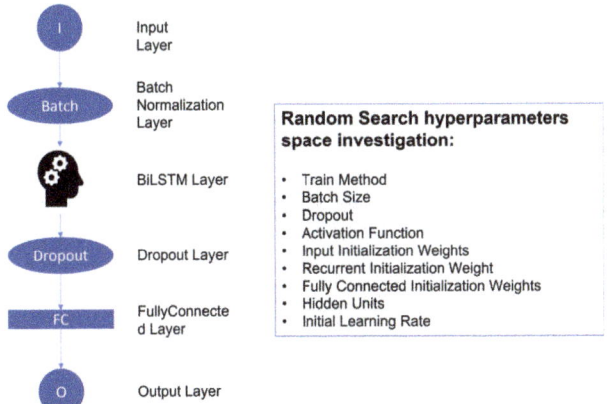

**Figure 3.** AI model graph based on Bi-LSTM network and hyperparameters investigated by the random-search optimization technique. The input layer normalizes data with the z-score method. A batch normalization layer normalizes a batch of data across all observations. The Bi-LSTM layer is defined by the number of hidden units (hidden state, correspondent to the number of information remembered between time steps), the activation function to update the cell and hidden state, and the weights initialization. The dropout layer randomly drops out input elements. The fully connected layer performs output of one dimension. The output layer computes the half-mean-squared-error loss for the regression task [57].

This study focused on the estimation of SOH based on charge cycles, which are examples of time sequences. The problem is therefore based on sequence-to-one regression networks and the loss function is the half-mean-squared-error shown in the Equation (2),

$$Loss = \frac{1}{2} \sum_{i=1}^{N} \frac{(\hat{y}_i - y_i)^2}{N}, \qquad (2)$$

where $N$ is the number of responses, $y_i$ is the target output, and $\hat{y}_i$ is the network's prediction for response $i$. Finally, an early stopping tecnhique was applied when the performance of the validation phase started to degrade in order to avoid overfitting on training dataset [60].

*2.3. Model Performance Evaluation*

Each model was evaluated and selected by taking into account different performance metrics for the SOH prediction results. Together with test data, the performance of all trained Bi-LSTM architectures was analysed based on:

- the RMSE considering the test dataset,
- the coefficient of determination $R^2$, and
- the customized regression accuracy (CRA) coefficient, which compared the predicted SOH, $S\hat{O}H$, with the corresponding measured value, $SOH$ through an identified threshold $thr$.

The specific performance evaluation of the neural networks and the selection of the best hyperparameter values are widely discussed and analysed in the Results and Discussion section.

*2.4. Variable SOC Windows during Partial Charging Events*

The time required to fully or partially charge the battery pack of an electric car is a crucial issue for most drivers. Depending on the charging power available from the grid, the battery pack charging process may take up to several hours. As a result, the current effort focused on determining the appropriate partial charging length as a trade-off in terms of accuracy and computational cost for on-board SOH estimation. Moreover, restraints in memory usage and data storage capacity is a widely known issue in modern on-board control units for passenger cars. Hence, reducing the data logged on the BMS could solve this problem. In particular, the shorter the length of partial charging data logged over time, the smaller the memory required and the computational cost for on-board data processing. Furthermore, the dataset sampling rate is a relevant aspect in memory use reduction. Due to the low dynamic range of the signal of interest, the sample rate for this activity has been set at 0.2 Hz.

The approach described in the previous section was developed in order to estimate the cell SOH by using only a portion of the data related to the overall battery charging process. In the first test scenario, before running the train–test split process regarding data, various lengths of partial charging segments over time were considered in the preprocessing step. In order to consider fixed portions of the 0–100% SOC window, the time lengths investigated were determined as a percentage. It should be noted that the more aged a cell is, the shorter the amount of data logged for a certain percentage of the SOC window due to capacity fading, as seen in Figure 4.

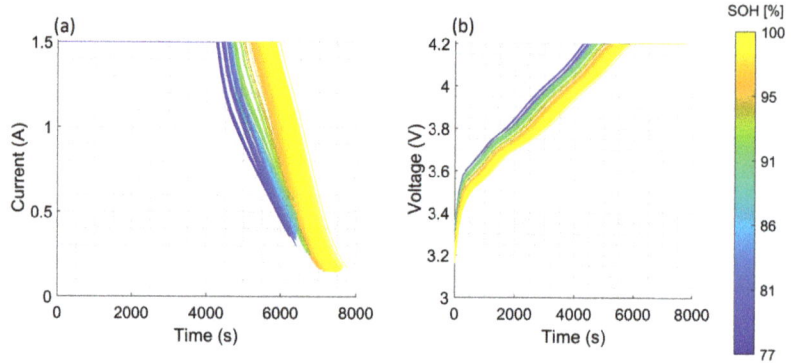

**Figure 4.** The acquired (**a**) current (A) and (**b**) voltage (V) are plotted as time series for each independent charging cycle.

Therefore, different data series of partial charging were expressed as a percentage of the total SOC domain and the retained SOC intervals were:
- 80%
- 60%
- 40%
- 20%.

The partial charge segments were collected for each single cycle over time considering a random starting point. The cut point was randomly chosen among a certain area of points to guarantee that the segments were mathcing the whole length data. If $n_k$ is the length of the $k_t h$ cycle in terms of samples data over the entire SOC domain, $L$ is the selected length as partial charging size related to the specific SOC window, the cut space $S$ from which the segment starting point was randomly selected can be defined in Equation (3):

$$0 \leq S \leq n_k - L. \tag{3}$$

Examples of retained partial charging segments, respectively related to SOC ranges of 80%, 60%, 40%, and 20% with respect to the entire SOC window, are highlighted in Figure 5.

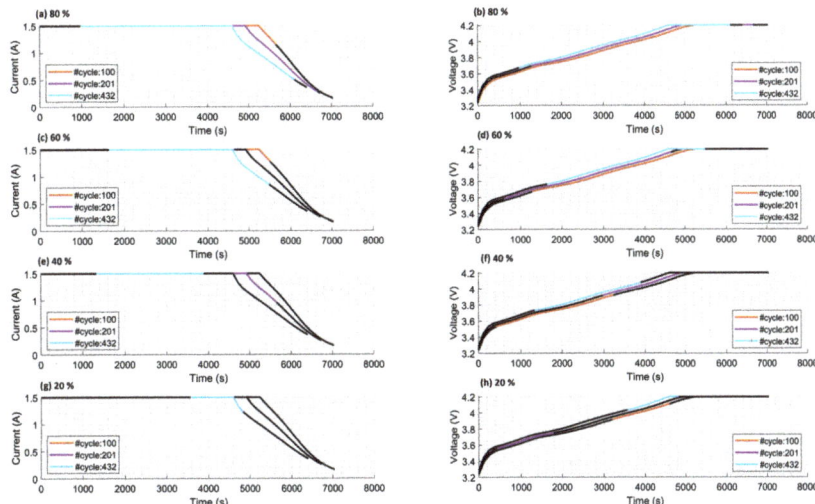

**Figure 5.** Operating current and voltage of few partial charging segments. (**a**,**b**) are operating current and voltage of the partial charging length equal to 80%. (**c**,**d**) are operating current and voltage of the partial charging length equal to 60%. (**e**,**f**) are operating current and voltage of the partial charging length equal to 40%. (**g**,**h**) are operating current and voltage of the partial charging length equal to 20%. In the graphs, the black lines is the full size of data (equal to 100% of length).

In order to proceed with the learning process for the Bi-LSTM neural networks, the dataset of each partial charge segment was randomly subdivided into training, validation, and test data [61]. Here, an 80–20% split was retained between the training and validation dataset on one hand, and the test dataset on the other. The features were the cell operating parameters, such as voltage (V), current (A), charged capacity in time (Ah), and cell temperature (°C). The target was the SOH values to be predicted by the model.

### 2.5. Best Partial Charging Length and Optimal SOC Window Identification for SOH Estimation

In this section, Bi-LSTM networks were trained, and the optimal topologies for each charging length considered were identified. The purpose of the investigation was to determine the optimal SOC range for estimating the remaining life of a cell during its charging process. Before analyzing the best SOC window for the SOH estimation, it was necessary to determine the optimal partial charging length $L_{opt}$. Indeed, a trade-off between prediction accuracy (RMSE, CRA), computational cost and memory use was analysed for the on-board SOH estimation by control units of Li-ion battery packs.

A sensitivity analysis was conducted over the best 1, 5, and 10 Bi-LSTM networks considering RMSE and CRA for each charging length. The computational costs for each considered charging length were investigated, retaining the time required to run the numerical models for cell SOH estimation. Moreover, the memory storage capacity was analysed based on the memory used by the models and the data logged. Finally, the trade-off-based optimal input length was found and employed for the best SOC window analysis. A complete explanation of sensitivity and trade-off analysis will be described in the Results and Discussion section.

Once the optimal partial length $L_{opt}$ was obtained, a new dataset was generated by cutting data at different starting points among the full size data of cycles. Particularly, the cut points were defined at each 10% step in the SOC window until reaching the last point, which guaranteed the contiguous size of data, i.e., while respecting the 0% and 100%

SOC limits. Hence, the SOC windows $SOC_{win}$ among the entire domain are shown in Equation (4) and expressed as a percentage of size data of a single charging cycle:

$$SOC_{L_{opt}} = [0; L_{opt}], [10; L_{opt} + 10], ..., [100 - L_{opt}; 100]. \tag{4}$$

For instance, if the optimal length $L_{opt}$ was observed to be 40%, then the $SOC_{L_{opt}}$ is reported in Equation (5):

$$SOC_{40} = [0; 40], [10; 50], [20; 60], [30; 70], [40; 80], [50; 90], [60; 100]. \tag{5}$$

Considering the example of dataset shown in Figure 5 for $L_{opt}$ equal to 40%, the related charging data are illustrated in Figure 6.

In this analysis, once $L_{opt}$ was determined, the same split of data was preserved between training, validation, and test. This allowed for a careful analysis of the findings pertaining to the selection of the threshold value *thr* utilized in the definition of the CRA, as stated in the Results and Discussion section.

As far as the learning process of the neural networks is concerned, the top 30 Bi-LSTM-trained networks from the previous section were used for a new learning process. However, the regression task, the feature definitions, and the target variable were identical. Finally, the optimal SOC range for capacity degradation estimation during charging events was determined.

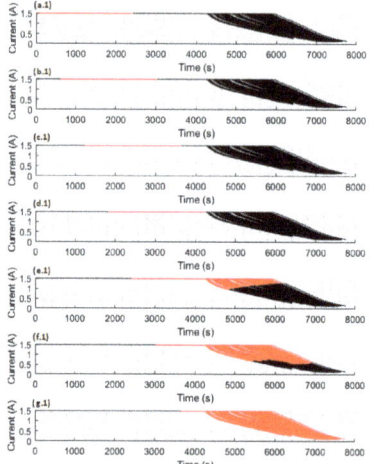
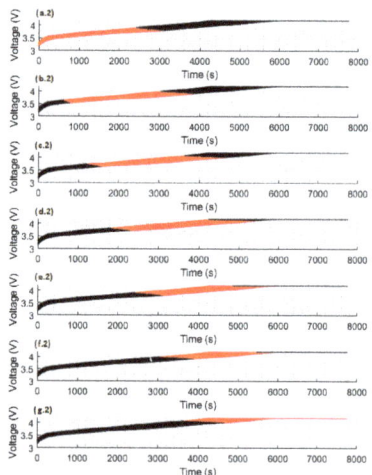

**Figure 6.** Fixed SOC window equal to 40% moving over the entire domain, for the generation of datasets. (**a.1,a.2**) are, respectively, current and voltage of SOC window [0,40]. (**b.1,b.2**) are, respectively, current and voltage of SOC window [10,50]. (**c.1,c.2**) are, respectively, current and voltage of SOC window [20,60]. (**d.1,d.2**) are, respectively, current and voltage of SOC window [30,70]. (**e.1,e.2**) are, respectively, current and voltage of SOC window [40,80]. (**f.1,f.2**) are, respectively, current and voltage of the SOC window [50,90]. (**g.1,g.2**) are, respectively, current and voltage of SOC window [60,100]. In the graph, curves for the entire cycles are plotted in black.

## 3. Results and Discussion

In this work, an AI-based SOH estimator was developed considering partial charging of a Li-ion cell to reduce the computational cost and memory occupancy for BMS applications. As already detailed in the Materials and Methods section, cycle aging tests were exploited for model developing, and through an optimization technique Bi-LSTM network architectures were established for their high performance in forcasting task and managing time-dependent data.

Before delving further into the numerical findings, a broad view of the performance metrics retained in analyzing the outcomes must be provided. The performance of the trained models was assessed by taking into account the estimation capabilities over the test dataset. In general, the model outputs were studied by comparing them with real targets in terms of the absolute error shown in Equation (6). The main metrics considered in order to evaluate the quality of the model predictions were RMSE reported in Equation (7). The coefficient of determination $R^2$ is reported in Equation (8) along with the customized regression accuracy (CRA). The CRA parameter can be defined by considering the problem to be similar with a classification task whereby the result was evaluated as correct if the value of the absolute error was lower then a specific threshold, as shown in Equation (9). We have

$$E_i = x_i - \hat{x}_i \tag{6}$$

$$RMSE = \sqrt{\frac{\sum_{i=1}^{n} E_i^2}{n}} \tag{7}$$

$$R^2 = 1 - \frac{\sum_{i=1}^{n} E_i^2}{\sum_{i=1}^{n}(x_i - \bar{x}_i)^2} \tag{8}$$

$$CRA = \frac{\sum_{i=1}^{n} T_i}{n} \times 100 \ \ with : \begin{cases} T_i = 1 & if \ |E_i| < threshold \\ T_i = 0 & if \ |E_i| > threshold \end{cases} \tag{9}$$

$x_i$ was the target value, $\hat{x}_i$ was the model output, $\bar{x}_i$ was the mean of the dataset label considering that each experimental test in the dataset had a number of samples equal to $n$, and $E_i$ is the residual between target and predicted values. All the results shown in this section are derived from the validation of the model on the testing data.

As seen in Figure 7, the *threshold* value defines the accuracy of the model prediction.

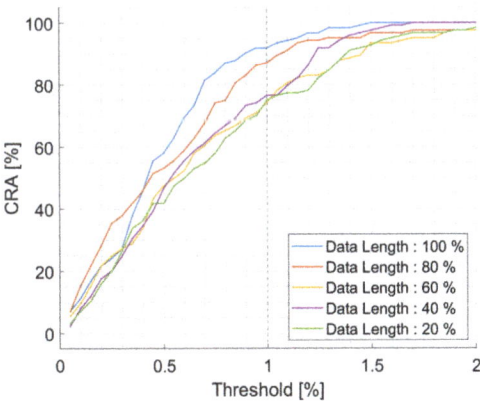

**Figure 7.** Sensitivity analysis of the neural network's accuracy depending on the threshold value as described in Equation (9).

In order to perform the sensitivity analysis, the best NN was determined for each charging length by minimizing the RMSE metric. Looking at Figure 7, the charging length equal to 40% is observed to have the most rapid growth and to be the only one reaching the 100% accuracy among the partial charging lengths. As far as the sensitivity analysis is concerned, a threshold of about 1% was chosen. Here, the threshold represents the tolerance of the estimated cell SOH compared with the related measurements. The 1% tolerance value appears to be consistent with the literature [62] because it has been demonstrated to limit the error in the cell SOH estimation within 2.2%. With the selected threshold value and for a partial length of 40%, the CRA reaches almost 80%. The results in Figure 7 are based on Analysis #1, where the training dataset for Bi-LSTM processing was created by randomly choosing several parts along the SOC domain in order to investigate which is the lower charging length and which still guarantees acceptable accuracy in cell SOH estimation. However, the overall accuracy of Analysis #1 is much lower than Analysis #3, where, in order to analyze which is the best SOC charging window for SOH estimation, each network was trained, respectively, with data from the same SOC window. Hence, in the latter analysis, the prediction results are far greater, and the threshold value may be drastically cut.

*3.1. Analysis #1: Variable SOC Windows for Partial Charging*

After the definition of the metrics involved in the analysis of the SOH estimation accuracy, the present section focuses on the study of the influence of each partial charging length over cell SOH forecasting. The trained model's prediction results, in terms of CRA and RMSE are shown (after the training process phase was performed by the random-search technique) in Figure 8.

Specifically, the top one, five, and 10 trained Bi-LSTM networks are presented for each percentage of charging length analyzed based on the performance standard deviations. The overall trend of CRA is directly proportional to the length of the partial charging considered. Increasing the charging length from 20% to 40%, the test CRA of the five best networks increases by approximately 7%. From the test $RMSE$ point of view, even if the boundary cases 20% and 100% are, respectively, the worst and the best options, the trend of the five best networks changes. Being an AI model based on data, a large number of observations are required to find and recognize some specific patterns, especially when random approaches are exploited for generalization purposes. However, the objective of the study is to understand and investigate whether high accuracy can be attained for the cell SOH estimation with only partial charging events. For instance, looking at Figure 8, it can be seen that an accuracy level of roughly 77% can be attained by a network setup by using an input of 40% of the SOC window during a charge phase. The accuracy definition pertains to the CRA with a threshold parameter value of 1%. Given the customized nature of this metric and the randomness associated with the selection of hyperparameters and training data for the Analysis #1, a CRA value of 77% is not optimal (higher values are reached for the analysis #3). However, the 40% data length achieves very good RMSE and R2 scores compared to the literature. Moreover, excluding the 100% length case corresponding to the full SOC domain, the 40% data length case has the lowest RMSE error. This indicates that an optimal Bi-LSTM configuration is not found for each charging length case by the random process approach, although it is theoretically possible that this could occur after several additional iterations. Finally, the significant result of the analysis shows that the SOH of a cell can be carefully detected by just monitoring 40% of the whole 0–100% SOC charge process.

In Figure 9, the regression task results for the best Bi-LSTM network per each input charging length are represented.

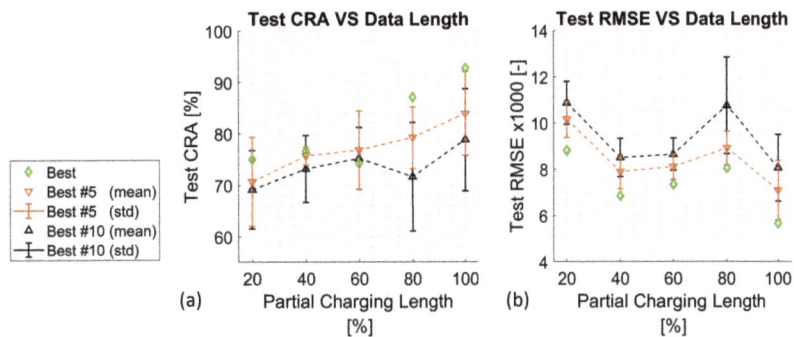

**Figure 8.** (**a**) CRA sensitivity analysis of the best one, five and 10 trained networks according to the RMSE on testing dataset. (**b**) RMSE sensitivity analysis of the best one, five and 10 trained neural networks according to the RMSE on testing dataset. For partial charging lengths, the minimum RMSE is equal to 0.0068 corresponding to 40% data length.

As summary results, Tables 3 and 4 resepectively report the validation performance of the developed models and the details of trained Bi-LSTM architectures for each SOC window length.

**Table 3.** Analysis #1: Best neural network regression statistics.

| Data Length [% of SOC] | 100 | 80 | 60 | 40 | 20 |
|---|---|---|---|---|---|
| m | 2.67 | 5.82 | 9.92 | 8.81 | 3.68 |
| q | 0.97 | 0.93 | 0.88 | 0.89 | 0.96 |
| Test RMSE $\times 1000$ | 5.65 | 8.04 | 7.33 | 6.80 | 8.93 |
| Test $R^2$ | 0.99 | 0.97 | 0.96 | 0.96 | 0.96 |

**Table 4.** Analysis #1: Best neural network training and architecture parameters.

| Data Length [% of SOC] | 100 | 80 | 60 | 40 | 20 |
|---|---|---|---|---|---|
| Hidden Layers | 1 | 1 | 1 | 1 | 1 |
| Hidden Neurons | 59 | 30 | 29 | 47 | 52 |
| State Activation Function | tanh | tanh | tanh | tanh | softsign |
| DropOut | 0.2 | 0.3 | 0.2 | 0.1 | 0.1 |
| Batch Size | 128 | 32 | 64 | 32 | 64 |
| Learning Rate | 0.0090 | 0.0089 | 0.0060 | 0.0069 | 0.0044 |
| Optimization Algorithm | sgdm | sgdm | adam | sgdm | adam |
| Training Epochs | 190 | 84 | 108 | 264 | 186 |

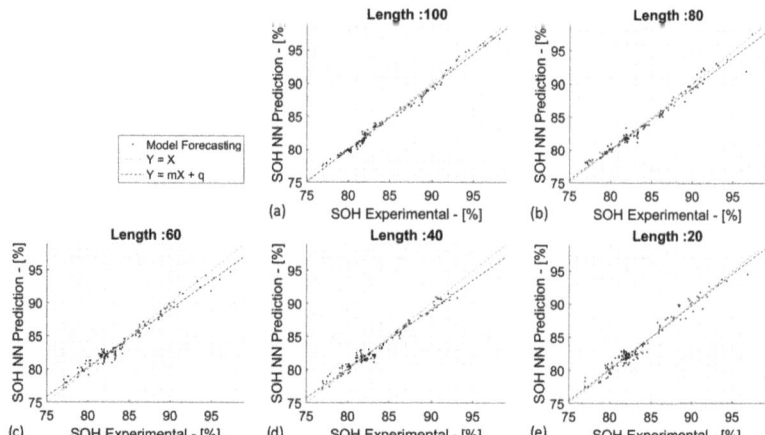

**Figure 9.** Best neural network regression performance. (**a**) refers to the full charging length equal to 100%, (**b**) refers to the partial charging length equal to 80%, (**c**) refers to the partial charging length equal to 60%, (**d**) refers to the partial charging length equal to 40%, (**e**) refers to the partial charging length equal to 20%. The black points represents the correlation points between predicted and target values. The green dashed line is the bisector, and the red dashed line is the fitting regression line. The regression parameters can be observed in Table 3. The predicted SOH points over the entire cycle aging test are those of test set for perfomance validation.

*3.2. Analysis #2: Computational Cost and Memory Occupancy for Best SOC Charging Window Length Identification*

In the on-board implementation of an SOH estimator, the computational power and memory occupancy are critical issues in current vehicle control units. In the present work, a profiling analysis was investigated as a performance metric together with $CRA$ and $RMSE$ in order to identify the best SOC window length for capacity fade monitoring. Hence, a profiling approach was developed to quantify the benefits of the proposed method in terms of computational costs and memory usage. In Figure 10, the computational performance required by the electronic control unit for the processing phase is shown.

The elapsed time in Figure 10 was computed by considering the average time for 10 runs among the best 30 neural networks for each charging dataset input length. The elapsed time seems to be almost linear. The computational time was processed through a laptop with Intel(R) Core (TM) i7-10510U CPU @ 1.80 GHz and 16 GB of RAM. The memory occupancy of stored data is plotted against input charging length and the figure clearly shows a linear behaviour. The longer the time series considered for the processing phase, the higher the space required by the memory. Finally, the memory used for the Bi-LSTM network size was computed as the required space to store the top 10 neural networks architectures for each data length. The main memory reduction factor is due to the dataset size reduction, which is clearly linear. On the other hand, the Bi-LSTM model sizes vary according to the same range (20 kB to 230 kB) for each input data length. As a consequence, the choice of the suitable SOC window length does not depend on the model dimension in this case.

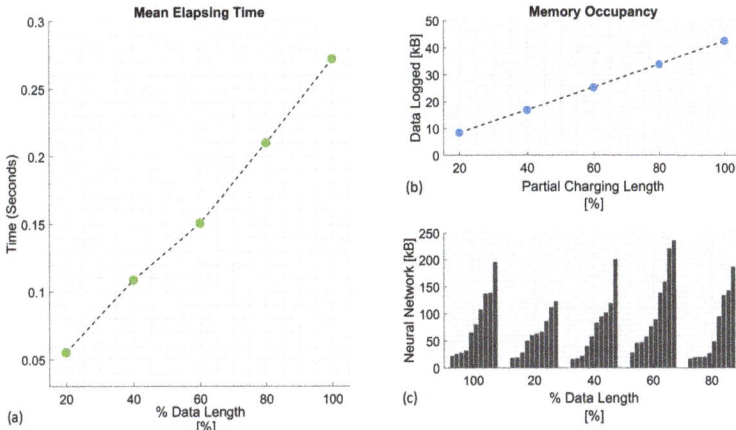

**Figure 10.** The plotted values are referred to the results shown in Analysis #1. (**a**) Time consumption during neural network prediction and depending on SOC window length. (**b**) Memory occupancy by logging data and depending on SOC window length. (**c**) Memory occupancy by Bi-LSTM networks and depending on SOC window length.

### 3.3. Analysis #3: Best SOC Window Identification for Optimal SOH Estimation

Considering the results obtained by Analysis #1 and Analysis #2, it can be assumed that the configuration with the 40% length of input data is the best SOC window length $L_{opt}$ in terms of cell SOH estimation capability and computational lightweighting trade-off. Therefore, the present section is focused on this charging length value.

In this part, we investigated what specific part of the charge process contains more information about the battery SOH, allowing a better estimation of the battery's remaining lifetime. The analysis results for $L_{opt} = 40\%$ are shown in Figure 11.

As shown in Figure 11, higher performance in terms of $CRA$ and $RMSE$ are obtained due to an accurate reporting of the partial charging start points.

The presented analyses report that the last charging SOC window (with a range between 60% and 100%) guarantees the highest $CRA$ and the lowest $RMSE$ on the cell SOH estimation. Hence, this range is considered to be the optimal SOC window for the cell SOH estimation. Considering the input data representation in Figure 6, the optimal SOC charge range of 60% to 100% corresponds to the constant voltage (CV) phase of the charging process. The CCCV tests are widely employed for the cell SOH estimation and assessing the battery performance while aging [63]. Specifically, the partial CV charging phase is proven to be the most suitable for SOH estimation holding more information and robustness about capacity fade [64]. However, instead of considering a single CV trace, the present work compared and analysed different partial charging lengths over the entire charging process domain. In fact, concerning the best Bi-LSTM neural network, it is important to highlight that the other portions of the domain achieve remarkable results with a CRA accuracy that consistently approaches 90%. Moreover, it is interesting that the RMSE value for the 0–40% trace is only slightly higher, i.e., 0.0054, than the best value of 0.0012 for the 60–100% trace. However, it should be noted that evaluating the top 15 or 25 neural network configurations displays higher differences between the SOC windows. This confirms the prevalence of the typical aging pattern in a certain SOC range, i.e., 60–100%.

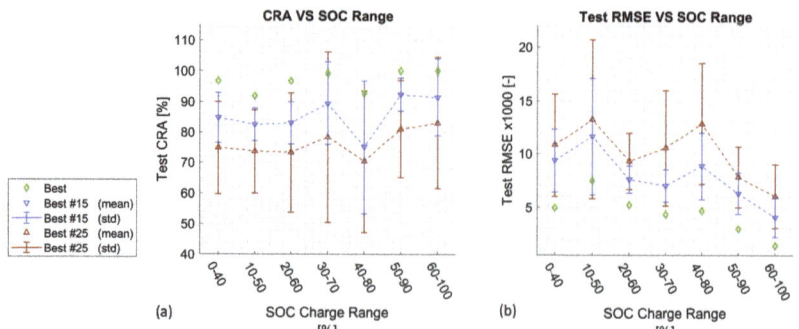

**Figure 11.** Sensitivity analysis of the best 15 and 25 neural networks according to the RMSE on the testing dataset for $L_{opt} = 40\%$. (**a**) CRA trend depending on the input SOC window selected. (**b**) Test RMSE ×1000 depending on the input SOC window selected.

In Figure 12, the regression results for the best Bi-LSTM network for each SOC charge range considered were represented, and we can observe that the 60–100% range case has the smallest deviation in the regression line by bisector, and it has the predicted points densely packed on the bisector.

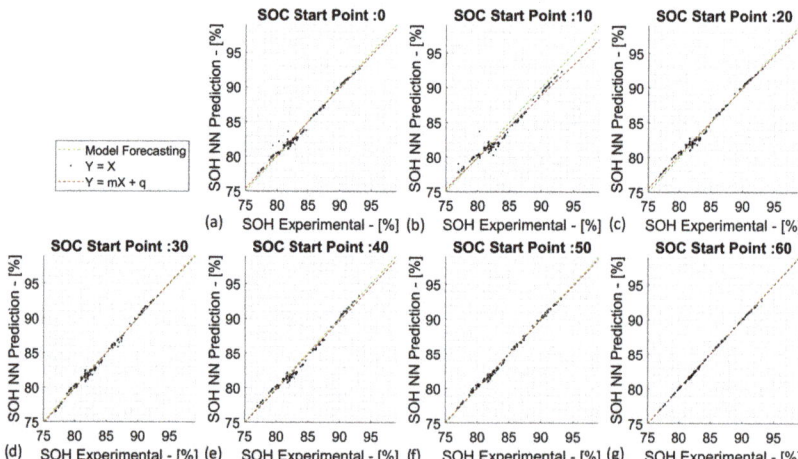

**Figure 12.** Best neural network regression performance. (**a**) refers to the SOC window [0,40], (**b**) refers to the SOC window [10,50], (**c**) refers to the SOC window [20,60], (**d**) refers to the SOC window [30,70], (**e**) refers to the SOC window [40,80], (**f**) refers to the SOC window [50,90], (**g**) refers to the SOC window [60,100]. The black points represents the correlation points between predicted and target values. The green dashed line is the bisector, and the red dashed line is the fitting regression line. The regression parameters can be observed in Table 5. The predicted SOH points over the entire cycle aging test are those of test set for perfomance validation.

As summary results, Tables 5 and 6 show, resepectively, the validation performance of the developed models and the details of the trained Bi-LSTM architectures for each SOC charge range analysed.

Table 5. Analysis #3: Best neural network regression statistics.

| Charge Segment SOC [%] | 0–40 | 10–50 | 20–60 | 30–70 | 40–80 | 50–90 | 60–100 |
|---|---|---|---|---|---|---|---|
| m | 0.96 | 0.90 | 0.96 | 1.02 | 0.98 | 0.98 | 0.99 |
| q | 3.29 | 7.99 | 3.82 | −1.59 | 1.34 | 1.35 | 0.30 |
| Test RMSE $\times 1000$ | 5.45 | 7.52 | 5.23 | 4.09 | 4.81 | 3.12 | 1.27 |
| Test $R^2$ | 0.98 | 0.95 | 0.98 | 0.99 | 0.98 | 0.99 | 0.99 |

Table 6. Analysis #3: Best neural network training and architecture parameters.

| Charge Segment SOC [%] | 0–40 | 10–50 | 20–60 | 30–70 | 40–80 | 50–90 | 60–100 |
|---|---|---|---|---|---|---|---|
| Hidden Layers | 1 | 1 | 1 | 1 | 1 | 1 | 1 |
| Hidden Neurons | 15 | 15 | 65 | 54 | 42 | 51 | 51 |
| State Activation Function | tanh | tanh | softsign | softsign | tanh | softsign | softsign |
| DropOut | 0.3 | 0.3 | 0.3 | 0.2 | 0.5 | 0.5 | 0.5 |
| Batch Size | 16 | 16 | 32 | 64 | 64 | 32 | 32 |
| Learning Rate | 0.0098 | 0.0098 | 0.0055 | 0.0099 | 0.0086 | 0.0083 | 0.0083 |
| Optimization Algorithm | sgdm | sgdm | rmsprop | sgdm | sgdm | sgdm | sgdm |
| Training Epochs | 92 | 30 | 43 | 155 | 103 | 91 | 62 |

*3.4. Best SOC Window: Training and Validation Information*

In this final section, training and validation performance details about the best Bi-LSTM network for the best SOC window of approximately 60–100% were discussed. As already explained, the performance in terms of $CRA$ and $RMSE$ of this trained network with the homogeneous dataset are much more analytically compared with Figures 8 and 9 from Analysis #1. The statistical metrics and Bi-LSTM architecture details are those shown in Tables 5 and 6. In Figure 13, the training history along the epochs are reported, comparing the loss function between the training dataset and validation dataset.

The learning process shows a trend that seems to be in line with good fit results, thus excluding overfitting and underfitting of a training phase. Finally, Figure 14 plots the SOH predicted points as those composing the test set among all aging cycles.

The results shown in the figure ensure promising performance in SOH estimation capability reaching 100% CRA and a low residual error for each cycle prediction, i.e., that the uncertainty of prediction is within 1%. However, because of the strong forecasting performance, it is feasible to achieve the same accuracy value of 100% in an area of uncertainty (threshold) decreased to 0.4% by analyzing the punctual error. For the sake of clarity, the CRA defines the percentage by which a model estimation may fall within a given uncertainty range of the target. The amplitude of the range is described by the threshold parameter value.

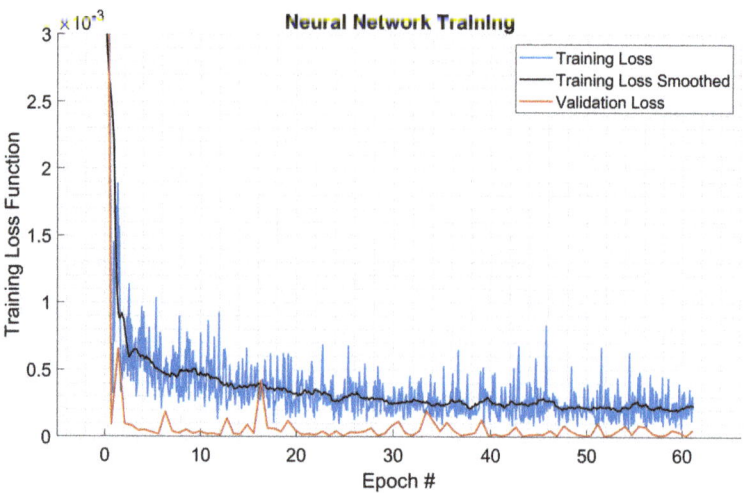

**Figure 13.** Training process along the epochs of Bi-LSTM related to the best SOC range.

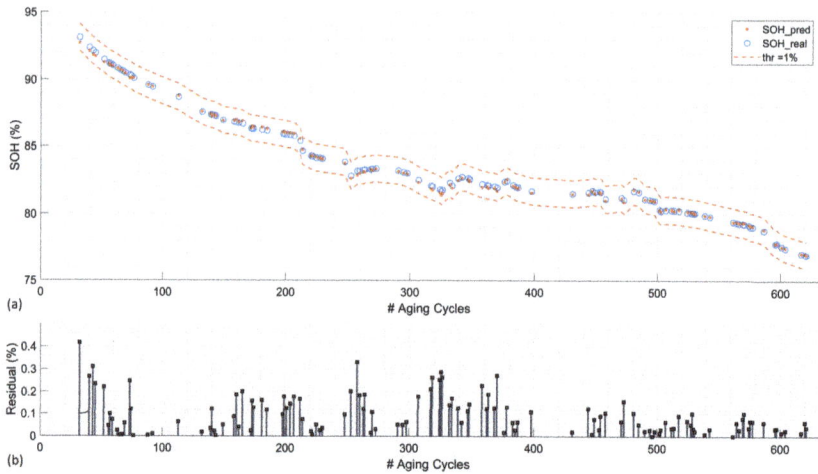

**Figure 14.** Neural network SOH estimation performance on testing dataset. # Aging Cycles is the number of cycles that one cell has cycled.

## 4. Conclusions

This study proposed a computationally lightweight methodology for the cell SOH estimation on board electric vehicles during partial charging processes. Several Bi-LSTM neural networks were trained, exploiting different datasets made of battery-charging data time series with varying lengths and random selection of the start point within the battery SOC domain. The proposed methodology identified the best SOC window length as a trade-off between the prediction accuracy and the computational cost for onboard SOH estimation. Moreover, the optimal SOC charge range which allows higher performance for the cell SOH estimation is identified. The proposed neural network considers time series made of the cell current, voltage, temperature, and the capacity charged while the output is the single regression value of the SOH. The case study retains an 18,650 cell with 3 Ah capacity and nickel–manganese–cobalt chemistry whereas the dataset is part of a collection of cycle aging tests performed by the Sandia Nation Laboratories. The results are consistent and show that the battery SOH can be predicted with the highest error of ±0.4%, monitoring just the last 40% of the SOC window (CV phase) of the total CCCV charge process by reducing the memory occupancy in the BMS for charge data logging and the computational time by a factor of about 2.3. Just a single cell operating condition has been considered in this study, consisting of charging and discharging at constant current and environmental temperature. In future work, the methodology could be extended to a wider range of working conditions, including additional temperatures, charging and discharging operations. Furthermore, the length of the charge segment to be monitored has been identified through a process that does not allow us to find the globally optimal solution. Therefore, additional improvements could be obtained by developing appropriate fine-tuning methodologies. An additional future development concerns the study of optimization of Bi-LSTM neural network architectures constrained to having smaller dimensions.

**Author Contributions:** Conceptualization, A.F. and T.A.G.; methodology, A.F. and T.A.G.; software, A.F. and T.A.G.; validation, A.F. and T.A.G.; formal analysis, A.F. and T.A.G.; investigation, A.F. and T.A.G.; resources, A.F. and T.A.G.; data curation, A.F. and T.A.G.; writing—original draft preparation, A.F. and T.A.G.; writing—review and editing, A.F., T.A.G., D.M., and P.G.A.; visualization, A.F. and T.A.G.; supervision, D.M. and P.G.A.; project administration, D.M. All authors have read and agreed to the published version of the manuscript.

**Funding:** This research received no external funding.

**Institutional Review Board Statement:** Not applicable.

**Informed Consent Statement:** Not applicable.

**Data Availability Statement:** Data available in a publicly accessible repository that does not issue DOIs. Publicly available datasets were analyzed in this study. This data can be found here: [https://www.batteryarchive.org/index.html].

**Conflicts of Interest:** The authors declare no conflict of interest.

## Abbreviations

The following abbreviations are used in this manuscript:

| | |
|---|---|
| SOH | State-of-Health |
| BMS | Battery Management System |
| AI | Artificial Intelligence |
| SOC | State of Charge |
| Bi-LSTM | Bidirectional Long Short-term Memory |
| CCCV | Constant Current Constant Voltage |
| RMSE | Root Mean Squared Error |
| CV | Constant Voltage |
| EOL | End of Life |
| ECM | Equivalent Circuit Model |
| EIS | Electrochemical Impedance Spectroscopy |
| PSO | Particle Swarm Optimizer |
| FNN | Feed-Forward Neural Netwrk |
| CNN | Convolutional Neural Network |
| NMC | Nickel Manganese Cobalt |
| CRA | Customized Regression Accuracy |
| thr | Threshold |
| $L_{opt}$ | Optimal length |
| $SOC_{win}$ | SOC window |
| $R^2$ | Coefficient of determination |
| CPU | Central Processing Unit |

## References

1. Van den Bossche, A. Light and Ultralight electric vehicles. In *4ième Conférence Internationale sur le Génie Electrique (CIGE 2010) (No. 2, pp. 3–9)*; Université de Bechar Algérie: Béchar, Algeria, 2010.
2. Blomgren, G.E. The Development and Future of Lithium Ion Batteries. *J. Electrochem. Soc.* **2016**, *164*, A5019. [CrossRef]
3. Keshan, H.; Thornburg, J.; Ustun, T.S. Comparison of lead-acid and lithium ion batteries for stationary storage in off-grid energy systems. In Proceedings of the 2016 4th IET Clean Energy and Technology Conference, Kuala Lumpur, Malaysia, 14–15 November 2019.
4. Vutetakis, D.; Timmons, J. A Comparison of Lithium-Ion and Lead-Acid Aircraft Batteries. *SAE Tech. Paper* **2008**, *1*, 2875.
5. See, K.W.; Wang, G.; Zhang, Y.; Wang, Y.; Meng, L.; Gu, X.; Zhang, N.; Lim, K.C.; Zhao, L.; Xie, B. Critical review and functional safety of a battery management system for large-scale lithium-ion battery pack technologies. *Int. J. Coal. Sci. Technol.* **2022** *9*, 36. [CrossRef]
6. Feng, X.; Ouyang, M.; Liu, X.; Lu, L.; Xia, Y.; He, X. Thermal runaway mechanism of lithium ion battery for electric vehicles: A review. *Energy Storage Mater.* **2017**, *10*, 246–267. [CrossRef]
7. Tran, M.K.; Mevawalla, A.; Aziz, A.; Panchal, S.; Xie, Y.; Fowler, M. A Review of Lithium-Ion Battery Thermal Runaway Modeling and Diagnosis Approaches. *Processes* **2022**, *10*, 1192. [CrossRef]
8. Leng, F.; Tan, C.; Pecht, M. Effect of Temperature on the Aging rate of Li Ion Battery Operating above Room Temperature. *Sci. Rep.* **2015**, *5*, 12967. [CrossRef]
9. Ma, S.; Jiang, M.; Tao, P.; Song, C.; Wu, J.; Wang, J.; Deng, T.; Shang, W. Temperature effect and thermal impact in lithium-ion batteries: A review. *Prog. Nat. Sci. Mater. Int.* **2018**, *28*, 6. [CrossRef]
10. Jaydeep, B. Effect of Temperature on Battery Life and Performance in Electric Vehicle. *Int. J. Sci. Res.* **2012**, *2*, 1–3.
11. Ouyang, D.; He, Y.; Weng, J.; Liu, J.; Chen, M.; Wang, J. Influence of low temperature conditions on lithium-ion batteries and the application of an insulation material. *RSC Adv.* **2019**, *9*, 9053–9066. [CrossRef]
12. Hossein, M.; Jason, H. Effects of overdischarge on performance and thermal stability of a Li-ion cell. *J. Power Sources* **2006**, *160*, 1395–1402.
13. Xu, B.; Oudalov, A.; Ulbig, A.; Andersson, G.; Kirschen, D.s. Modeling of Lithium-Ion Battery Degradation for Cell Life Assessment. *IEEE Trans. Smart Grid* **2016**, *99*, 1. [CrossRef]
14. Raj, T.; Wang, A.A.; Monroe, C.W.; Howey, D.A. Investigation of Path-Dependent Degradation in Lithium-Ion Batteries. *Eur. Chem. Soc. Publ.* **2020**, *3*, 12. [CrossRef]
15. Barcellona, S.; Colnago, S.; Dotelli, G.; Latorrata, S.; Piegari, L. Aging effect on the variation of Li-ion battery resistance as function of temperature and state of charge. *J. Energy Storage* **2022**, *50*, 104658. [CrossRef]
16. Shahjalal, M.; Roy, P.K.; Shams, T.; Fly, A.; Chowdhury, J.I.; Ahmed, R.; Liu, K. A review on second-life of Li-ion batteries: Prospects, challenges, and issues. *Energy* **2022**, *241*, 122881. [CrossRef]
17. Shu, X.; Shen, S.; Shen, J.; Zhang, Y.; Li, G.; Chen, Z.; Liu, Y. State of health prediction of lithium-ion batteries based on machine learning: Advances and perspectives. *iScience* **2021**, *24*, 11. [CrossRef] [PubMed]
18. Lin, C.; Tang, A.; Wang, W. A Review of SOH Estimation Methods in Lithium-ion Batteries for Electric Vehicle Applications. *Energy Procedia* **2015**, *75*, 1920–1925. [CrossRef]
19. Berecibar, M.; Gandiaga, I.; Villarreal, I.; Omar, N.; Van Mierlo, J.; Van den Bossche, P. Critical review of state of health estimation methods of Li-ion batteries for real applications. *Renew. Sustain. Energy Rev.* **2016**, *56*, 572–587. [CrossRef]
20. Noura, N.; Boulon, L.; Jemeï, S. A Review of Battery State of Health Estimation Methods: Hybrid Electric Vehicle Challenges. *World Electr. Veh. J.* **2020**, *11*, 66. [CrossRef]
21. Han, H.; Xu, H.; Yuan, Z.; Shen, Y. A new SOH prediction model for lithium-ion battery for electric vehicles. In Proceedings of the 2014 17th International Conference on Electrical Machines and Systems (ICEMS), Hangzhou, China, 22–25 October 2014.
22. Anselma, P.G.; Kollmeyer, P.; Lempert, J.; Zhao, Z.; Belingardi, G.; Emadi, A. Battery state-of-health sensitive energy management of hybrid electric vehicles: Lifetime prediction and ageing experimental validation. *Appl. Energy* **2021**, *285*, 116440. [CrossRef]
23. Falai, A.; Giuliacci, T.A.; Misul, D.; Paolieri, G.; Anselma, P.G. Modeling and On-Road Testing of an Electric Two-Wheeler towards Range Prediction and BMS Integration. *Energies* **2022**, *15*, 2431. [CrossRef]
24. Amir, S.; Gulzar, M.; Tarar, M.O.; Naqvi, I.H.; Zaffar, N.A.; Pecht, M.G. Dynamic Equivalent Circuit Model to Estimate State-of-Health of Lithium-Ion Batteries. *IEEE Access* **2022**, *10*, 18279–18288. [CrossRef]
25. Galeotti, M.; Cinà, L.; Giammanco, C.; Cordiner, S.; Di Carlo, A. Performance analysis and SOH (state of health) evaluation of lithium polymer batteries through electrochemical impedance spectroscopy. *Energy* **2015**, *89*, 678–686. [CrossRef]
26. Kieran, M.; Hemtej, G.; Kevin, M.R.; Tadhg, K. Review—Use of impedance spectroscopy for the estimation of Li-ion battery state of charge, state of health and internal temperature. *J. Electrochem. Soc.* **2021**, *168*, 080517.
27. Sihvo, J.; Roinila, T.; Stroe, D.I. SOH analysis of Li-ion battery based on ECM parameters and broadband impedance measurements. In Proceedings of the IECON 2020 the 46th Annual Conference of the IEEE Industrial Electronics Society, Singapore, 18–21 October 2020.
28. Preetpal, S.; Che, C.; Cher, T.; Shyh-Chin, H. Semi-Empirical Capacity Fading Model for SoH Estimation of Li-Ion Batteries. *Appl. Sci.* **2019**, *9*, 3012.

29. Huanyang, H.; Jinhao, M.; Yuhong, W.; Lei, C.; Jichang, P.; Ji, W.; Qian, X.; Tianqi, L.; Remus, T. An Enhanced Data-Driven Model for Lithium-Ion Battery State-of-Health Estimation with Optimized Features and Prior Knowledge. *Automot. Innov.* **2022**, *5*, 134–145. [CrossRef]
30. Huang, J.; Wang, S.; Xu, W.; Shi, W.; Fernandez, C. A Novel Autoregressive Rainflow—Integrated Moving Average Modeling Method for the Accurate State of Health Prediction of Lithium-Ion Batteries. *Processes* **2021**, *9*, 795. [CrossRef]
31. Gao, K.; Xu, J.; Li, Z.; Cai, Z.; Jiang, D.; Zeng, A. A Novel Remaining Useful Life Prediction Method for Capacity Diving Lithium-Ion Batteries. *ACS Omega* **2022**, *7*, 26701–26714. [CrossRef]
32. Pang, B.; Chen, L.; Dong, Z. Data-Driven Degradation Modeling and SOH Prediction of Li-Ion Batteries. *Energies* **2022**, *15*, 5580. [CrossRef]
33. Azis, N.A.; Joelianto, E.; Widyotriatmo, A. State of Charge (SoC) and State of Health (SoH) Estimation of Lithium-Ion Battery Using Dual Extended Kalman Filter Based on Polynomial Battery Model. In Proceedings of the 2019 6th International Conference on Instrumentation, Control, and Automation (ICA), Bandung, Indonesia, 31 July–2 August 2019.
34. Li, R.; Li, W.; Zhang, H.; Zhou, Y.; Tian, W. On-Line Estimation Method of Lithium-Ion Battery Health Status Based on PSO-SVM. *Front. Energy Res.* **2021**, *9*, 401. [CrossRef]
35. Bian, Z.; Ma, Y. An Improved Particle Filter Method to Estimate State of Health of Lithium-Ion Battery. *IFAC-PapersOnLine* **2021**, *54*, 344–349. [CrossRef]
36. Chinedu, O.; Nagarajan, R. Statistical Characterization of the State-of-Health of Lithium-Ion Batteries with Weibull Distribution Function—A Consideration of Random Effect Model in Charge Capacity Decay Estimation. *Batteries* **2012**, *3*, 32.
37. Sui, X.; He, S.; Vilsen, S.B.; Meng, J.; Teodorescu, R.; Stroe, D. A review of non-probabilistic machine learning-based state of health estimation techniques for Lithium-ion battery. *Appl. Energy* **2021**, *300*, 117346. [CrossRef]
38. Bao, Z.; Jiang, J.; Zhu, C.; Gao, M. A New Hybrid Neural Network Method for State-of-Health Estimation of Lithium-Ion Battery. *Energies* **2022**, *15*, 4399. [CrossRef]
39. Zhou, J.; He, Z.; Gao, M.; Liu, Y. Battery state of health estimation using the generalized regression neural network. In Proceedings of the 2015 8th International Congress on Image and Signal Processing (CISP), Shenyang, China, 14–16 October 2015.
40. Jiantao, Q.; Feng, L.; Yuxiang, M.; Jiaming, F. A Neural-Network-based Method for RUL Prediction and SOH Monitoring of Lithium-Ion Battery. *IEEE Access* **2019**, *1*, 1.
41. Fan, Y.; Wu, H.; Chen, W.; Jiang, Z.; Huang, X.; Chen, S.-Z. A Data Augmentation Method to Optimize Neural Networks for Predicting SOH of Lithium Batteries. In Proceedings of the International Conference on Robotics Automation and Intelligent Control (ICRAIC 2021), Wuhan, China, 26–28 November 2021.
42. Jo, S.; Jung, S.; Roh, T. Battery State-of-Health Estimation Using Machine Learning and Preprocessing with Relative State-of-Charge. *Energies* **2021**, *14*, 7206. [CrossRef]
43. Morello, R.; Di Rienzo, R.; Roncella, R.; Saletti, R.; Schwarz, R.; Lorentz, V.R.; Hoedemaekers, E.R.G.; Rosca, B.; Baronti, F. Advances in Li-Ion Battery Management for Electric Vehicles. In Proceedings of the IECON 2018—44th Annual Conference of the IEEE Industrial Electronics Society, Washington, DC, USA, 21–23 October 2018.
44. Gabbar, H.A.; Othman, A.M.; Abdussami, M.R. Review of Battery Management Systems (BMS) Development and Industrial Standards. *Technologies* **2021**, *9*, 28. [CrossRef]
45. Yang, S.; Zhang, Z.; Cao, R.; Wang, M.; Cheng, H.; Zhang, L.; Jiang, Y.; Li, Y.; Chen, B.; Ling, H.; et al. Implementation for a cloud battery management system based on the CHAIN framework. *Energy AI* **2021**, *5*, 100088. [CrossRef]
46. Park, K.; Choi, Y.; Choi, W.J.; Ryu, H.-Y.; Kim, H. LSTM-Based Battery Remaining Useful Life Prediction With Multi-Channel Charging Profiles. *IEEE Access* **2020**, *8*, 20786–20798. [CrossRef]
47. Cinomona, B.; Chung, C.; Tsai, M.-C. Long Short-Term Memory Approach to Estimate Battery Remaining Useful Life Using Partial Data. *IEEE Access* **2020**, *8*, 165419–165431. [CrossRef]
48. Gong, D.; Gao, Y.; Kou, Y.; Wang, Y. State of health estimation for lithium-ion battery based on energy features. *Energy* **2022**, *257*, 124812. [CrossRef]
49. Sun, S.; Sun, J.; Wang, Z.; Zhou, Z.; Cai, W. Prediction of Battery SOH by CNN-Bi-LSTM Network Fused with Attention Mechanism. *Energies* **2022**, *15*, 4428. [CrossRef]
50. Pham, T.; Truong, L.; Nguyen, M.; Garg, A.; Gao, L.; Quan, T. Sequence-in-Sequence Learning for SOH Estimation of Lithium-Ion Battery. In Proceedings of the 11th International Conference on Electronics, Communications and Networks (CECNet), Xiamen, China, 18–21 November 2021.
51. Dos Reis, G.; Strange, C.; Yadav, M.; Li, S. Lithium-ion battery data and where to find it. *Energy&AI* **2021**, *5*, 100081.
52. Sandia National Lab. Data for Degradation of Commercial Lithium-Ion Cells as a Function of Chemistry and Cycling Conditions. 2020. Available online: https://www.batteryarchive.org/snl_study.html (accessed on 10 June 2022).
53. Medium. A Comparison of Grid Search and Randomized Search Using Scikit Learn. 2019. Available online: https://medium.com/@peterworcester_29377/a-comparison-of-grid-search-and-randomized-search-using-scikit-learn-29823179bc85 (accessed on 8 July 2022).
54. Khan, N.; Ullah, F.U.M.; Ullah, A.; Lee, M.Y.; Baik, S.W. Batteries State of Health Estimation via Efficient Neural Networks With Multiple Channel Charging Profiles. *IEEE Access* **2020**, *9*, 7797–7813. [CrossRef]
55. Siami-Namini, S.S.; Tavakoli, N.T.; Siami Namin, A.S.N. The Performance of LSTM and Bi-LSTM in Forecasting Time Series. In Proceedings of the IEEE International Conference on Big Data, Los Angeles, CA, USA, 9–12 December 2019.

56. Machine Learning Mastery. Hyperparameter Optimization with Random Search and Grid Search. 2020. Available online: https://machinelearningmastery.com/hyperparameter-optimization-with-random-search-and-grid-search/ (accessed on 15 August 2022).
57. Mathworks. Deep Learning with Time Series and Sequence Data. Available online: https://www.mathworks.com/help/deeplearning/deep-learning-with-time-series-sequences-and-text.html (accessed on 15 August 2022).
58. Machine Learning Mastery. A Gentle Introduction to Dropout for Regularizing Deep Neural Networks. 2018. Available online: https://machinelearningmastery.com/dropout-for-regularizing-deep-neural-networks/ (accessed on 15 August 2022).
59. Sebastian Ruder. An Overview of Gradient Descent Optimization Algorithms. 2016. Available online: https://ruder.io/optimizing-gradient-descent/ (accessed on 15 August 2022).
60. Machine Learning Mastery. A Gentle Introduction to Early Stopping to Avoid Overtraining Neural Networks. 2018. Available online: https://machinelearningmastery.com/early-stopping-to-avoid-overtraining-neural-network-models/ (accessed on 20 August 2022).
61. Machine Learning Mastery. Train-Test Split for Evaluating Machine Learning Algorithms. 2020. Available online: https://machinelearningmastery.com/train-test-split-for-evaluating-machine-learning-algorithms/ (accessed on 20 August 2022).
62. Shi, M.; Xu, J.; Lin, C.; Mei, X. A fast state-of-health estimation method using single linear feature for lithium-ion batteries. *Energy* **2022**, *256*, 124652. [CrossRef]
63. Bin, X.; Bing, X.; Luoshi, L. State of Health Estimation for Lithium-Ion Batteries Based on the Constant Current–Constant Voltage Charging Curve. *Electronics* **2020**, *9*, 1279.
64. Ruan, H.; He, H.; Wei, Z.; Quan, Z.; Li, Y. State of Health Estimation of Lithium-ion Battery Based on Constant-Voltage Charging Reconstruction. *IEEE J. Emerg. Sel. Top. Power Electron.* **2021**. [CrossRef]

*Review*

# A Review of Lithium-Ion Battery Capacity Estimation Methods for Onboard Battery Management Systems: Recent Progress and Perspectives

Jichang Peng [1], Jinhao Meng [2,*], Dan Chen [2], Haitao Liu [1], Sipeng Hao [1], Xin Sui [3] and Xinghao Du [2]

[1] Smart Grid Research Institute, Nanjing Institute of Technology, Nanjing 211167, China
[2] College of Electrical Engineering, Sichuan University, Chengdu 610065, China
[3] Department of Energy Technology, Aalborg University, 9220 Aalborg, Denmark
* Correspondence: scmjh2008@163.com

**Abstract:** With the widespread use of Lithium-ion (Li-ion) batteries in Electric Vehicles (EVs), Hybrid EVs and Renewable Energy Systems (RESs), much attention has been given to Battery Management System (BMSs). By monitoring the terminal voltage, current and temperature, BMS can evaluate the status of the Li-ion batteries and manage the operation of cells in a battery pack, which is fundamental for the high efficiency operation of EVs and smart grids. Battery capacity estimation is one of the key functions in the BMS, and battery capacity indicates the maximum storage capability of a battery which is essential for the battery State-of-Charge (SOC) estimation and lifespan management. This paper mainly focusses on a review of capacity estimation methods for BMS in EVs and RES and provides practical and feasible advice for capacity estimation with onboard BMSs. In this work, the mechanisms of Li-ion batteries capacity degradation are analyzed first, and then the recent processes for capacity estimation in BMSs are reviewed, including the direct measurement method, analysis-based method, SOC-based method and data-driven method. After a comprehensive review and comparison, the future prospective of onboard capacity estimation is also discussed. This paper aims to help design and choose a suitable capacity estimation method for BMS application, which can benefit the lifespan management of Li-ion batteries in EVs and RESs.

**Keywords:** lithium-ion battery; battery management system; capacity estimation; electric vehicle; battery degradation

## 1. Introduction

On the background of energy crisis and global warming, applications such as renewable energy systems and new energy vehicles (Electric Vehicles (EVs) and Hybrid EVs) have become a necessary way of saving energy and decreasing carbon emission [1,2]. As the key component in the power supply of the EVs and Renewable Energy Systems(RESs) [3–5], the energy management of the battery pack directly affects its performance in various operation conditions [6,7]. Due to its high energy density, long service life, no memory effect, etc. [8,9], the Lithium-ion (Li-ion) battery has become a first choice for EVs and RESs [10]. For example, lithium iron phosphate (LFP) has a 90~140 Wh/kg energy density and up to 2000 life cycles, which usually consists of LiFePO$_4$ cathode and graphite anode. In addition, Li-ion battery chemistries also include lithium Nickel Manganese Cobalt oxide (NMC) and lithium Nickel Cobalt Aluminum oxide (NCA) with a higher energy density (140~250 Wh/kg) [11]. Recently, battery manufacturers have also developed new products with relatively superior performance, such as the blade battery (LFP) from BYD which has good thermal safety characteristics through nail penetration tests [12]. Thanks to its excellent properties, the scope of Li-ion batteries has also expanded to various areas like robots, Automated Guided Vehicles (AGVs) and consumer electronics. Especially, with the concept of low carbon, Li-ion batteries will play an important role in the future. According

to Research and Markets research data in Statista [13], the global lithium-ion battery scales to about 185 GWh in 2020, and the market is expected to grow to 950 GWh in 2026 as shown in Figure 1.

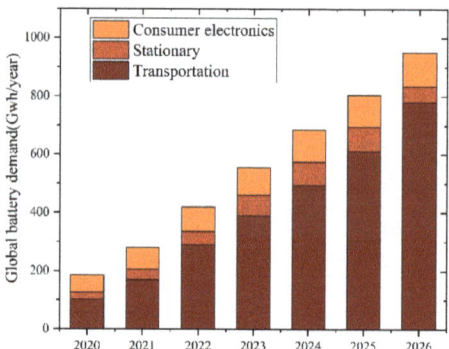

**Figure 1.** Global battery demand 2020–2026.

A typical structure of the Battery Energy Storage System (BESS) is illustrated in Figure 2, which mainly includes battery cells, Battery Management System (BMS), Power Conversion System (PCS), etc. Among all the components, BMS is responsible for the safety operation of the cells in the BESS. The functions of BMS include state estimation, voltage/temperature monitoring and fault diagnosis and warning. One key parameter here is the battery capacity representing the maximum Ah throughput at present. In essence, the battery capacity is the number and energy of the electrons inside the electrodes [14,15]. One consensus is that the Li-ion battery capacity will fade with battery degradation, which could be influenced by numerous external factors in operation conditions. Although the degradation of Li-ion battery can be briefly divided into two modes: Loss of Active Materials in electrodes (LAM), and Loss of Lithium Inventory (LLI), it is difficult to distinguish the aging modes in reality. However, the capacity of an Li-ion battery is critical for the energy management decision marking of BMS. For example, the battery State of Charge (SOC) represents current energy left, which is a ratio of the present Ah amount to its capacity [16]. It is impossible to obtain an accurate SOC without knowing the battery capacity. Once a precise SOC is received, BMS can choose when to charge or discharge each cell. In order to avoid the overuse of the Li-ion battery, its capacity should also be clearly defined. Otherwise, safety hazards, such as failure and thermal runaway [17,18], may exist when the Li-ion battery reaches its End-Of-Life (EOL) [19]. Capacity is also a fundamental index for the secondary use of the Li-ion battery [20,21]. In general, the battery capacity is especially important for the lifespan management of the cells by BMS [22,23].

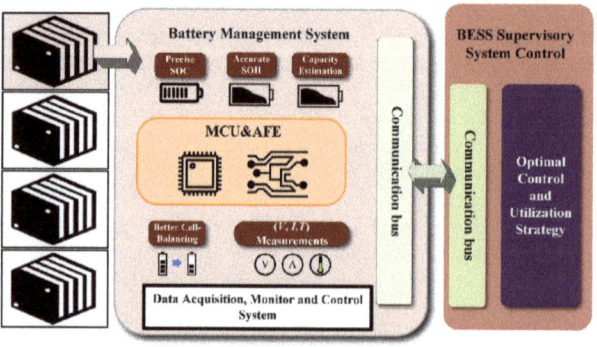

**Figure 2.** Structure of the battery energy storage system.

Battery capacity is usually regarded as the indicator of its lifespan, and it is believed to reach its EOL once the battery capacity reaches 80% of its initial value [24]. An accurate capacity can improve the accuracy of SOC estimation, thus enabling the users to perform charging operations and battery maintenance prompt. A slightly changed capacity will gradual deteriorate the battery's electrical and thermal characteristics and further lead to other severe safety issues [25]. However, a series of barriers hinder an accurately measurement of the Li-ion battery's capacity. One primary fact is the capacity of Li-ion battery is related to current rate and temperature [26,27] considering the effect of electrode kinetics. Then, it is easy to understand that the Li-ion battery's capacity greatly influences the working conditions of the battery pack, which increase the difficulties of obtaining an accurate battery capacity. Another critical factor is the limitation from BMS, the computing power of the microprocessor is limited due to the cost [28]. It can be deduced that onboard implementable battery capacity estimation algorithms are still needed for most EV applications [29,30]. One expectation is that the fast development of Internet-of-Things (IoT) and artificial intelligence can improve the capacity estimation techniques for BMS [31,32].

Great efforts have been made to obtain an accurate battery capacity in the literature, as shown in Figure 3. The points shown in the graph are the phrases that appear more than 20 times. The results of the high-frequency word analysis show a strong correlation between the battery capacity and the EVs. After analyzing the results, battery capacity is often used as an additional result for SOC estimation, or as a representation of energy and working efficiency. With the current market expansion and safety requirements, the battery capacity has become extremely important for battery health. From the analysis rules that brighter the node color means a more recent research period, the study of battery capacity has totally become a hot area with the keywords related to battery health all existing in brighter color. In the past, most works related to BMS focused on battery SOC [33,34]. With the wide application of EVs and RESs, the battery State-Of-Health (SOH), capacity, safety and Remaining Useful Life (RUL) are becoming the points of discussion. We have to mention that more than 500 articles have been investigated from 2016 to 2021; all related to battery capacity.

However, we also realized that there is a limited number of reviews on capacity estimation, especially for online implementable capacity estimation in BMS. Ref. [35] covers almost all the battery states including SOH, SOC, State-Of-Power (SOP), State-Of-Energy (SOE), etc., and selects the current research hotspots for discussion and evaluation. It is oriented towards the BMS and summaries the features of various states, but the capacity estimation methods are not well addressed. Refs. [30,36] provide a discussion of the classification of existing capacity estimation methods. Although the principles for the classification are different, they both discuss, in detail, the research methods, but [36] is more focused on the model-based method for Li-ion battery SOH estimation, and [30] published in 2015 has not covered any discussions about machine learning based methods. Most of the existing reviews on battery capacity estimation focus on the generalization of existing methods and do not distinguish between their application conditions or scenarios. The current booming market of EVs also requires the practicality of onboard BMS. It is found that there is a lack of a summary of the existing knowledge for onboard capacity estimation. Therefore, this work overviews and compares the current battery capacity estimation methods suitable for onboard BMS. The characteristics of various capacity estimation are reviewed and discussed in this paper.

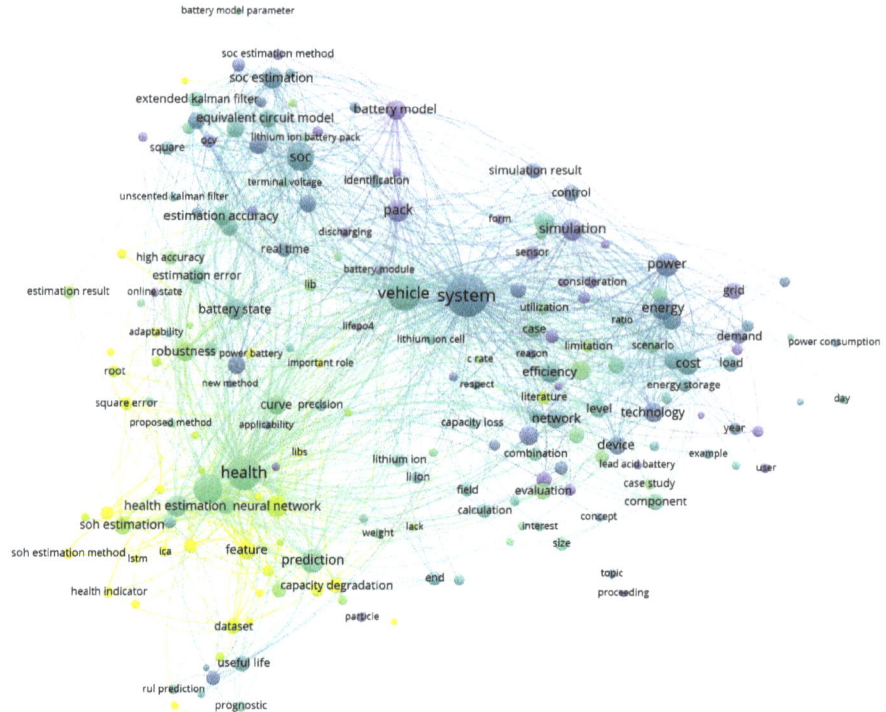

**Figure 3.** High-frequency keyword co-occurrence network for battery capacity on Scopus from 2016 to 2021.

The rest of this paper will be structured as follows: Section 2 briefly analyzes the battery degradation mechanisms. Section 3 reviews the existing methods for onboard battery capacity estimation. Discussion and perspectives are expressed in Section 4. Section 5 is the conclusion of this work.

## 2. Li-Ion Battery Degradation Mechanism Analysis

An Li-ion battery mainly contains the lithium metal oxide as the cathode, and graphite as the anode material at present. A separator exists between the two electrodes for insulation, which only allows the pass of Li-ions, and the electrons can only exchange through external circuits. Additionally, an electrolyte is also needed to assist the transfer of Li-ion. Thus, it is clear that the Li-ions exchange from the electrodes during battery charging and discharging [37]. For EVs, the reduction of the battery capacity results in less energy available, which directly reflects the performance degradation of the battery pack. The capacity loss of the battery is a non-linear process containing complex aging mechanism. However, the aging mechanism of batteries cannot be precisely described, especially for the decay rules of cycle life. To conveniently analyze the battery degradation, recent research usually divides the battery aging into two main forms: calendar aging and cycling aging [38–41].

Calendar aging refers to the capacity loss during storage, which is mainly influenced by high temperature and SOC [42–44]. Five aging cases are set in [43] for the calendar aging of 15 Li-ion batteries for a period between 24 and 36 months. The test results clearly prove there is a non-linear battery degradation during calendar aging, and the fading rate of the Li-ion battery is accelerated by increased storage temperature and SOC. Among all the influencing factors, high storage temperature is believed to be the most critical factor for battery calendar degradation [45,46]. LLI is the main reason for calendar aging with high temperature [15].

However, cycling aging is always accompanied by calendar aging in an actual application, which makes it complicated to clarify the degradation procedure. Cycling aging is the main reason for battery aging in BESS; Belt et al. [47] have tested 107 commercial cells and the results show that the charge-depleting by cycling aging is far more than the calendar one. Cycling aging is closely related to the charging and discharging process of Li-ion batteries. Chemical reactions are essential for the process of Li-ion movement between the electrodes [48]. Thus, the investigation of battery cycling aging needs to consider the current rate, Depth of Discharge (DoD) and SOC, etc. [49]. During cycling aging, the distribution of the current density, SOC and temperature is not consistent inside the cell as illustrated in [50]. The inhomogeneities of distribution in the cathode material will further induce mechanical force to fatigue the electrodes, and thus accelerate the battery cycling degradation.

It is clear that the Li-ion battery degradation is the coupling of multiple factors, as shown in Figure 4. The degradation modes of Li-ion battery are also LAM and LLI as previously described [51]. The thickening of a Solid Electrolyte Interface (SEI) and lithium plating in the graphite anode will both consume the lithium inventory, and thus results in the LLI of an Li-ion battery. No chemical reaction is ideal without any losses and generates some extra products. Thus, the reactions during cycling and storage cause the LAM of both cathode and anode. Both LAM and LLI are observed by the incremental capacity and differential voltage curves in LFP, LMO (Lithium Manganese Oxide) and LTO (Lithium Titanium Oxide) batteries according to the results in [52]. Inappropriate temperature ranges, overcharging or discharging, and high SOC (>80%) are the external factors which can speed up the battery aging process [53–55]. Complicated degradation behaviors of the battery make it difficult to clarify all the details in theory.

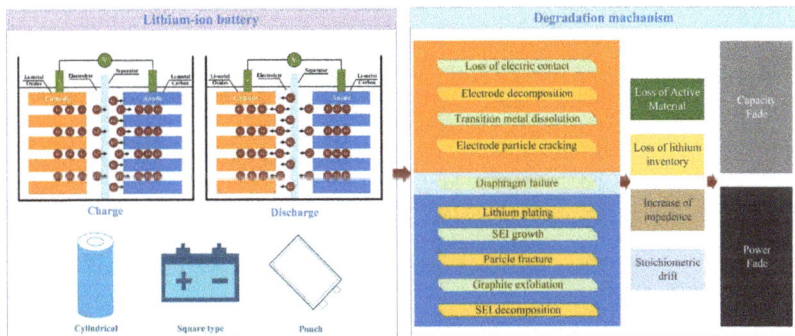

**Figure 4.** The effect of factors on the battery capacity degradation.

Here, we also illustrate the degradation measurement results of two commercial Li-ion batteries under calendar aging and cycling aging in Figure 5. With a higher storage temperature, the capacity of a battery in T = 45 °C decreases faster than 30 °C. It is also clear that the Li-ion battery charged with a larger current (2.7 C) degrades faster than with a lower charging C-rate (1.3 C). A Neware battery tester is used to cycle the battery, more details about the cycling aging setting can refer to [56].

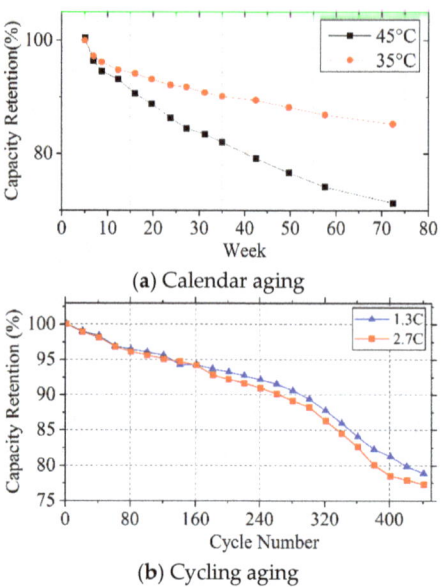

**Figure 5.** Li-ion battery degradation. (**a**) 36.9 Ah Li-ion batteries stored at SOC = 100% in thermostat (T = 35 °C, T = 45 °C), the capacities are measured every four weeks; (**b**) 1.5 Ah NMC based 18,650 Li-ion batteries are charged with 1.3 C and 2.7 C, and discharged by 5 C. The capacities are measured every 20 cycles.

## 3. Review of Capacity Estimation Methods

Considering the complexity of battery degradation, it is still challenging for the BMS to accurately predict the battery capacity onboard. Thus, researchers have made significant efforts to solve this problem. This section will brief introduces the battery capacity estimation methods in the literature. We mainly divide the methods into direct measurement methods, analysis-based methods, SOC-based methods and data-driven methods, whose principle and current processes will be detailed in the following subsection.

A.  Direct Measurement Method

The most straightforward way to receive the battery capacity is to accumulate the charge during its cycling period [57]. Direct measurement methods need a full charge or discharge of the battery under a specific condition. Current various standards from International Electrotechnical Commission (IEC) [58], International Organization for Standardization (ISO) [59] and Institute of Electrical and Electronics Engineers Standards Association (IEEE-SA) [60] have been proposed for testing the Li-ion battery capacity in a standard condition. For example, ref. [58] defines a $1/3\ I_t$ constant discharging current for EV and $1\ I_t$ discharging currént for HEV, for the purpose of measuring the battery capacity. As for the capacity measurement in [59], 1 C current is recommended for discharging the high power battery and C/3 is used for measuring high energy battery. It is not difficult to realize that the measured battery capacity may not be the same for different C-rates and temperature settings in those standards. In addition, the test procedure is rather strict compared with the working environment of the battery pack in a real application. [58] needs the battery soaked at a predefined temperature for at least 12 h to ensure thermal stabilization, which requires the cell temperature changes lower than 1 °C in 1 h time interval. The current and voltage measurement accuracy should be less than +/− 1%, and the time is measured less than +/− 0.1% in [59]. Thus, it is not practical to always meet the above requirements in a battery application, which limits these test methods to laboratory tests as references.

Another concern is that in reality, the BESS cannot always fully charge or discharge in various load conditions. Direct measurement methods cannot give a result if the battery is partially charged or discharged, which often happens in real cases. To clarify this point, an SOC profile of BESS for primary frequency regulation lasting one week [31] is shown below in Figure 6. Mostly, the SOC of the BESS varies within 40–60%, which confirms the unrealistic implement of direct measurement methods in a real application. It is noted that fully charging or discharging the battery is also quiet time-consuming [61].

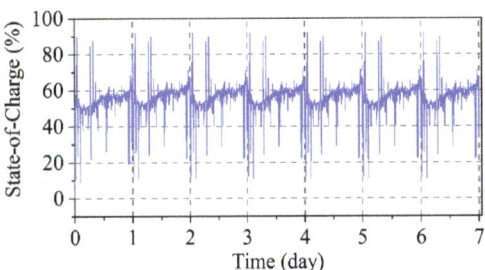

**Figure 6.** SOC profile of BESS for the primary frequency regulation of grid.

For convenience, an option is to measure the internal resistance to reflect the battery capacity. The battery internal resistance can be directly measured by applying a current pulse to the battery [62] as shown in Figure 7. Usually, the current pulse lasts a few seconds, and then the internal resistance can be calculated by the following Equation,

$$R_{bat} = \frac{\Delta U}{\Delta I} = \frac{U_A - U_C}{I_A - I_C} = \frac{U_B - U_D}{I_B - I_D} \quad (1)$$

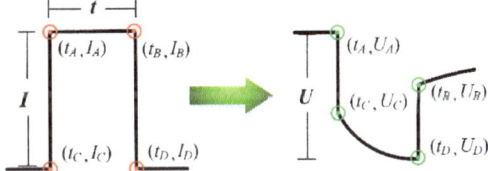

**Figure 7.** The DC internal resistance measurement for Li-ion battery.

Unfortunately, the internal resistance is more related to the power fade of the battery, which does not always exhibit a linear relationship with capacity fade. The capacity degradation is related to lithium corrosion at the anode, while the power fade is related to SEI growth and LAM [63]. Moreover, the internal resistance measurement is also affected by C-rate, temperature and SOC, and the internal resistance is quite small in the milliohm range [43,64,65]. Thus, some uncertainties may exist if only internal resistance is used for calculation. Direct measurement methods are strictly performed by charging and discharging of the battery in laboratories. As for onboard BMS implementation, the practical application requires estimation methods that can be done with limited complexity.

Therefore, more advanced methods are needed to estimate the battery capacity by processing the current, voltage, temperature and mechanical stress. Those existing methods include analysis-based methods, SOC-based methods and data-driven methods, which will be introduced in the following subsections.

B. Analysis-Based Methods

For indirect methods, the voltage, current and temperature can be recorded by sensors, and then used to estimate the capacity. In this work, we mainly introduce five kinds of analysis-based methods with IC (Incremental curve) curve, DV (Differential voltage) curve, DT (Differential thermal) curve, mechanical stress and Electrochemical Impedance Spectroscopy (EIS) as shown Figure 8. At present, more attention is paid to the Li-ion battery capacity. The capacity, which limits the available energy, is the key indicator for State-Of-Health (SOH), which is defined as the ratio of current maximum capacity to its initial capacity [66].

$$SOH = \frac{Q_{present}}{Q_{initial}} \quad (2)$$

where $Q_{present}$ denotes the current capacity and $Q_{initial}$ is the nominal capacity. Thus, we will not distinguish capacity estimation and SOH estimation in the following explanations.

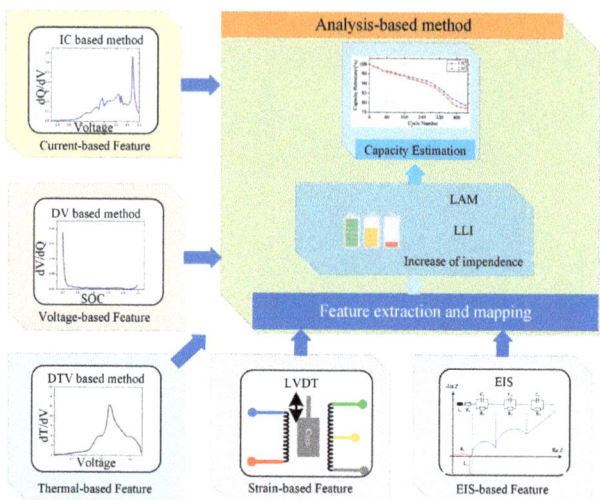

**Figure 8.** Schematic of analysis-based method.

(1) Incremental curve analysis method

IC curve analysis method focuses on the variation of capacity with voltage, which is expressed as,

$$IC = \frac{dQ}{dV} \quad (3)$$

In the IC curve, the $dQ$ can be easily obtained by Coulomb counting of the current. Since noise always exists in current and voltage measurements, a filter is usually needed to smooth the IC curve [52,67,68]. Usually, a low-pass filter [69], Savitzky–Golay filter [70,71] and Kalman filter [72] have been used to process the IC curve for reducing the noise sensitivity. A two dimensional Luenberger–Gaussian-moving-average filter is designed in [68] to enhance the IC curve extraction. Once a smooth IC curve is obtained, the mechanisms of the Li-ion battery can be analyzed accordingly. The peaks and valleys in the IC curve are related to the voltage plateau of the battery. In [52], the degradations of three peaks in the IC curve are related to LAM and LLI, and the small shift of the IC curve indicates a slight increasement of the battery internal resistance. Then, the shape variation of the IC curve, especially, the peak and valley changes can be used for analyzing the capacity degradation trend and capacity estimation. The analysis procedure of the IC curve is summarized as Figure 8.

Since the insertion and extraction of Li-ions can change the phase transformation of the material in the electrodes, the IC curve is an effective tool to reflect the battery degradation. The LLI and LAM of six LFP batteries are quantitatively analyzed by the IC curve in [73] for the battery health diagnosis, which utilizes the heigh, area, shape and position of the five peaks in the IC curve. A regression model can be then easily established by using the variations of the IC peak. A linear regression model characterized by ordinary least squares is used for battery health estimation in [73]. Ref. [74] also uses the IC peak as the feature for battery capacity estimation, which chooses the grey relational analysis as the estimator and the maximum error is claimed less than 4%. Utilizing the IC peak and the related area, the capacity of the retired battery is also evaluated in [75]. A IC curve based mode is proposed in [76] to describe the phase transition behavior of active material of a Li-ion battery, which is used for battery capacity estimation and later verified on LFP, NMC, LTO chemistries. The proposed model has also claimed to reduce noise from sampling and measurements, which is suitable for onboard BMS.

The IC curve mainly extracts the variation of the voltage with Li-ion battery degradation, and thus can be used for aging mechanism analysis. However, a very low current rate is needed to obtain the IC curve for battery diagnosis, such as 1/10 C in [45]. Analyzing the battery degradation with IC curve also requires specialized knowledge of the electrochemical reactions inside the battery.

(2) Differential voltage curve analysis method

DV curve [77] is quite like IC curve in analysis, which is described as,

$$DV = \frac{dV}{dQ} \quad (4)$$

From Equation (4), we understand that the DV curve can also be obtained from the current and voltage measurements during charge or discharge. The basic idea to analyze the degradation of a battery cathode and anode is based on the variation of the DV curve [78]. The battery capacity can be also deduced from the trend of the variation in DV curves. Despite the requirement of a smooth filter, some publications also choose DV curve as a tool for capacity estimation. Ref. [79] measures the DV curve of a half-cell, and investigates the DV curves of positive and negative electrodes, respectively. Features of DV curve related to LAM and LLI are used for battery degradation estimation. The DV curve changes of a LTO battery is discussed in [80] to analyze the aging mechanism during a number of 1080 cycling tests. The aging modes of positive and negative electrodes are expressed by the peak and valley of DV curves, which can be further used as the reference for the degradation information extraction from a Li-ion battery. A comparison of IC and DV curves is studied in [52] for the aging mechanism identification of five commercial Li-ion batteries, and experimental results prove the good consistency of IC and DV curve in battery degradation mode analysis. In short, using a DV curve for battery capacity estimation is similar to an IC curve; both utilize the variation of the curve's shape to analyze the aging mechanisms and then extract features as the input of a regression model for capacity estimation. The characteristics of the DV curve can also refer to the IC curve in the previous section. We have to mention that one good advantage of those methods is the use of direct measurements from BMS which is low cost to implement. A common challenge is to deal with the measurement noise in reality [81].

(3) Differential thermal analysis method

Considering that the general measurements from BMS contain current, voltage and temperature, the differential thermal voltammetry [82] is chosen to diagnosis the Li-ion battery degradation as,

$$DT = \frac{dT}{dV} \quad (5)$$

There always is heat generation during the battery operation, which is divided to reversible and irreversible heat. The reversible heat is generated by the entropy change of the electrochemical reaction in electrodes, while irreversible heat is mainly the ohmic resistance heat, polarization heat and the side reactions heat [83]. Those thermal characteristics are also charged with battery degradation. Therefore, it is possible to use the temperature variation during charge or discharge as the indictor for battery capacity estimation [67,84,85]. Here, we plot the DT curve of the Oxford dataset [86] as shown in Figure 9, which indicates a variation on the peak and valley with the increase of the cycling number.

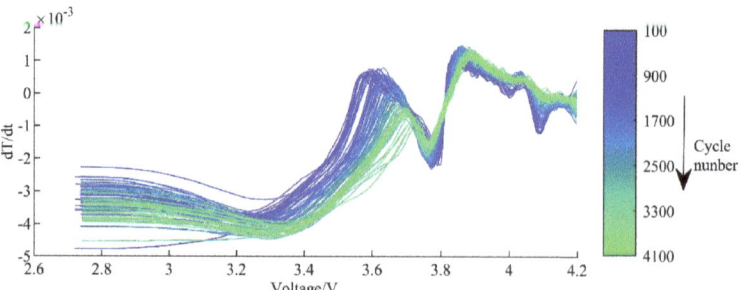

**Figure 9.** DT curve of the Li-ion battery cell in Oxford dataset.

The DT curve of LCO and NCA batteries are analyzed in [85], and the peak and valley of the DT curve can be used for capacity estimation. The position of the peak and valley in the DT curve is selected as the features for a Gaussian Process Regression (GPR)-based data-driven estimator in [87]. Both IC and DT curves are analyzed and compared, which shows that the MAE of DT curve is 19.50% less than IC curve on Oxford and NASA dataset [88]. Ref. [89] extracts the temperature variation during a 1 C constant current charge process, the time interval of two temperature cooling areas are selected as the feature for capacity estimation on both calendar aging and cycling aging.

Compared with IC and DV curves, the DT curve can be suitable for high current charge or discharge. However, the DT curve is also sensitive to measurement noise. The DT curve can be used for aging mechanisms analysis because it is essential the heat generation during the electrochemical reactions. In addition, the connections of the DT curve to battery aging still need to be investigated.

(4) Mechanical stress analysis method

The volume of the particle will change during the charge or discharge of an Li-ion battery. Taking an LFP battery as an example, the LFP particle expands by 6.77% during lithiation [90], while the intercalation of Li-ion in the anode leads to 12% changes in the volume of graphite [15]. Thus, it is reasonable to use the mechanical stress for battery performance analysis [91–93].

It is proved in [94] that the thickness of the Li-ion battery varies by 54.5 μm after the first discharge and the changes of the thickness is 13.5 μm after ten full cycles for a 60-Ah NMC Li-ion battery. It is clear that the mechanical stress of the Li-ion battery will change both with SOC and battery aging. The swell of a commercial 5-Ah Li-ion battery is investigated in [95], where the relationship between the battery swelling and C-rate, SOC and temperature is analyzed. Especially, the authors claim that the $ds/dQ$ can be used for identifying the phase transition in the negative electrode, and further acts as the tool for aging estimation. Ref. [96] proposes a force-based incremental capacity analysis method for Li-ion battery capacity fading estimation, which detects the expansion force of a MNC cell from a HEV battery pack. The experimental results have proven that the proposed method is better than IC curve in signal-to-noise ratio. A high relevance of the second derivative of strain and IC curve are found in [97], where the strain of the electrode can be used for the state estimation as IC curve. The results also prove the second derivative of strain is less sensitive to C-rate compared with IC method, which is expected to be more suitable for real battery-based applications.

Although mechanical stress is believed to be an effective way for battery detection, the main limitation is the requirement of a specially designed device for swell measurement such as displacement sensor, pressure sensor, test fixture, etc. In this way, the cost is increased and also its implementation to a battery pack must be carefully considered for not destroying its original design.

(5) EIS analysis method

EIS is an effective tool with high sensitivity to the electrochemical reactions inside the Li-ion battery; it has been used for battery modelling [98,99], SOC and temperature estimation [100], and also battery degradation diagnosis [101,102]. Generally, EIS is measured in the battery equilibrium state with small current or voltage injection in a frequency range between mHz and kHz [103]. The real and imaginary parts of impedance measurement is selected as the input of a Gaussian Process Regression (GPR) model for Li-ion battery capacity and RUL estimation, in which the variation of EIS with battery degradation is shown in Figure 10. The EIS curve turns from blue to red with battery aging. Ref. [104] combines EIS with ultrasonic time-of-flight analysis to investigate the electrochemical characteristics and structure variation inside a Li-ion battery when the cycle number of a battery increases. Refs. [105,106] extract the parameters of ECM from EIS curve, and then analyze the connections between the parameters and battery aging for estimating the battery capacity. [107] uses S transform to a fast calculation of battery impedance, and the zero-crossing point of real impedance R0 is chosen for battery capacity estimation.

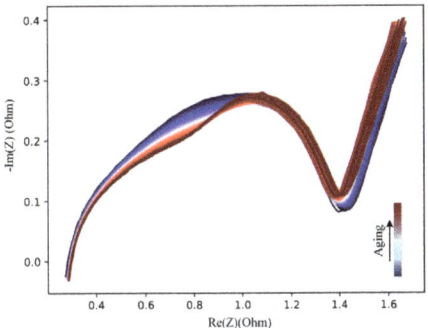

**Figure 10.** The variation of EIS with Li-ion battery degradation with data from [101].

Battery EIS has a strong potential to reflect the electrochemical reactions in the frequency domain, which is expected to have great potential for onboard BMS application. However, most EIS related battery degradation analysis are based on the commercial electrochemical workstation which is accurate yet expensive. Considering the volume and weight of the electrochemical workstations, they are difficult to be directly used in an EV environment. Therefore, two kinds of solutions have been proposed recently to address the issue of EIS measurement with BMS and machine drive electronics. One method is to combine with the onboard chargers [108] or DC–DC converters [109,110]. An onboard charger is integrated with a Dual Active Bridge (DAB) converter for charging and EIS measurement of the battery pack in EV [108]. The second method is using low power measurement module in a BMS for small AC signal injection. [111] uses Single-Cell Supervisor (SCS) designed by NXP Semiconductors injecting the AC current to the battery, and the battery impedance can be calculated by measuring the voltage response. However, there are still very limited examples on hardware design of EIS measurement [112] considering the cost of current BMS. Another concern is that the measurement of EIS is easily affected by noise which hinders a reliable usage of EIS on battery capacity estimation.

C. SOC-based method

An SOC-based methods can be divided into SOC indirect estimation and the SOC observer-based method, which is illustrated in Figure 11. SOC indirect estimation calculates the battery capacity through a period of Coulomb counting and SOC variation, which usually estimates the battery in a short time scale online. The SOC observer-based method directly estimates the battery capacity utilizing an observer based on battery Equivalent Circuit Model (ECM) model, which uses only current and voltage as the input, and SOC and capacity can be estimated synchronously.

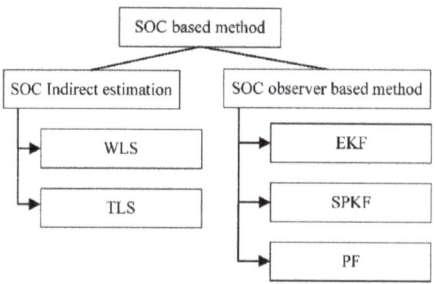

**Figure 11.** SOC-based method.

(1) SOC indirect estimation

SOC indirect method is essentially based on the coulomb counting equations, which is expressed by the following equation,

$$SOC(t_2) = SOC(t_1) + \frac{1}{Q}\int_{t_1}^{t_2} \frac{\eta i(t)}{3600} dt \qquad (6)$$

where $\eta$ is the coulomb efficiency, $i(t)$ is the current, and $Q$ is the capacity. From Equation (5), the battery capacity can be calculated once the SOC variation is known.

LS is a commonly used method for parameter estimation of a linear model, which offers a mathematical model to fit the experimental data with a minimum residual sum of squares errors [113]. According to Equation (6), the capacity can be solved by linear regression with SOC and current information.

Weighted Least Squares (WLS) is calculated by considering the weights of the data based on an ordinary LS [114]. Based on the linear function structure $y = Qx$, where $y$ is measurement, and $x$ is an independent variable. In this case, the model can be expressed as $Y = y - \Delta y = \tilde{Q}X$ as shown in Figure 12a, where Y is the measurement vector and X is the independent variable vector, and the measurement errors $\Delta y$ is considered as the weightings. Ref. [115] uses the errors from the observations as the weights to calculate a fitting equation. Capacity is set as a time-varying parameter and then solved by a recursive approximation [116].

(a) WLS  (b) TLS

**Figure 12.** Data errors on two LS-based method.

Ref. [117] proposes a capacity estimation algorithm based on LS methods for PHEV and EVs. With the given current signal, the OCV is calculated by LS and the relationship between SOC and OCV is mapped. Then capacity is obtained by the iteration process. Wei et al. [118] propose an SOC and SOH estimation based on two LS estimators. Under the condition that no pre-determined parameters are necessary, OCV is derived by LS estimator, and then the capacity is converted to linear fitting problem to solve according to the mapping relationship between OCV and SOC.

Total Least Squares (TLS): During the calculation of ordinary LS, it assumes that the input data are accurate, but in actual operation process, the input and output data are both influenced by measurement noise. Presented in terms of data fitting, WLS only accounts for $\Delta y$, while Total-Least-Squares (TLS) accounts for both $\Delta y$ and $\Delta x$. Therefore, TLS is introduced for capacity estimation, considering the disturbances from both input and output [119,120]. In this way, the model can be expressed as $Y = y - \Delta y = \tilde{Q}X = \tilde{Q}(x - \Delta x)$ in Figure 12b, where $\Delta y$ is the measurement errors and $\Delta x$ is the input data errors. The TLS problem can be solved by a singular value decomposition of the matrix, but the multiplier of the singular value decomposition of the n × n matrix is $6N^3$ [121]. After analyzing the calculation of TLS, Rhode et al. [122] find that it is difficult to derive an analytical solution to the matrix and therefore they propose a recursive form to solve the TLS problem and satisfactory estimation results are finally obtained. Ref. [123] uses the constraint Rayleigh quotient as the cost function in the TLS calculation process, which greatly reduces the complexity of the TLS.

(2) SOC observer-based method

The SOC observer-based method attempts to estimate the battery capacity according to the SOC estimation results in a dual time scale framework.

As for the SOC observer-based method, the capacity estimation is developed based on ECMs. The ECMs describe the external characteristics of the Li-ion battery using resistance, capacitance, and voltage source as shown in Figure 13 [124]. It is believed that the complexity of ECMs is suitable for online BMS applications [125]. Depending on different numbers n of the RC networks, which describe the dynamic characteristic including the polarization characteristics and diffusion effects [126], the Rint model ($n = 0$), Thevenin model ($n = 1$), and dual-polarization model ($n = 2$) are proposed, respectively [127].

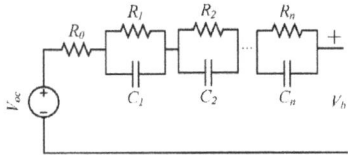

**Figure 13.** ECM with $n$-RC.

In existing studies, capacity is often considered as one of the parameters to be estimated in parallel with another battery state such as SOC [128]. This is also known as joint estimation [129]. Based on the coulomb counting Equation (2), SOC estimation and capacity estimation are coupled. In the joint estimation of SOC–SOH, capacity as a dynamic parameter is treated as an extended state of the filter, and then parametric filtering is performed [130,131]. For filters, there are two core components, prediction, and correction. Predicting the state from the previous moment and correcting the result based on the observations, the prediction and correction are continuously recursive to complete the estimation of the state. Figure 14 is a schematic diagram of the filtering process using the Kalman filter as an example [132,133]. The equations for a stochastic linear discrete system are described as:

$$x_k = Ax_{k-1} + Bu_k + \omega_k \tag{7}$$

$$y_k = Hx_k + v_k \tag{8}$$

where $x$ is the state vector, $A$ is the state transfer matrix, $u$ is the state control vector, $B$ is the control variable matrix, $y$ is the measurement vector, $H$ is the transformation matrix from the state vector to the measurement vector, w and v are both noises obeying a Gaussian distribution, and $P$ is the covariance matrix.

Extended Kalman Filter (EKF): KF is somehow limited to linear systems [134]. In ref. [135], the researcher linearizes the non-linear OCV–SOC curves into seven segments to meet the requirements of KF for a linear system model. However, in the case of long-term battery capacity decrease, as mentioned in Section 1, the degradation process is of much nonlinearity. Thus the improved KF-based EKF is a better choice for achieving battery capacity estimation [136]. Ref. [137] proposes a co-estimation of multiple battery states using the correlation of parameters in the state space. Based on the EKF for SOC estimation and cumulative charge, the battery capacity is solved simultaneously. Ref. [138] proposes a dual filter of state and parameter based on EKF to achieve a simultaneous estimation of SOC and SOH. The parametric filtering is used to update the capacity online to improve the accuracy of SOC estimation.

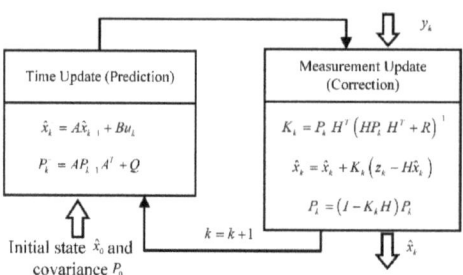

**Figure 14.** Filtering process of Kalman filter.

Sigma Point Kalman Filter (SPKF): For those situations where the degree of nonlinearity is higher, SPKF is a better choice. SPKF is an improvement of KF without linearization steps or computing the derivatives [139]. The mean and covariance are obtained by a set of weighted points passing through the non-linear function. According to the pattern of choosing the sigma points, SPKF is divided into Unscented Kalman Filter (UKF) [140] and Central Difference Kalman Filter (CDKF) [141]. Compared with EKF, SPKF avoids the necessity of complex differential processes and has a better covariance approximation. However, under more frequent current fluctuations conditions, its accuracy and stability still cannot be guaranteed [142]. Ref. [143] proposes a UKF-based dual filtering framework based on the coupling of SOC and capacity as shown in the coulomb Equation (6). Real-time capacity is updated as recursive filtering of SOC state and capacity parameter proceeds. Ref. [144] presents a joint estimation for multiple critical states along with an autoregressive equivalent circuit model. The state space-coupling model is solved using UKF which has good dynamic tracking properties for multiple states.

Particle Filter (PF): PF is commonly seen in tracking the capacity in the full lifespan of a battery [145–147]. By employing the Monte Carlo sampling techniques, PF offers the possibility of dealing with any type of distribution by a proper group of particles or samples approximating the respective probability density functions [148]. Ref [149] uses an optimized dynamic single exponential model to describe the degradation of capacity and particle filtering to complete the optimal state solution. The performance of single-step, multi-step and long-term capacity prediction based on particle filtering is analyzed in detail.

It is noted that the time scales of state and parameters changes are different [150]. Typically, the time scale for SOC changes is much smaller. In [151], the time scale of SOC is set as 1s, while capacity is 60s. Thus, the two variables are filtered separately by independent filters. In addition, in the common joint SOC–SOH (denoted by capacity) estimation, the coupled nature of the state space equations requires a high degree of accuracy for both estimators. Otherwise, the cross-interference between the two will become a difficult problem for the joint estimation. i.e., SOC and capacity uncertainties can interfere with each other in the process of information exchange. Ref. [152] proposed the decoupling of parameter identification and state estimation. The capacity identification estimator is fully decoupled from the SOC state estimator. Different time scales are adopted to further improve the accuracy of the results.

As we can find for the SOC-based method, accurate SOC is the important prerequisite for capacity estimation. However, SOC itself is not an easy measurable status for the Li-ion battery, which limits the usage of SOC-based methods in reality.

D. Data-driven method

With the fast development of IoT and artificial intelligence, the daily operation measurement of the battery system is easy to be recorded to a cloud platform which could be further used for cloud to edge estimation. The data-driven approach is characterized by a reliance on a large amount of dataset to make decisions and does not require a specific battery model. In a data-driven approach, a model can be used to map the data as long as a sufficiently representative sample is available, without the need to pre-determine a definitive model in advance. In the case of batteries, the operation measurement that we can collect may contain aging information. The degree of aging can be reflected by certain characteristics during the charging and discharging process, and the data-driven approach constructs an approximate model to match the true aging situation with this information. Figure 15a is a schematic diagram of the data-driven approach application procedure, which mainly includes three processes: data collection & preprocessing, offline training and online estimation. The main purpose of data collection is measuring the voltage, current and temperature during the operation of the battery pack. Then, the data-driven model can be trained offline with high computing power processor, and the trained model is later implemented in a BMS for online estimation. In this type of method, the keys lie in the processing of the data, the extraction of key features and model training, the main data stream of those processes is shown in Figure 15b.

The existing data-driven method is introduced in the following subsections.

Neural Network (NN): The basic NN is a three-layer structure network including an input layer, hidden layer and output layer. The input layer neuron can be regarded as extracting capacity-related features [153]. It is critical to choose a suitable indicator. Refs. [154,155] use NN to investigate the battery capacity. From the perspective of feature extraction, the former uses discharge voltage, while the latter adds also the temperature effects. The weight coefficients from the input layer to the hidden layer or from the hidden layer to the output layer need to be obtained after training a large number of samples [105]. Ref. [156] chooses to train a generalized regression NN with the battery's constant current charging time to estimate SOH. The instantaneous discharging voltage drop and the amount of Ah-throughout for a certain depth of discharge are captured as features. Ref. [157] uses the Broad Learning System (BLS) to process historical capacity data and generate feature nodes as the input layer of the neural network. This method does not require an in-depth study of the battery aging mechanism, but it also requires at least 25% of the historical capacity data.

Support Vector Machine (SVM): Support Vector Machine (SVM) is another technique. The core is to divide the data set in a hyperplane [158,159] so that the geometric interval between each data point can be maximized in the hyperplane. It can be transformed into an objective function under constraints to solve the optimization problem. Ref. [160] studies the relationship between the electrolyte concentration and voltage with the battery capacity. The non-linear relationship is then fitted by SVM. A Least Squares Support Vector Machine (LSSVM) is used in [161], with charging voltage, discharging current, temperature, and cycle times as inputs, and the residual sum of squares error is selected as the cost function to calculate the capacity retention rate. Ref. [162] uses Particle Swarm Optimization (PSO) to find the suitable hyper-parameters for the SVM kernel function and trains the impedance values as the features to complete the battery SOH estimation.

Bayesian learning method: Bayesian methods solve the posterior information with assumed prior probabilities to infer the unknown parameters [163]. There are a variety of data-driven methods that use their associated theory, such as the Relevance Vector Machine (RVM), which provides an output of posterior probabilities based on a Bayesian approach. Compared to SVM, it eliminates the need for model selection, but it often requires more training time. The literature uses empirical modal decomposition for battery capacity data, and sets up a multi-start prediction matrix to train RVM. It reduces the stochastic uncertainty associated with the starting point of a single prediction and parameter settings. GPR is derived from the Bayesian framework [164], and uses the Gaussian process prior knowledge to perform regression analysis on the data. Ref. [165] uses voltage segments in short periods during constant current operation as the input of GPR for capacity estimation. The non-parametric regression properties of the GPR technique allow the estimation to be adapted to the complexity of the data.

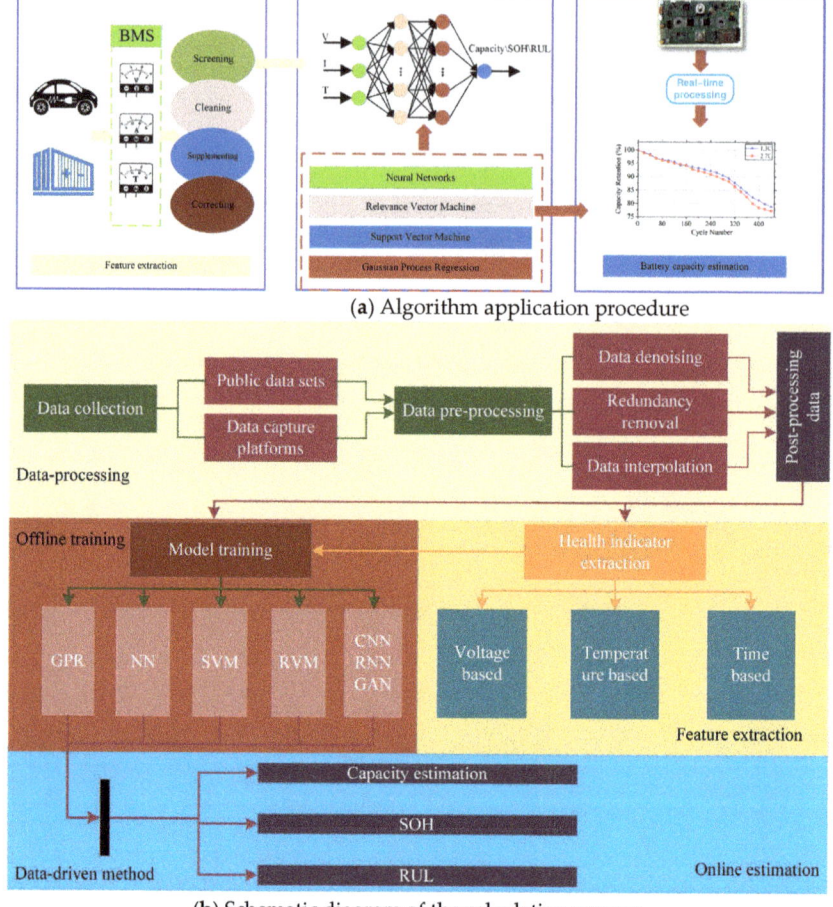

(a) Algorithm application procedure

(b) Schematic diagram of the calculation process

Figure 15. The basic process of the data-driven approach.

Deep learning method: Deep learning utilizes multiple hidden layers in the network [32], which can reflect more complex mapping between the features and battery health. Methods, such as Convolutional Neural Network (CNN) [166,167], Recurrent Neural Network (RNN) [168,169] and Long Short-Term Memory (LSTM) [170,171] have been used for battery SOH estimation recently, and have shown promising performance in estimation accuracy. Ref. [167] takes advantage of CNN and Transformers for accurate SOH estimation of Li-ion batteries, which utilizes the attention mechanism to extract more important features from the original measurement. A differential evolution grey wolf optimizer is used in [171] to tune the hyperparameters of LSTM for an accurate battery health estimation. A hybrid of gate recurrent unit and CNN is shown to estimate the Li-ion battery SOH in [172], which utilizes voltage, current and temperature as the input of the network.

The implementation of data-driven methods relies on the validity of the data and a complex training process. The advantage is that the model can be adapted to the data through training, but this also means that a large sampling and training dataset is required to achieve an accurate estimation.

## 4. Discussion

From the previous description, there have been a large number of studies on battery capacity estimation. We realize that some methods require specific implementation conditions. This paper discusses current battery capacity estimation methods for online BMS implementation, which are briefly divided into: direct measurement methods, analysis-based methods, SOC-based methods and data-driven methods. Since direct measurement methods are mostly limited to laboratory tests as a reference, the other three kinds of methods are compared with pros and cons in Table 1.

**Table 1.** A comparison of capacity estimation methods.

| Methods | Examples & Relevant References | Estimation Error | Strength | Drawback |
|---|---|---|---|---|
| Analysis-based method | IC curve [67–76] | Max relevant error [4%] RMSE [0.0066–0.0605] | Reflect the chemical characteristics of the battery, Simple model structure | Noise sensitivity |
| | DV cure [77–81] | Max relevant error [3%] | | |
| | DT cure [82–89] | Max relevant error [5.9%] RMSE [0.0027–0.0251] | | |
| | Mechanical stress [90–97] | Max relevant error [12%] | | |
| | EIS [98–112] | Max relavant error [2.2%] RMSE [0.0098–0.0116] | | |
| SOC-based method | WLS [114–118] | Max relevant error [1%] | Less computation, Easy for online imply | Difficult to cope with complex non-linear problems |
| | TLS [119–123] | Max relevant error [0.15%] | | |
| | EKF [134–138] | Max relevant error [0.5%] RMSE [0.0306–0.0599] | Closed-loop error management, Real-time dynamic tracking, Effective to handle the noise, Non-linear systems applicable | Complex model and parameter building process, High dependency on models, |
| | SPKF(UKF) [139–144] | RMSE [0.002–0.1275] | | |
| | PF and their variants [145–149] | Max relevant error [0.4%] RMSE [0.0019] | | |
| Data-driven method | NN [153–157] | RMSE [0.0121–0.0223] | No need to focus on internal mechanisms, simple model building, high adaptive capability, Powerful approximating ability, Non-linear systems applicable | High level of data dependency, offline training needed, Large computation effort, Over-fitting |
| | SVM [158–162] | RMSE [0.03–0.07] | | |
| | Bayesian method [163–165] | Max relevant error [3%] RMSE [0.0041–0.0068] | | |
| | Deep learning method [166–172] | Max relevant error [5%] RMSE [0.0032–0.0653] | | |

Analysis-based methods utilize the electric, thermal and strain characteristics of an Li-ion battery during charge or discharge, which always need a high precise sensor for measurement. The relationship between the analyzed physical quantity and battery degradation is closely related to measurement conditions and the materials of the battery chemistry; such factors limit the usage of the analysis-based method. In addition, professional knowledge is required to use curves for degradation mechanism analysis. IC and DV curves need to measure the voltage and current with high accuracy and low C-rate. All the analysis-based methods should be processed by a specially designed filter for receiving a smooth curve. The DT curve and mechanism strain are affected by various factors, such as the thermal management and structural design of the battery pack. The positions of the thermal and strain sensors in the battery module are still challenging, which limit their current usage in BMS products. Further investigations of the DT curve and mechanical stress are still needed to clarify the variation of the thermal and mechanical characteristics with battery degradation. EIS can detect the electrochemical kinetics inside the Li-ion battery with a small AC signal injection, which can reflect the battery degradation degree by the battery impedances, such as Ohmic resistance, polarization resistance, and SEI film resistance. With a broadband signal injection, the status of Li-ion battery can be analyzed in the frequency domain. Recently, some chips (DNB1168 [173]) have been made for EIS measurement of Li-ion batteries, which push the application of EIS to BMS a promising solution for the next generation BESS.

SOC-based methods rely on an accurate battery model. Although ECMs have been proven to have a good balance of accuracy and complexity, they are still far from been satisfied usage in a BMS. If the battery model is not reliable, the SOC estimation error will later affect the capacity estimation. In addition, the SOC-based methods usually obtain the results by iterations, which consumes the computing resources of BMS hardware. Thus, the application of SOC-based method will increase the cost of BMS. How to improve the computing efficiency and modeling accuracy still needs more further studies.

Data-driven methods treat the Li-ion battery as a black box without the need to deal with the complicated degradation mechanisms of the battery. However, quiet few data-driven models have been used in the current BMS. One reason is that the performance of the data-driven method is closely related to the feature and quality of the measurement dataset. There is a lack of open source datasets from a real BESS application for training the data-driven model. In reality, it is also difficult to obtain a labelled dataset, and measuring all the needed dataset from experimental testbench is costly. Another point is that the data-driven method lacks interpretability, and the credibility of the estimation results may be doubtful for real applications. The training of the data-driven model is closely related to a proper setting of the training procedure, while the training process is also time consuming. However, with the fast development of IoT and AI, data-driven methods will probably play a critical role in the future BMSs.

One key point we have to mention here is the materials of the electrodes might affect the applications of the capacity estimation methods for an application. For example, the NMC and LFP based battery characterize by a different OCV–SOC profile. The flat voltage curve of LFP battery corresponding to SOC may influence the capacity estimation accuracy. Thus, special attentions have to be paid when extracting features from the flat voltage curve or using OCV as the input of the method. IC curve can reflect the phase transitions of the electrodes, and probably can be used for feature extraction instead of a flat voltage curve for LFP battery.

We believe that the future onboard capacity estimation framework will be a hybrid of data-driven methods and other techniques as shown in Figure 16. Analysis-based methods can provide health features for the data-driven model. In this way, using an analysis-based method only needs a data processing procedure, and professional knowledge from the Li-ion battery degradation can be ignored. An SOC-based method can also be running in the BMS terminal, and can later collaborate with data-driven method through a fusion mechanism.

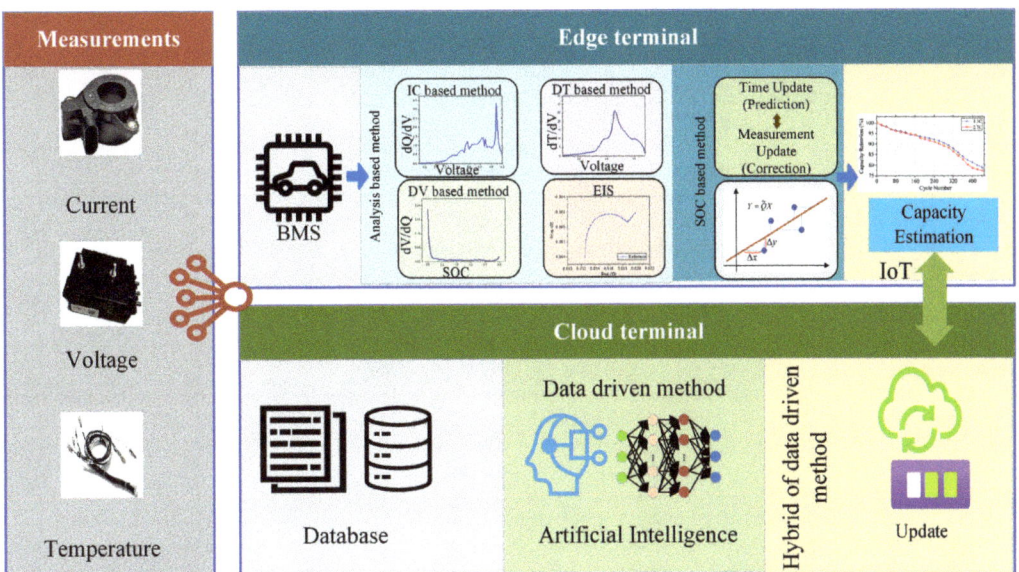

**Figure 16.** The prospective of future onboard capacity estimation framework.

## 5. Conclusions

This paper discusses a variety of methods for onboard BMS capacity estimation, which are based on different principles. These methods are divided into four main categories, direct measurement methods, analysis-based methods, SOC-based methods and data-driven methods. With emphasis on the onboard BMS implementable methods, the characteristics of each method are reviewed and discussed. Analysis-based methods with IC/DV/DT curves and mechanical strain are suitable for Li-ion battery degradation mechanism investigation. IC and DV curves are easier to be applied to a BMS application since no more sensor or measurement devices are needed in the battery packs. EIS is a promising solution for onboard BMS usage in the near future with an update of the hardware. SOC-based methods rely on the accuracy of the battery model and the iterative process requires more computing resources from the microprocessor in a BMS. With the development of the cloud-to-edge technique, data-driven methods will play an important role in the next-generation BESS.

From the methods discussed in this work, we hope to have summarized the recent progress in the battery capacity estimation area. In the future, a hybrid of various methods could be a more practical solution for real BMS applications, especially, a combination of data-driven and analysis-based methods.

**Author Contributions:** Conceptualization, J.P., J.M.; methodology, J.P., J.M.; software, J.P., J.M. and D.C.; validation, X.D., D.C. and X.S.; formal analysis, J.P., X.D.; investigation, J.M., D.C.; writing—original draft preparation, J.P., J.M.; writing—review and editing, J.P., J.M.; supervision, S.H., H.L. All authors have read and agreed to the published version of the manuscript.

**Funding:** This work is supported by the University-level Research Foundation of Nanjing Institute of Technology under grant YKJ2019114, the Open Research Fund of Jiangsu Collaborative Innovation Center for Smart Distribution Network under grant XTCX202005, and the National Natural Science Foundation of China under Grant 52107229.

**Data Availability Statement:** Not applicable.

**Conflicts of Interest:** The authors declare no conflict of interest.

## References

1. Tang, C.Y.; Chen, P.T.; Jheng, J.H. Bidirectional Power Flow Control and Hybrid Charging Strategies for Three-Phase PV Power and Energy Storage Systems. *IEEE Trans. Power Electron.* **2021**, *36*, 12710–12720. [CrossRef]
2. Ku, T.T.; Li, C.S. Implementation of Battery Energy Storage System for an Island Microgrid with High PV Penetration. *IEEE Trans. Ind. Appl.* **2021**, *57*, 3416–3424. [CrossRef]
3. Matthew, T.L.; Suthar, B.; Northrop, P.W.C.; De, S.; Santhanagopalan, S.; Subramanian, V.R. Battery Energy Storage System (BESS) and Battery Management System (BMS) for Grid-Scale Applications. *Proc. IEEE* **2014**, *102*, 1014–1030.
4. Bharatee, A.; Ray, P.K.; Ghosh, A. A Power Management Scheme for Grid-Connected PV Integrated with Hybrid Energy Storage System. *J. Mod. Power Syst. Clean Energy* **2022**, *10*, 954–963. [CrossRef]
5. Wan, T.; Tao, Y.; Qiu, J.; Lai, S. Data-Driven Hierarchical Optimal Allocation of Battery Energy Storage System. *IEEE Trans. Sustain. Energy* **2021**, *12*, 2097–2109. [CrossRef]
6. Rezvanizaniani, S.M.; Liu, Z.; Chen, Y.; Lee, J. Review and Recent Advances in Battery Health Monitoring and Prognostics Technologies for Electric Vehicle (EV) Safety and Mobility. *J. Power Sources* **2014**, *256*, 110–124. [CrossRef]
7. Wang, J.; Kang, L.; Liu, Y. Optimal Scheduling for Electric Bus Fleets Based on Dynamic Programming Approach by Considering Battery Capacity Fade. *Renew. Sustain. Energy Rev.* **2020**, *130*, 109978. [CrossRef]
8. Zhuang, W.; Lu, S.; Lu, H. Progress in Materials for Lithium-Ion Power Batteries. In Proceedings of the 2014 International Conference on Intelligent Green Building and Smart Grid (IGBSG), Taipei, Taiwan, 23–25 April 2014. [CrossRef]
9. Weiss, H.; Winkler, T.; Ziegerhofer, H. Large Lithium-Ion Battery-Powered Electric Vehicles—From Idea to Reality. In Proceedings of the 2018 ELEKTRO, Mikulov, Czech Republic, 21–23 May 2018; pp. 1–5.
10. Hannan, M.A.; Hoque, M.M.; Hussain, A.; Yusof, Y.; Ker, P.J. State-of-the-Art and Energy Management System of Lithium-Ion Batteries in Electric Vehicle Applications: Issues and Recommendations. *IEEE Access* **2018**, *6*, 19362–19378. [CrossRef]
11. Zubi, G.; Dufo-López, R.; Carvalho, M.; Pasaoglu, G. The Lithium-Ion Battery: State of the Art and Future Perspectives. *Renew. Sustain. Energy Rev.* **2018**, *89*, 292–308. [CrossRef]
12. BYD Blade Battery: Everything You Should Know. Available online: https://topelectricsuv.com/news/byd/byd-blade-battery-update/ (accessed on 1 October 2022).
13. Statista Projected Global Battery Demand from 2020 to 2030, by Application (in Gigawatt Hours). Available online: https://www.statista.com/statistics/ (accessed on 1 October 2022).
14. Plett, G.L. *Battery Management Systems, Volume II: Equivalent-Circuit Methods*; Artech House: Norwood, MA, USA, 2020; Volume II, ISBN 9781630810276.
15. Wang, L.; Qiu, J.; Wang, X.; Chen, L.; Cao, G.; Wang, J.; Zhang, H.; He, X. Insights for Understanding Multiscale Degradation of LiFePO4 Cathodes. *eScience* **2022**, *2*, 125–137. [CrossRef]
16. Meng, J.; Luo, G.; Ricco, M.; Swierczynski, M.; Stroe, D.-I.; Teodorescu, R. Overview of Lithium-Ion Battery Modeling Methods for State-of-Charge Estimation in Electrical Vehicles. *Appl. Sci.* **2018**, *8*, 659. [CrossRef]
17. Feng, X.; Ren, D.; He, X.; Ouyang, M. Mitigating Thermal Runaway of Lithium-Ion Batteries. *Joule* **2020**, *4*, 743–770. [CrossRef]
18. Feng, X.; Ouyang, M.; Liu, X.; Lu, L.; Xia, Y.; He, X. Thermal Runaway Mechanism of Lithium Ion Battery for Electric Vehicles: A Review. *Energy Storage Mater.* **2018**, *10*, 246–267. [CrossRef]
19. Monden, Y.; Mizutani, M.; Yamazaki, S.; Kobayashi, T. Charging and Discharging Control of a Hybrid Battery Energy Storage System Using Different Battery Types in Order to Avoid Degradation. In Proceedings of the 2021 IEEE International Future Energy Electronics Conference (IFEEC), Taipei, Taiwan, 16–19 November 2021; pp. 1–6. [CrossRef]
20. Zhao, Y.; Pohl, O.; Bhatt, A.I.; Collis, G.E.; Mahon, P.J.; Rüther, T.; Hollenkamp, A.F. A Review on Battery Market Trends, Second-Life Reuse, and Recycling. *Sustain. Chem.* **2021**, *2*, 167–205. [CrossRef]
21. Martinez-Laserna, E.; Gandiaga, I.; Sarasketa-Zabala, E.; Badeda, J.; Stroe, D.I.; Swierczynski, M.; Goikoetxea, A. Battery Second Life: Hype, Hope or Reality? A Critical Review of the State of the Art. *Renew. Sustain. Energy Rev.* **2018**, *93*, 701–718. [CrossRef]
22. Dai, H.; Jiang, B.; Hu, X.; Lin, X.; Wei, X.; Pecht, M. Advanced Battery Management Strategies for a Sustainable Energy Future: Multilayer Design Concepts and Research Trends. *Renew. Sustain. Energy Rev.* **2021**, *138*, 110480. [CrossRef]
23. Hossain Lipu, M.S.; Hannan, M.A.; Karim, T.F.; Hussain, A.; Saad, M.H.M.; Ayob, A.; Miah, M.S.; Indra Mahlia, T.M. Intelligent Algorithms and Control Strategies for Battery Management System in Electric Vehicles: Progress, Challenges and Future Outlook. *J. Clean. Prod.* **2021**, *292*, 126044. [CrossRef]
24. Schuster, S.F.; Bach, T.; Fleder, E.; Müller, J.; Brand, M.; Sextl, G.; Jossen, A. Nonlinear Aging Characteristics of Lithium-Ion Cells under Different Operational Conditions. *J. Energy Storage* **2015**, *1*, 44–53. [CrossRef]
25. Deyab, M.A.; Mohsen, Q. Improved Battery Capacity and Cycle Life in Iron-Air Batteries with Ionic Liquid. *Renew. Sustain. Energy Rev.* **2021**, *139*, 110729. [CrossRef]
26. Stroe, D.; Świerczyński, M.; Stan, A.; Teodorescu, R.; Andreasen, S.J. Accelerated Lifetime Testing Methodology for Lifetime Estimation of Lithium-Ion Batteries Used in Augmented Wind Power Plants. *IEEE Trans. Ind. Appl.* **2014**, *50*, 4006–4017. [CrossRef]
27. Meng, J.; Cai, L.; Stroe, D.-I.; Ma, J.; Luo, G.; Teodorescu, R. An Optimized Ensemble Learning Framework for Lithium-Ion Battery State of Health Estimation in Energy Storage System. *Energy* **2020**, *206*, 118140. [CrossRef]
28. Lu, L.; Han, X.; Li, J.; Hua, J.; Ouyang, M. A Review on the Key Issues for Lithium-Ion Battery Management in Electric Vehicles. *J. Power Sources* **2013**, *226*, 272–288. [CrossRef]

29. Xiong, R.; Zhang, Y.; Wang, J.; He, H.; Peng, S.; Pecht, M. Lithium-Ion Battery Health Prognosis Based on a Real Battery Management System Used in Electric Vehicles. *IEEE Trans. Veh. Technol.* **2019**, *68*, 4110–4121. [CrossRef]
30. Farmann, A.; Waag, W.; Marongiu, A.; Sauer, D.U. Critical Review of On-Board Capacity Estimation Techniques for Lithium-Ion Batteries in Electric and Hybrid Electric Vehicles. *J. Power Sources* **2015**, *281*, 114–130. [CrossRef]
31. Cai, L.; Meng, J.; Stroe, D.-I.; Luo, G.; Teodorescu, R. An Evolutionary Framework for Lithium-Ion Battery State of Health Estimation. *J. Power Sources* **2019**, *412*, 615–622. [CrossRef]
32. Sui, X.; He, S.; Vilsen, S.B.; Meng, J.; Teodorescu, R.; Stroe, D.I. A Review of Non-Probabilistic Machine Learning-Based State of Health Estimation Techniques for Lithium-Ion Battery. *Appl. Energy* **2021**, *300*, 117346. [CrossRef]
33. Mohamed, A.T. A Review Paper on Batteries Charging Systems with the State of Charge Determination Techniques. *IET Conf. Publ.* **2019**, *2019*, 1–6. [CrossRef]
34. Qays, M.O.; Buswig, Y.; Hossain, M.L.; Abu-Siada, A. Recent Progress and Future Trends on State of Charge Estimation Methods to Improve Battery-Storage Efficiency: A Review. *CSEE J. Power Energy Syst.* **2019**, *8*, 105–114. [CrossRef]
35. Wang, Y.; Tian, J.; Sun, Z.; Wang, L.; Xu, R.; Li, M.; Chen, Z. A Comprehensive Review of Battery Modeling and State Estimation Approaches for Advanced Battery Management Systems. *Renew. Sustain. Energy Rev.* **2020**, *131*, 110015. [CrossRef]
36. Xiong, R.; Li, L.; Tian, J. Towards a Smarter Battery Management System: A Critical Review on Battery State of Health Monitoring Methods. *J. Power Sources* **2018**, *405*, 18–29. [CrossRef]
37. Xiong, R.; Pan, Y.; Shen, W.; Li, H.; Sun, F. Lithium-Ion Battery Aging Mechanisms and Diagnosis Method for Automotive Applications: Recent Advances and Perspectives. *Renew. Sustain. Energy Rev.* **2020**, *131*, 110048. [CrossRef]
38. Corno, M.; Pozzato, G. Active Adaptive Battery Aging Management for Electric Vehicles. *IEEE Trans. Veh. Technol.* **2020**, *69*, 258–269. [CrossRef]
39. Shi, M.; Shi, X.; Li, Z.; Wang, X. Study on the Aging Characteristics of Li-Ion Battery Based on the Electro-Thermal and Aging Joint Simulation Platform. In Proceedings of the 2020 IEEE/IAS Industrial and Commercial Power System Asia (I&CPS Asia), Weihai, China, 13–15 July 2020; pp. 257–261.
40. Meyer, C.; Bockholt, H.; Haselrieder, W.; Kwade, A. Characterization of the Calendering Process for Compaction of Electrodes for Lithium-Ion Batteries. *J. Mater. Process. Technol.* **2017**, *249*, 172–178. [CrossRef]
41. Schreiner, D.; Klinger, A.; Reinhart, G. Modeling of the Calendering Process for Lithium-Ion Batteries with DEM Simulation. *Procedia CIRP* **2020**, *93*, 149–155. [CrossRef]
42. Mocera, F.; Soma, A.; Clerici, D. Study of Aging Mechanisms in Lithium-Ion Batteries for Working Vehicle Applications. In Proceedings of the 2020 Fifteenth International Conference on Ecological Vehicles and Renewable Energies (EVER), Monte-Carlo, Monaco, 10–12 September 2020.
43. Stroe, D.I.; Swierczynski, M.; Kær, S.K.; Teodorescu, R. Degradation Behavior of Lithium-Ion Batteries During Calendar Ageing—The Case of the Internal Resistance Increase. *Proc. IEEE Trans. Ind. Appl.* **2018**, *54*, 517–525. [CrossRef]
44. Dufek, E.J.; Tanim, T.R.; Chen, B.-R. Sangwook Kim Battery Calendar Aging and Machine Learning. *Joule* **2022**, *6*, 1363–1367. [CrossRef]
45. Dubarry, M.; Liaw, B.Y.; Chen, M.-S.; Chyan, S.-S.; Han, K.-C.; Sie, W.-T.; Wu, S.-H. Identifying Battery Aging Mechanisms in Large Format Li Ion Cells. *J. Power Sources* **2011**, *196*, 3420–3425. [CrossRef]
46. Mendoza-Hernandez, O.S.; Hosono, E.; Asakura, D.; Matsuda, H.; Shironita, S.; Umeda, M.; Sone, Y. Impact of Calendar Degradation on the Performance of LiFePO$_4$—Graphite Li-Ion Cells during Charge-Discharge Cycling at $-5$ °C. *J. Electrochem. Soc.* **2019**, *166*, A3525–A3530. [CrossRef]
47. Belt, J.; Utgikar, V.; Bloom, I. Calendar and PHEV Cycle Life Aging of High-Energy, Lithium-Ion Cells Containing Blended Spinel and Layered-Oxide Cathodes. *J. Power Sources* **2011**, *196*, 10213–10221. [CrossRef]
48. Nikitina, V.A. Advanced Electrochemical Analysis of Metal-Ion Battery Materials for Rationalizing and Improving Battery Performance. *Curr. Opin. Electrochem.* **2021**, *29*, 100768. [CrossRef]
49. Guo, J.; Li, Y.; Meng, J.; Pedersen, K.; Gurevich, L.; Stroe, D.-I. Understanding the Mechanism of Capacity Increase during Early Cycling of Commercial NMC/Graphite Lithium-Ion Batteries. *J. Energy Chem.* **2022**, *74*, 34–44. [CrossRef]
50. Fleckenstein, M.; Bohlen, O.; Roscher, M.A.; Bäker, B. Current Density and State of Charge Inhomogeneities in Li-Ion Battery Cells with LiFePO$_4$ as Cathode Material Due to Temperature Gradients. *J. Power Sources* **2011**, *196*, 4769–4778. [CrossRef]
51. Lyu, C.; Zhao, Y.; Luo, W.; Wang, L. Aging Mechanism Analysis and Its Impact on Capacity Loss of Lithium Ion Batteries. In Proceedings of the 2019 14th IEEE Conference on Industrial Electronics and Applications (ICIEA), Xi'an, China, 19–21 June 2019; pp. 2148–2153.
52. Han, X.; Ouyang, M.; Lu, L.; Li, J.; Zheng, Y.; Li, Z. A Comparative Study of Commercial Lithium Ion Battery Cycle Life in Electrical Vehicle: Aging Mechanism Identification. *J. Power Sources* **2014**, *251*, 38–54. [CrossRef]
53. Barré, A.; Deguilhem, B.; Grolleau, S.; Gérard, M.; Suard, F.; Riu, D. A Review on Lithium-Ion Battery Ageing Mechanisms and Estimations for Automotive Applications. *J. Power Sources* **2013**, *241*, 680–689. [CrossRef]
54. Chen, M.; Zhang, L.; Yu, F.; Zhou, L. An Aging Experimental Study of Li-Ion Batteries for Marine Energy Power Station Application. In Proceedings of the 2019 Prognostics and System Health Management Conference (PHM-Qingdao), Qingdao, China, 25–27 October 2019. [CrossRef]
55. Tian, J.; Xiong, R.; Shen, W. State-of-Health Estimation Based on Differential Temperature for Lithium Ion Batteries. *IEEE Trans. Power Electron.* **2020**, *35*, 10363–10373. [CrossRef]

56. Huang, H.; Meng, J.; Wang, Y.; Feng, F.; Cai, L.; Peng, J.; Liu, T. A Comprehensively Optimized Lithium-Ion Battery State-of-Health Estimator Based on Local Coulomb Counting Curve. *Appl. Energy* **2022**, *322*, 119469. [CrossRef]
57. Sataloff, R.T.; Johns, M.M.; Kost, K.M. Fundamentals and Applications of Lithium-Ion Batteries in Electric Drive Vehicles. In *Geriatric Otolaryngology*; Thieme Medical Publishers: New York, NY, USA, 2015; ISBN 9781626239777.
58. IEC-62660-2; Secondary Lithium-Ion Cells for the Propulsion of Electric Road Vehicles—Part 2: Reliability and Abuse Testing 2018. International Electrotechnical Commission: Geneva, Switzerland, 2018.
59. ISO 12405-3; ISO Electrically Propelled Road Vehicles—Test Specification for Lithium-Ion Traction Battery Packs and Systems—Part 3: Safety Performance Requirements. ISO: Geneva, Switzerland, 2018.
60. IEEE Std 450-2020 (Revision IEEE Std 450-2010); IEEE Recommended Practice for Maintenance, Testing, and Replacement of Vented Lead-Acid Batteries for Stationary Applications. IEEE: New York, NY, USA, 2021; pp. 1–71.
61. Meng, J.; Cai, L.; Stroe, D.-I.; Luo, G.; Sui, X.; Teodorescu, R. Lithium-Ion Battery State-of-Health Estimation in Electric Vehicle Using Optimized Partial Charging Voltage Profiles. *Energy* **2019**, *185*, 1054–1062. [CrossRef]
62. Meng, J.; Cai, L.; Luo, G.; Stroe, D.-I.; Teodorescu, R. Lithium-Ion Battery State of Health Estimation with Short-Term Current Pulse Test and Support Vector Machine. *Microelectron. Reliab.* **2018**, *88–90*, 1216–1220. [CrossRef]
63. Belt, J.R.; Ho, C.D.; Motloch, C.G.; Miller, T.J.; Duong, T.Q. A Capacity and Power Fade Study of Li-Ion Cells during Life Cycle Testing. *J. Power Sources* **2003**, *123*, 241–246. [CrossRef]
64. Wen, A.; Meng, J.; Peng, J.; Cai, L.; Xiao, Q. Online Parameter Identification of the Lithium-Ion Battery with Refined Instrumental Variable Estimation. *Complexity* **2020**, *2020*, 8854618. [CrossRef]
65. Du, X.; Meng, J.; Zhang, Y.; Huang, X.; Wang, S.; Liu, P.; Liu, T. An Information Appraisal Procedure: Endows Reliable Online Parameter Identification to Lithium-Ion Battery Model. *IEEE Trans. Ind. Electron.* **2022**, *69*, 5889–5899. [CrossRef]
66. Xiong, R.; Sun, F.; He, H. Model-Based Health Condition Monitoring Method for Multi-Cell Series-Connected Battery Pack. In Proceedings of the 2016 IEEE Transportation Electrification Conference and Expo (ITEC), Dearborn, MI, USA, 27–29 June 2016; pp. 1–5. [CrossRef]
67. Lin, M.; Wu, D.; Meng, J.; Wu, J.; Wu, H. A Multi-Feature-Based Multi-Model Fusion Method for State of Health Estimation of Lithium-Ion Batteries. *J. Power Sources* **2022**, *518*, 230774. [CrossRef]
68. Tang, X.; Liu, K.; Lu, J.; Liu, B.; Wang, X.; Gao, F. Battery Incremental Capacity Curve Extraction by a Two-Dimensional Luenberger–Gaussian-Moving-Average Filter. *Appl. Energy* **2020**, *280*, 115895. [CrossRef]
69. Maures, M.; Zhang, Y.; Martin, C.; Delétage, J.-Y.; Vinassa, J.-M.; Briat, O. Impact of Temperature on Calendar Ageing of Lithium-Ion Battery Using Incremental Capacity Analysis. *Microelectron. Reliab.* **2019**, *100–101*, 113364. [CrossRef]
70. Feng, X.; Weng, C.; He, X.; Wang, L.; Ren, D.; Lu, L.; Han, X.; Ouyang, M. Incremental Capacity Analysis on Commercial Lithium-Ion Batteries Using Support Vector Regression: A Parametric Study. *Energies* **2018**, *11*, 2323. [CrossRef]
71. Li, L.; Li, Y.; Cui, W.; Chen, Z.; Wang, D.; Zhou, B.; Hong, D. A Novel Health Indicator for Online Health Estimation of Lithium-Ion Batteries Using Partial Incremental Capacity and Dynamic Voltage Warping. *J. Power Sources* **2022**, *545*, 231961. [CrossRef]
72. Tang, X.; Zou, C.; Yao, K.; Chen, G.; Liu, B.; He, Z.; Gao, F. A Fast Estimation Algorithm for Lithium-Ion Battery State of Health. *J. Power Sources* **2018**, *396*, 453–458. [CrossRef]
73. Jiang, Y.; Jiang, J.; Zhang, C.; Zhang, W.; Gao, Y.; Li, N. State of Health Estimation of Second-Life LiFePO4 Batteries for Energy Storage Applications. *J. Clean. Prod.* **2018**, *205*, 754–762. [CrossRef]
74. Li, X.; Wang, Z.; Zhang, L.; Zou, C.; Dorrell, D.D. State-of-Health Estimation for Li-Ion Batteries by Combing the Incremental Capacity Analysis Method with Grey Relational Analysis. *J. Power Sources* **2019**, *410–411*, 106–114. [CrossRef]
75. Ma, H.; Deng, Y.; Liu, W.W.; Li, T.; Zhang, H. State of Health Estimation of Retired Battery for Echelon Utilization Based on Charging Curve. *Procedia CIRP* **2022**, *105*, 458–463. [CrossRef]
76. Li, X.; Jiang, J.; Wang, L.Y.; Chen, D.; Zhang, Y.; Zhang, C. A Capacity Model Based on Charging Process for State of Health Estimation of Lithium Ion Batteries. *Appl. Energy* **2016**, *177*, 537–543. [CrossRef]
77. Bloom, I.; Jansen, A.N.; Abraham, D.P.; Knuth, J.; Jones, S.A.; Battaglia, V.S.; Henriksen, G.L. Differential Voltage Analyses of High-Power, Lithium-Ion Cells: 1. Technique and Application. *J. Power Sources* **2005**, *139*, 295–303. [CrossRef]
78. Han, X.; Feng, X.; Ouyang, M.; Lu, L.; Li, J.; Zheng, Y.; Li, Z. A Comparative Study of Charging Voltage Curve Analysis and State of Health Estimation of Lithium-Ion Batteries in Electric Vehicle. *Automot. Innov.* **2019**, *2*, 263–275. [CrossRef]
79. Honkura, K.; Takahashi, K.; Horiba, T. Capacity-Fading Prediction of Lithium-Ion Batteries Based on Discharge Curves Analysis. *J. Power Sources* **2011**, *196*, 10141–10147. [CrossRef]
80. Han, X.; Ouyang, M.; Lu, L.; Li, J. Cycle Life of Commercial Lithium-Ion Batteries with Lithium Titanium Oxide Anodes in Electric Vehicles. *Energies* **2014**, *7*, 4895–4909. [CrossRef]
81. Liang, T.; Song, L.; Shi, K. On-Board Incremental Capacity/Differential Voltage Curves Acquisition for State of Health Monitoring of Lithium-Ion Batteries. In Proceedings of the 2018 IEEE International Conference on Applied System Invention (ICASI), Tokyo, Japan, 13–17 April 2018; pp. 976–979.
82. Wu, B.; Yufit, V.; Merla, Y.; Martinez-Botas, R.F.; Brandon, N.P.; Offer, G.J. Differential Thermal Voltammetry for Tracking of Degradation in Lithium-Ion Batteries. *J. Power Sources* **2015**, *273*, 495–501. [CrossRef]
83. Zhang, Q.; Wei, F.; Zhang, P.; Dong, R.; Li, J.; Li, P.; Jia, Q.; Liu, Y.; Mao, J.; Shao, G. Research on the Reversible and Irreversible Heat Generation of LiNi$_{1-x-y}$Co$_x$Mn$_y$O$_2$-Based Lithium-Ion Batteries. *Fire Technol.* **2022**. [CrossRef]

84. Maher, K.; Yazami, R. A Study of Lithium Ion Batteries Cycle Aging by Thermodynamics Techniques. *J. Power Sources* **2014**, *247*, 527–533. [CrossRef]
85. Yang, J.; Cai, Y.; Mi, C. Lithium-Ion Battery Capacity Estimation Based on Battery Surface Temperature Change under Constant-Current Charge Scenario. *Energy* **2022**, *241*, 122879. [CrossRef]
86. Birkl, C. *Diagnosis and Prognosis of Degradation in Lithium-Ion Batteries*; University of Oxford: Oxford, UK, 2017.
87. Wang, Z.; Yuan, C.; Li, X. Lithium Battery State-of-Health Estimation via Differential Thermal Voltammetry with Gaussian Process Regression. *IEEE Trans. Transp. Electrif.* **2021**, *7*, 16–25. [CrossRef]
88. Saha, B.; Goebel, K. Battery Data Set. Available online: https://www.nasa.gov/intelligent-systems-division (accessed on 1 October 2022).
89. Wu, Y.; Jossen, A. Entropy-Induced Temperature Variation as a New Indicator for State of Health Estimation of Lithium-Ion Cells. *Electrochim. Acta* **2018**, *276*, 370–376. [CrossRef]
90. Zhang, W.-J. Structure and Performance of LiFePO4 Cathode Materials: A Review. *J. Power Sources* **2011**, *196*, 2962–2970. [CrossRef]
91. Oh, K.-Y.; Epureanu, B.I.; Siegel, J.B.; Stefanopoulou, A.G. Phenomenological Force and Swelling Models for Rechargeable Lithium-Ion Battery Cells. *J. Power Sources* **2016**, *310*, 118–129. [CrossRef]
92. Li, R.; Li, W.; Singh, A.; Ren, D.; Hou, Z.; Ouyang, M. Effect of External Pressure and Internal Stress on Battery Performance and Lifespan. *Energy Storage Mater.* **2022**, *52*, 395–429. [CrossRef]
93. Jeong, J.; Kwak, E.; Kim, J.; Oh, K.-Y. Novel Active Management of Compressive Pressure on a Lithium-Ion Battery Using a Phase Transition Actuator. *Energy Rep.* **2022**, *8*, 10762–10775. [CrossRef]
94. Deich, T.; Hahn, S.L.; Both, S.; Birke, K.P.; Bund, A. Validation of an Actively-Controlled Pneumatic Press to Simulate Automotive Module Stiffness for Mechanically Representative Lithium-Ion Cell Aging. *J. Energy Storage* **2020**, *28*, 101192. [CrossRef]
95. Oh, K.Y.; Siegel, J.B.; Secondo, L.; Kim, S.U.; Samad, N.A.; Qin, J.; Anderson, D.; Garikipati, K.; Knobloch, A.; Epureanu, B.I.; et al. Rate Dependence of Swelling in Lithium-Ion Cells. *J. Power Sources* **2014**, *267*, 197–202. [CrossRef]
96. Samad, N.A.; Kim, Y.; Siegel, J.B.; Stefanopoulou, A.G. Battery Capacity Fading Estimation Using a Force-Based Incremental Capacity Analysis. *J. Electrochem. Soc.* **2016**, *163*, A1584–A1594. [CrossRef]
97. Schiffer, Z.J.; Cannarella, J.; Arnold, C.B. Strain Derivatives for Practical Charge Rate Characterization of Lithium Ion Electrodes. *J. Electrochem. Soc.* **2016**, *163*, A427–A433. [CrossRef]
98. Deng, Z.; Zhang, Z.; Lai, Y.; Liu, J.; Li, J.; Liu, Y. Electrochemical Impedance Spectroscopy Study of a Lithium/Sulfur Battery: Modeling and Analysis of Capacity Fading. *J. Electrochem. Soc.* **2013**, *160*, A553–A558. [CrossRef]
99. Capkova, D.; Knap, V.; Fedorkova, A.S.; Stroe, D.-I. Analysis of 3.4 Ah Lithium-Sulfur Pouch Cells by Electrochemical Impedance Spectroscopy. *J. Energy Chem.* **2022**, *72*, 318–325. [CrossRef]
100. Du, X.; Meng, J.; Peng, J.; Zhang, Y.; Liu, T.; Teodorescu, R. Sensorless Temperature Estimation of Lithium-Ion Battery Based on Broadband Impedance Measurements. *IEEE Trans. Power Electron.* **2022**, *37*, 10101–10105. [CrossRef]
101. Zhang, Y.; Tang, Q.; Zhang, Y.; Wang, J.; Stimming, U.; Lee, A.A. Identifying Degradation Patterns of Lithium Ion Batteries from Impedance Spectroscopy Using Machine Learning. *Nat. Commun.* **2020**, *11*, 1706. [CrossRef]
102. Jiang, B.; Zhu, J.; Wang, X.; Wei, X.; Shang, W.; Dai, H. A Comparative Study of Different Features Extracted from Electrochemical Impedance Spectroscopy in State of Health Estimation for Lithium-Ion Batteries. *Appl. Energy* **2022**, *322*, 119502. [CrossRef]
103. Meddings, N.; Heinrich, M.; Overney, F.; Lee, J.-S.; Ruiz, V.; Napolitano, E.; Seitz, S.; Hinds, G.; Raccichini, R.; Gaberšček, M.; et al. Application of Electrochemical Impedance Spectroscopy to Commercial Li-Ion Cells: A Review. *J. Power Sources* **2020**, *480*, 228742. [CrossRef]
104. Knehr, K.W.; Hodson, T.; Bommier, C.; Davies, G.; Kim, A.; Steingart, D.A. Understanding Full-Cell Evolution and Non-Chemical Electrode Crosstalk of Li-Ion Batteries. *Joule* **2018**, *2*, 1146–1159. [CrossRef]
105. Khan, N.; Ullah, F.U.M.; Afnan; Ullah, A.; Lee, M.Y.; Baik, S.W. Batteries State of Health Estimation via Efficient Neural Networks with Multiple Channel Charging Profiles. *IEEE Access* **2021**, *9*, 7797–7813. [CrossRef]
106. Sihvo, J.; Roinila, T.; Stroe, D.I. SOH Analysis of Li-Ion Battery Based on ECM Parameters and Broadband Impedance Measurements. In Proceedings of the IECON Proceedings (Industrial Electronics Conference), Singapore, 18–21 October 2020; pp. 1923–1928.
107. Wang, X.; Kou, Y.; Wang, B.; Jiang, Z.; Wei, X.; Dai, H. Fast Calculation of Broadband Battery Impedance Spectra Based on S Transform of Step Disturbance and Response. *IEEE Trans. Transp. Electrif.* **2022**, *8*, 3659–3672. [CrossRef]
108. Wang, X.; Wei, X.; Chen, Q.; Dai, H. A Novel System for Measuring Alternating Current Impedance Spectra of Series-Connected Lithium-Ion Batteries with a High-Power Dual Active Bridge Converter and Distributed Sampling Units. *IEEE Trans. Ind. Electron.* **2021**, *68*, 7380–7390. [CrossRef]
109. Huang, W.; Abu Qahouq, J.A. An Online Battery Impedance Measurement Method Using DC–DC Power Converter Control. *IEEE Trans. Ind. Electron.* **2014**, *61*, 5987–5995. [CrossRef]
110. Qahouq, J.A.A.; Xia, Z. Single-Perturbation-Cycle Online Battery Impedance Spectrum Measurement Method with Closed-Loop Control of Power Converter. *IEEE Trans. Ind. Electron.* **2017**, *64*, 7019–7029. [CrossRef]

111. Raijmakers, L.H.J.; Shivakumar, K.M.; Donkers, M.C.F.; Lammers, M.J.G.; Bergveld, H.J. Crosstalk Interferences on Impedance Measurements in Battery Packs**This Work Has Received Financial Support from the Dutch Ministry of Economic Affairs under the Grant A Green Deal in Energy Materials (ADEM) and from the Horizon 2020 Programme of the European Union under the Grant Integrated Components for Complexity Control in Affordable Electrified Cars (3Ccar-662192). *IFAC-PapersOnLine* **2016**, *49*, 42–47. [CrossRef]
112. Wei, X.; Wang, X.; Dai, H. Practical On-Board Measurement of Lithium Ion Battery Impedance Based on Distributed Voltage and Current Sampling. *Energies* **2018**, *11*, 64. [CrossRef]
113. He, C.; Feng, Z. Research on Parmeters of Acquisition Model Based on Non-Linear Least Squares Method. In Proceedings of the 2011 International Conference on Remote Sensing, Environment and Transportation Engineering, Nanjing, China, 24–26 June 2011; pp. 1499–1502. [CrossRef]
114. Toh, K.A.; Eng, H.L. Between Classification-Error Approximation and Weighted Least-Squares Learning. *IEEE Trans. Pattern Anal. Mach. Intell.* **2008**, *30*, 658–669. [CrossRef] [PubMed]
115. Plett, G.L. Recursive Approximate Weighted Total Least Squares Estimation of Battery Cell Total Capacity. *J. Power Sources* **2011**, *196*, 2319–2331. [CrossRef]
116. Kim, E.; Member, S.; Kim, K.; Member, S. Distance Estimation with Weighted Least Squaresfor Mobile Beacon-Based Localization in Wireless Sensor Networks. *Signal Process.* **2010**, *17*, 559–562.
117. Tang, X.; Mao, X.; Lin, J.; Koch, B. Capacity Estimation for Li-Ion Batteries. In Proceedings of the 2011 American Control Conference, San Francisco, CA, USA, 29 June 2011–1 July 2011; pp. 947–952. [CrossRef]
118. Wei, J.; Chen, C. State of Charge and Health Estimation for Lithium-Ion Batteries Using Recursive Least Squares. In Proceedings of the 2020 5th International Conference on Advanced Robotics and Mechatronics (ICARM), Shenzhen, China, 18–21 December 2020; pp. 686–689. [CrossRef]
119. Arablouei, R.; Dogancay, K. Linearly-Constrained Recursive Total Least-Squares Algorithm. *IEEE Signal Process. Lett.* **2012**, *19*, 821–824. [CrossRef]
120. Wei, Z.; Zou, C.; Leng, F.; Soong, B.H.; Tseng, K.J. Online Model Identification and State-of-Charge Estimate for Lithium-Ion Battery with a Recursive Total Least Squares-Based Observer. *IEEE Trans. Ind. Electron.* **2018**, *65*, 1336–1346. [CrossRef]
121. Feng, D.Z.; Bao, Z.; Jiao, L.C. Total Least Mean Squares Algorithm. *IEEE Trans. Signal Process.* **1998**, *46*, 2122–2130. [CrossRef]
122. Rhode, S.; Usevich, K.; Markovsky, I.; Gauterin, F. A Recursive Restricted Total Least-Squares Algorithm. *IEEE Trans. Signal Process.* **2014**, *62*, 5652–5662. [CrossRef]
123. Kim, T.; Wang, Y.; Sahinoglu, Z.; Wada, T.; Hara, S.; Qiao, W. A Rayleigh Quotient-Based Recursive Total-Least-Squares Online Maximum Capacity Estimation for Lithium-Ion Batteries. *IEEE Trans. Energy Convers.* **2015**, *30*, 842–851. [CrossRef]
124. Lai, X.; Zheng, Y.; Sun, T. A Comparative Study of Different Equivalent Circuit Models for Estimating State-of-Charge of Lithium-Ion Batteries. *Electrochim. Acta* **2018**, *259*, 566–577. [CrossRef]
125. Meng, J.; Ricco, M.; Luo, G.; Swierczynski, M.; Stroe, D.I.; Stroe, A.I.; Teodorescu, R. An Overview and Comparison of Online Implementable SOC Estimation Methods for Lithium-Ion Battery. *IEEE Trans. Ind. Appl.* **2018**, *54*, 1583–1591. [CrossRef]
126. Xia, Z.; Abu Qahouq, J.A. Evaluation of Parameter Variations of Equivalent Circuit Model of Lithium-Ion Battery under Different SOH Conditions. In Proceedings of the 2020 IEEE Energy Conversion Congress and Exposition (ECCE), Detroit, MI, USA, 11–15 October 2020; pp. 1519–1523. [CrossRef]
127. Kharisma, M.D.; Ridwan, M.; Ilmiawan, A.F.; Ario Nurman, F.; Rizal, S. Modeling and Simulation of Lithium-Ion Battery Pack Using Modified Battery Cell Model. In Proceedings of the 2019 6th International Conference on Electric Vehicular Technology (ICEVT), Bali, Indonesia, 18–21 November 2019; pp. 25–30. [CrossRef]
128. Plett, G.L. Extended Kalman Filtering for Battery Management Systems of LiPB-Based HEV Battery Packs—Part 3. State and Parameter Estimation. *J. Power Sources* **2004**, *134*, 277–292. [CrossRef]
129. Yu, Q.; Xiong, R.; Yang, R.; Pecht, M.G. Online Capacity Estimation for Lithium-Ion Batteries through Joint Estimation Method. *Appl. Energy* **2019**, *255*, 113817. [CrossRef]
130. Plett, G.L. Extended Kalman Filtering for Battery Management Systems of LiPB-Based HEV Battery Packs—Part 2. Modeling and Identification. *J. Power Sources* **2004**, *134*, 262–276. [CrossRef]
131. Chen, C.; Xiong, R.; Shen, W. A Lithium-Ion Battery-in-the-Loop Approach to Test and Validate Multiscale Dual H Infinity Filters for State-of-Charge and Capacity Estimation. *IEEE Trans. Power Electron.* **2018**, *33*, 332–342. [CrossRef]
132. Wang, H.; Leng, J. A Brief Review on the Development of Kalman Filter. In Proceedings of the 2018 Chinese Control and Decision Conference (CCDC), Shenyang, China, 9–11 June 2018; pp. 694–699. [CrossRef]
133. Xu, L.; Wang, J.; Chen, Q. Kalman Filtering State of Charge Estimation for Battery Management System Based on a Stochastic Fuzzy Neural Network Battery Model. *Energy Convers. Manag.* **2012**, *53*, 33–39. [CrossRef]
134. Rahmoun, A.; Biechl, H.; Rosin, A. SOC Estimation for Li-Ion Batteries Based on Equivalent Circuit Diagrams and the Application of a Kalman Filter. In Proceedings of the 2012 Electric Power Quality and Supply Reliability, Tartu, Estonia, 11–13 June 2012; pp. 273–276. [CrossRef]
135. Yu, Z.; Huai, R.; Xiao, L. State-of-Charge Estimation for Lithium-Ion Batteries Using a Kalman Filter Based on Local Linearization. *Energies* **2015**, *8*, 7854–7873. [CrossRef]

136. Sedighfar, A.; Moniri, M.R. Comparison of Three Well-Known Filters for the Battery State of Health Estimation Application. In Proceedings of the 2018 4th International Conference on Frontiers of Signal Processing (ICFSP), Poitiers, France, 24–27 September 2018; pp. 164–168. [CrossRef]
137. Shen, P.; Ouyang, M.; Lu, L.; Li, J.; Feng, X. The Co-Estimation of State of Charge, State of Health, and State of Function for Lithium-Ion Batteries in Electric Vehicles. *IEEE Trans. Veh. Technol.* **2018**, *67*, 92–103. [CrossRef]
138. Qian, K.F.; Liu, X.T. Hybrid Optimization Strategy for Lithium-Ion Battery's State of Charge/Health Using Joint of Dual Kalman Filter and Modified Sine-Cosine Algorithm. *J. Energy Storage* **2021**, *44*, 103319. [CrossRef]
139. Plett, G.L. Sigma-Point Kalman Filtering for Battery Management Systems of LiPB-Based HEV Battery Packs. Part 1: Introduction and State Estimation. *J. Power Sources* **2006**, *161*, 1356–1368. [CrossRef]
140. Xiong, R.; Mu, H. Accurate State of Charge Estimation for Lithium-Ion Battery Using Dual Uncsented Kalman Filters. In Proceedings of the 2017 Chinese Automation Congress (CAC), Jinan, China, 20–22 October 2017; pp. 5484–5487. [CrossRef]
141. Wang, L.; Wang, L.; Liao, C.; Liu, J. Sigma-Point Kalman Filter Application on Estimating Battery SOC. In Proceedings of the 2009 5th IEEE Vehicle Power and Propulsion Conference, VPPC '09, Dearborn, MI, USA, 7–10 September 2009; pp. 1592–1595. [CrossRef]
142. Sangwan, V.; Kumar, R.; Rathore, A.K. State-of-Charge Estimation for Li-Ion Battery Using Extended Kalman Filter (EKF) and Central Difference Kalman Filter (CDKF). In Proceedings of the 2017 IEEE Industry Applications Society Annual Meeting, Cincinnati, OH, USA, 1–5 October 2017; pp. 1–6. [CrossRef]
143. Ma, L.; Xu, Y.; Zhang, H.; Yang, F.; Wang, X.; Li, C. Co-Estimation of State of Charge and State of Health for Lithium-Ion Batteries Based on Fractional-Order Model with Multi-Innovations Unscented Kalman Filter Method. *J. Energy Storage* **2022**, *52*, 104904. [CrossRef]
144. Liu, F.; Shao, C.; Su, W.; Liu, Y. Online Joint Estimator of Key States for Battery Based on a New Equivalent Circuit Model. *J. Energy Storage* **2022**, *52*, 104780. [CrossRef]
145. Xing, Y.; Ma, E.W.M.; Tsui, K.L.; Pecht, M. A Case Study on Battery Life Prediction Using Particle Filtering. In Proceedings of the IEEE 2012 Prognostics and System Health Management Conference (PHM-2012 Beijing), Beijing, China, 23–25 May 2012. [CrossRef]
146. Cong, X.; Zhang, C.; Jiang, J.; Zhang, W.; Jiang, Y. A Hybrid Method for the Prediction of the Remaining Useful Life of Lithium-Ion Batteries with Accelerated Capacity Degradation. *IEEE Trans. Veh. Technol.* **2020**, *69*, 12775–12785. [CrossRef]
147. Hao, X.; Wu, J. Online State Estimation Using Particles Filters of Lithium-Ion Polymer Battery Packs for Electric Vehicle. In Proceedings of the 2015 IEEE International Conference on Systems, Man, and Cybernetics, Hong Kong, China, 9–12 October 2015; pp. 783–788. [CrossRef]
148. Omariba, Z.B.; Zhang, L.; Sun, D. Remaining useful life prediction of electric vehicle lithium-ion battery based on particle filter method. In Proceedings of the IEEE 3rd International Conference on Big Data Analysis (ICBDA), Shanghai, China, 9–12 March 2018; pp. 412–416. [CrossRef]
149. Cai, L.; Lin, J.; Liao, X. A Data-Driven Method for State of Health Prediction of Lithium-Ion Batteries in a Unified Framework. *J. Energy Storage* **2022**, *51*, 104371. [CrossRef]
150. Liu, X.; Jin, Y.; Zeng, S.; Chen, X.; Feng, Y.; Liu, S.; Liu, H. Online Identification of Power Battery Parameters for Electric Vehicles Using a Decoupling Multiple Forgetting Factors Recursive Least Squares Method. *CSEE J. Power Energy Syst.* **2020**, *6*, 735–742. [CrossRef]
151. Xiong, R.; Sun, F.; Chen, Z.; He, H. A Data-Driven Multi-Scale Extended Kalman Filtering Based Parameter and State Estimation Approach of Lithium-Ion Olymer Battery in Electric Vehicles. *Appl. Energy* **2014**, *113*, 463–476. [CrossRef]
152. Wei, Z.; Zhao, J.; Ji, D.; Tseng, K.J. A Multi-Timescale Estimator for Battery State of Charge and Capacity Dual Estimation Based on an Online Identified Model. *Appl. Energy* **2017**, *204*, 1264–1274. [CrossRef]
153. Zhao, F.; Li, P.; Li, Y.; Li, Y. The Li-Ion Battery State of Charge Prediction of Electric Vehicle Using Deep Neural Network. In Proceedings of the 2019 Chinese Control and Decision Conference (CCDC), Nanchang, China, 3–5 June 2019; pp. 773–777. [CrossRef]
154. Chen, C.-R.; Huang, K.-H. The Estimation of the Capacity of Lead-Acid Storage Battery Using Artificial Neural Networks. In Proceedings of the 2006 IEEE International Conference on Systems, Man and Cybernetics, Taipei, Taiwan, 8–11 October 2006; pp. 1575–1579.
155. Sarvi, M.; Adeli, S. A Neural Network Method for Estimation of Battery Available Capacity. In Proceedings of the 45th International Universities Power Engineering Conference UPEC, Cardiff, UK, 31 August–3 September 2010; pp. 28–32.
156. Zhou, J.; He, Z.; Gao, M.; Liu, Y. Battery State of Health Estimation Using the Generalized Regression Neural Network. In Proceedings of the 2015 8th International Congress on Image and Signal Processing (CISP), Shenyang, China, 14–16 October 2015; pp. 1396–1400. [CrossRef]
157. Zhao, S.; Zhang, C.; Wang, Y. Lithium-Ion Battery Capacity and Remaining Useful Life Prediction Using Board Learning System and Long Short-Term Memory Neural Network. *J. Energy Storage* **2022**, *52*, 104901. [CrossRef]
158. Lei, X.; Chan, C.C.; Liu, K.; Ma, L. Pruning LS-SVM Based Battery Model for Electric Vehicles. In Proceedings of the Third International Conference on Natural Computation (ICNC 2007), Haikou, China, 24–27 August 2007; Volume 3, pp. 333–337.
159. Yan, Q. SOC Prediction of Power Battery Based on SVM. In Proceedings of the 2020 Chinese Control and Decision Conference (CCDC), Hefei, China, 22–24 August 2020; pp. 2425–2429.

160. Zhao, M.; Chen, Y.; Luo, B.; Zhong, L.; Xing, R.L. Research of Battery Capacity Fiber On-Line Intelligent Testing Technology Based on SVM. In Proceedings of the 2008 7th World Congress on Intelligent Control and Automation, Chongqing, China, 25–27 June 2008; pp. 3067–3070. [CrossRef]
161. Chu, L.; Zhou, F.; Guo, J. Investigation of Cycle Life of Li-Ion Power Battery Pack Based on LV-SVM. In Proceedings of the 2011 International Conference on Mechatronic Science, Electric Engineering and Computer (MEC), Jilin, China, 19–22 August 2011; pp. 1602–1605.
162. Xie, J.; Li, W.; Hu, Y. Aviation Lead-Acid Battery State-of-Health Assessment Using PSO-SVM Technique. In Proceedings of the 2014 IEEE 5th International Conference on Software Engineering and Service Science, Beijing, China, 27–29 June 2014; pp. 344–347.
163. He, Z.; Gao, M.; Ma, G.; Liu, Y.; Chen, S. Online State-of-Health Estimation of Lithium-Ion Batteries Using Dynamic Bayesian Networks. *J. Power Sources* **2014**, *267*, 576–583. [CrossRef]
164. Liu, K.; Shang, Y.; Ouyang, Q.; Widanage, W.D. A Data-Driven Approach with Uncertainty Quantification for Predicting Future Capacities and Remaining Useful Life of Lithium-Ion Battery. *IEEE Trans. Ind. Electron.* **2021**, *68*, 3170–3180. [CrossRef]
165. Richardson, R.R.; Birkl, C.R.; Osborne, M.A.; Howey, D.A. Gaussian Process Regression for in Situ Capacity Estimation of Lithium-Ion Batteries. *IEEE Trans. Ind. Informatics* **2019**, *15*, 127–138. [CrossRef]
166. Sun, S.; Sun, J.; Wang, Z.; Zhou, Z.; Cai, W. Prediction of Battery SOH by CNN-BiLSTM Network Fused with Attention Mechanism. *Energies* **2022**, *15*, 4428. [CrossRef]
167. Gu, X.; See, K.W.; Li, P.; Shan, K.; Wang, Y.; Zhao, L.; Lim, K.C.; Zhang, N. A Novel State-of-Health Estimation for the Lithium-Ion Battery Using a Convolutional Neural Network and Transformer Model. *Energy* **2022**, *262*, 125501. [CrossRef]
168. You, G.-W.; Park, S.; Oh, D. Diagnosis of Electric Vehicle Batteries Using Recurrent Neural Networks. *IEEE Trans. Ind. Electron.* **2017**, *64*, 4885–4893. [CrossRef]
169. Raman, M.; Champa, V.; Prema, V. State of Health Estimation of Lithium Ion Batteries Using Recurrent Neural Network and Its Variants. In Proceedings of the 2021 IEEE International Conference on Electronics, Computing and Communication Technologies (CONECCT), Bangalore, India, 9–11 July 2021; pp. 1–6.
170. Zhang, J.; Hou, J.; Zhang, Z. Online State-of-Health Estimation for the Lithium-Ion Battery Based on An LSTM Neural Network with Attention Mechanism. In Proceedings of the 2020 Chinese Control and Decision Conference (CCDC), Hefei, China, 22–24 August 2020; pp. 1334–1339.
171. Ma, Y.; Shan, C.; Gao, J.; Chen, H. A Novel Method for State of Health Estimation of Lithium-Ion Batteries Based on Improved LSTM and Health Indicators Extraction. *Energy* **2022**, *251*, 123973. [CrossRef]
172. Fan, Y.; Xiao, F.; Li, C.; Yang, G.; Tang, X. A Novel Deep Learning Framework for State of Health Estimation of Lithium-Ion Battery. *J. Energy Storage* **2020**, *32*, 101741. [CrossRef]
173. Single Cell Supervisor (Linx) DNB1168 Datang NXP Semiconductors. Available online: https://www.datangnxp.com/en/details/products/43 (accessed on 7 October 2022).

Article

# A State of Charge Estimation Approach for Lithium-Ion Batteries Based on the Optimized Metabolic EGM(1,1) Algorithm

Qiang Sun [1,2,*], Shasha Wang [3], Shuang Gao [2], Haiying Lv [1,*], Jianghao Liu [1], Li Wang [1], Jifei Du [4] and Kexin Wei [2]

1. College of Engineering and Technology, Tianjin Agricultural University, Tianjin 300392, China
2. Key Laboratory of the Ministry of Education on Smart Power Grids, Tianjin University, Tianjin 300072, China
3. Research and Development Centre, Beijing Electric Vehicle Co., Ltd., Beijing 100176, China
4. National Active Distribution Network Technology Research Center (NANTEC), Beijing Jiaotong University, Beijing 100044, China
* Correspondence: sunqiang@tju.edu.cn (Q.S.); tjauee2018@tjau.edu.cn (H.L.)

**Abstract:** The accurate estimation of the state of charge (SOC) for lithium-ion batteries' performance prediction and durability evaluation is of paramount importance, which is significant to ensure reliability and stability for electric vehicles. The SOC estimation approaches based on big data collection and offline adjustment could result in imprecision for SOC estimation under various driving conditions at different temperatures. In the traditional GM(1,1), the initialization condition and the identifying parameter could not be changed as soon as they are confirmed. Aiming at the requirements of battery SOC estimation with non-linear characteristics of a dynamic battery system, the paper presents a method of battery state estimation based on Metabolic Even GM(1,1) to expand battery state data and introduce temperature factors in the estimation process to make SOC estimation more accurate. The latest information data used in the optimized rolling model is introduced through the data cycle updating. The experimental results show that the optimized MEGM(1,1) effectively considers the influence of initial data, and has higher accuracy than the traditional GM(1,1) model in the application of data expansion. Furthermore, it could effectively solve the problem of incomplete battery information and battery capacity fluctuation, and the dynamic performance is satisfactory to meet the requirements of fast convergence. The SOC estimation based on the presented strategy for power batteries at different temperatures could reach the goal of the overall error within 1% under CLTC conditions with well robustness and accuracy.

**Keywords:** lithium-ion battery; metabolic even grey model; parameter identification; state of charge estimation

## 1. Introduction

The air pollution and energy shortages caused by automobile exhaust emissions have become increasingly prominent, along with the rapid development of the automobile industry. On account of high energy density, long life cycle, environmental protection, and pollution-free characteristics, lithium-ion batteries are becoming more and more important as power batteries for electric vehicles (EV) [1–5]. The attenuation of the battery during use is accompanied by changes in parameters such as capacity and impedance, which directly affect the reliability of battery operation. In order to ensure power, economy, and safety operation for EVs, battery state assessment is particularly important. Conventional battery state prediction methods mainly predict the battery state by detecting parameters such as battery charge and discharge rate, battery life, and open circuit voltage (OCV). Due to the complicated relationship between these parameters and the state of charge (SOC), traditional experimental methods have low prediction accuracy and low reliability [6–13]. Owing to the time-varying and non-linear characteristics of the battery system, the high-precision prediction of the battery state is undoubtedly an arduous task.

The significance of accurate battery modeling is that it could express the internal and external battery characteristics for different structures and different driving conditions, which can be of much help in the optimal development of lithium-ion batteries. The equivalent circuit model (ECM) and electrochemical model (EM) approaches are widely employed to perform the characteristics of lithium-ion batteries for SOC estimation [14–18]. The EM describes the battery's dynamic characteristics based on differential iterations, which could be much more complicated. As a result, the EM method is usually adopted in battery design applications. In addition, current battery models rarely consider the effect of temperature on model parameters. With regard to the ECM with electrical components such as resistance and capacitor, this physical model is widely employed in battery management systems (BMS), since it could offer trade-off solutions with complexity and accuracy. Much research has shown the ECM is featured in SOC estimation, and the model is capable of exactly describing the physical and chemical characteristics of lithium-ion batteries, and satisfying the BMS requirement with low calculation cost [19–23]. Furthermore, the Extended Kalman Filter (EKF) algorithm depends on the battery state space model and employs a recursive iteration method to linearize the battery SOC [24,25]. Its accuracy is significantly influenced by the model's preciseness. Although the Unscented Kalman Filter (UKF) algorithm employs statistical linearization to reduce error and calculation, the SOC estimation accuracy still fluctuates along with the unit model [26–30]. Model-based methods could illustrate the physical and chemical characteristics of the battery, but the correctness of the parameters relies on the accuracy and robustness of the battery model [31–35]. In recent years, battery forecasting has developed towards non-modeling. The methods based on data-driven and statistical analysis mainly include the time sequence method, support vector machine method, and Markov method [36–39]. The sequential method can describe the periodic law of data changes, but it is unrealistic to record the data of various working conditions during the entire battery life cycle, and it is difficult to establish the learning model [40–43]. Although the support vector machine method has high accuracy, the amount of training data is large and time-consuming. The Markov method can reflect the periodic change characteristics of data and has strong randomness, but it is difficult to determine the situational state set.

Grey System Theory is widely used in various fields, and successfully solves the problem of incomplete information prediction. Grey Model-GM(1,1) is an approach that employs a small amount of incomplete data samples to establish a grey prediction model and then describes the development trend in a long-term predictive manner [44,45]. The advantages of rapid prediction have been widely accepted. With the extension of battery life, the parameters and performance of the battery also change correspondingly, and there're obvious uncertainties in estimating the SOC value with raw data. The traditional Even Grey Model-EGM(1,1) predicts that the cumulative error is relatively large and cannot reflect the periodic changes in the data [46–48]. The Metabolic Even Grey Model-MEGM(1,1) method is based on the latest data generated by the system and adopts the original data in the rolling deletion system to establish a new data sequence, thereby establishing a lithium-ion battery state estimation model, which can reflect the characteristics of the latest data in real time. Especially with the accumulation of system variables, when the battery system has parameter perturbation or sudden change, the algorithm based on MEGM(1,1) iterative model can achieve accurate SOC estimation of lithium battery.

In recent years, for the sake of lacking its own driving cycles, China directly cited the *new European driving cycle* (NEDC) driving standard, and it has significant influence in promoting the development of the automotive industry of China [49]. Nevertheless, along with the change in vehicle ownership, road structure and traffic condition in China, the endurance range of EVs under NEDC operating conditions has a large deviation from the actual driving situation in China. Furthermore, since the NEDC does not comply with the characteristics of the actual driving behavior of vehicles in China, it could not actually denote the real application effects of energy-saving and emission-reduction technologies, such as the idle start–stop and brake energy regeneration technology [50].

In addition, the *world-harmonized light-duty vehicle test cycle* (WLTC) was developed employing the actual driving behavior data collected through six regions, including America, Japan, India, Korea, the EU, and Switzerland. However, it is lacking Chinese data acquisition [49,51]. With different levels of congestion, WLTC driving condition better reflects the characteristics of fast and slow vehicle speeds, but the idling ratio and average speed of the two main operating conditions under WLTC condition are quite different from the actual operating conditions in China [51]. Thus, for China, the WLTC driving condition could not provide a suitable solution to solve the problem during the NEDC driving cycles mentioned previously. In order to comply with the real and much more critical driving conditions, the local test cycle *China Light-duty Vehicle Test Cycle* (CLTC) for endurance certification of EVs was announced by the *Ministry of Industry and Information Technology* (MIIT) of China.

For the modeling of the battery pack, the data-driven approach based on Gaussian process regression is proposed to put forward a feasible solution with non-linear approximation, nonparametric modeling, and probabilistic prediction [52]. As the battery pack includes hundreds of cells in series and parallel, inconsistencies among the cells will be difficult to build an accurate physical model for their behavior performance. As a result, the widely employed model-based approaches are unsuitable for SOC estimation of battery packs [53,54]. In the paper, for the sake of accurate SOC estimation of the lithium-ion battery cell, the characteristics of the battery with varied ambient temperatures are experimentally studied. The *Thevenin* equivalent circuit model with the sixth-order polynomial of OCV–SOC function relation is derived by exploiting the physical characteristics of the lithium-ion battery cell. Furthermore, along with the mechanism analysis of the traditional GM(1,1) approach is analyzed in detail, the optimized MEGM(1,1) algorithm is put forward based on the presented ECM. The estimation accuracy of the employed approach is explored via an experimental platform for the lithium-ion battery, which works under the CLTC driving condition with five different temperatures. All the experimental and theoretical results, compared with the traditional GM(1,1) estimation method, illustrate that the optimized MEGM(1,1)-based SOC estimation approach is with fast convergence, good robustness, and well accuracy for the critical driving condition.

## 2. Lithium Ion Batteries' Modeling

There are various kinds of battery modeling methods, among which the *Thevenin* model for a battery has the characteristics of simple operation and it could illustrate the steady and transient characteristics of batteries. Therefore, this paper employs this modeling approach to establish the state space function of the battery. In addition, the *Thevenin* topology for lithium-ion battery is displayed in Figure 1, where $V_{ocv}$ is the electromotive force of the battery, $R_1$ and $R_2$ are defined as ohmic internal resistor and polarization resistor, and $C$ is the polarization capacitor, which is connected in parallel with the polarization resistance $R_2$.

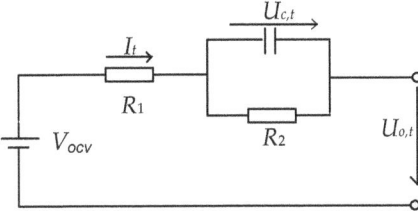

**Figure 1.** *Thevenin* topology for lithium-ion battery.

By means of the *Kirchhoff* and *Thevenin* principles, the expression for batteries' equivalent topology could be expressed in the following forms:

$$V_{ocv} = U_{o,t} + R_1 I_t + U_{c,t} \tag{1}$$

$$I_t = \frac{U_{c,t}}{R_2} + C\frac{dU_{c,t}}{dt} \tag{2}$$

The notation $V_{ocv}$ denotes the battery's open circuit voltage, and OCV reflects the relationship function with respect to SOC as $V_{ocv} = F(S_t)$. $U_{o,t}$ is the terminal voltage of the battery, and $U_{c,t}$ is the polarization voltage. Thus, SOC could be derived through the current integration method.

$$S_t = S_{t_0} - \frac{1}{Q_0}\int_{t_0}^{t} \eta I_t dt \tag{3}$$

where $\eta$ is the battery's discharge efficiency, $Q_0$ is battery capacity, and $S_t$ is the SOC value of the battery at the moment $t$. The state space equation could be represented by means of integrating and discretizing the formulas consequently, as:

$$\begin{cases} \begin{bmatrix} S_{t+\Delta t} \\ U_{c,t+\Delta t} \end{bmatrix} = \begin{bmatrix} 1 & 0 \\ 0 & \exp(-\frac{\Delta t}{R_2 C}) \end{bmatrix} \begin{bmatrix} S_t \\ U_{c,t} \end{bmatrix} + \begin{bmatrix} -\frac{\eta \Delta t}{Q_0} \\ R_2(1 - \exp(-\frac{\Delta t}{R_2 C})) \end{bmatrix} I_t + \begin{bmatrix} w_{1,t} \\ w_{2,t} \end{bmatrix} \\ U_{o,t} = F(S_t) - R_1 I_t - U_{c,t} + v_t \end{cases} \tag{4}$$

where $x_t = [S_t, U_{c,t}]^T$ is defined as the state variable, $U_{o,t}$ is the observation variable, $I_t$ is the control variable, $w_t = [w_{1,t}, w_{2,t}]^T$ is systematic noise, $v_t$ is observation noise. Consequently, the coefficient matrix of the battery's state space model can be represented as:

$$\begin{cases} A_t = \begin{bmatrix} 1 & 0 \\ 0 & \exp(-\frac{\Delta t}{R_2 C}) \end{bmatrix} \\ B_t = \begin{bmatrix} -\frac{\eta \Delta t}{Q_0} \\ R_2(1 - \exp(-\frac{\Delta t}{R_2 C})) \end{bmatrix} \\ C_t = [\frac{\partial F(S_t)}{\partial S_t}|_{S_t = S_t^-}, -1] \end{cases} \tag{5}$$

The OCV of the battery is not completely equivalent to the electromotive force due to the hysteresis effect, but when the battery is fully left standing, the two values are very close, and it is difficult to directly measure the real electromotive force. Therefore, in actual research, the OCV of the battery is usually used to describe its electromotive force under the SOC value. The OCV–SOC correspondence is the key performance parameter of the battery. It is often employed to illustrate the working state of the battery and perform SOC calibration with the necessary parameters for establishing the battery model.

Before performing simulation and estimation for the battery's SOC, parameter identification is performed on the *Thevenin* equivalent circuit topology, and discharge test along with pulse power features is also conducted. The CB2P0 type cell of LiFePO$_4$ lithium-ion battery is employed for an experiment, and the specifications are displayed in Table 1.

**Table 1.** Specifications of CB2P0 lithium-ion battery.

| Parameters | Values |
|---|---|
| Nominal Capacity (Ah) | 30 |
| Rated Voltage (V) | 3.2 |
| Charge Cutoff Voltage (V) | 3.65 |
| Discharge Cutoff Voltage (V) | 2.5 |
| Charge Temperature (°C) | 0~55 |
| Discharge Temperature (°C) | −25~55 |

The accurate SOC estimation for lithium-ion batteries' performance prediction and durability evaluation is of paramount importance, which is critical to ensure the reliability and stability of electric vehicles. The identification mechanism for the battery's parameters is mainly based on its physical behavior under different working temperatures. For the sake of accurate battery parameter identification, the pulse discharge test is conducted in this study during the range of −25 °C to 55 °C. According to the requirement of *ISO12405-2: 2012 for lithium-ion traction battery packs and system applications*, the battery is discharged at the current of C/3 to the cutoff voltage of 2.5V. The characteristics curve between the battery's SOC and OCV is illustrated in Figure 2.

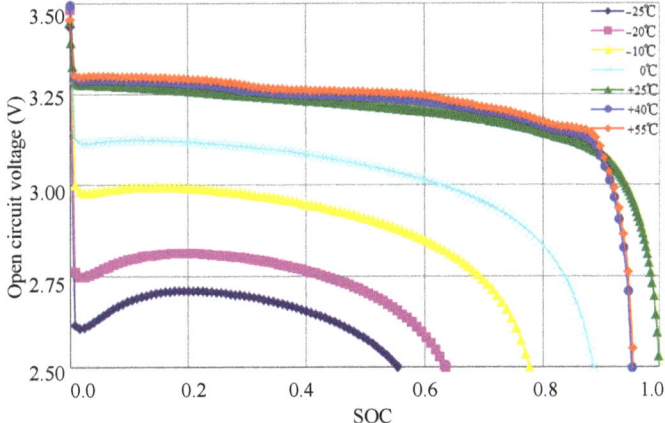

**Figure 2.** Characteristics curves on OCV and SOC.

As shown in Figure 2, under the condition of a small rate discharge, the discharge capacity can basically reach the rated value at room temperature. The performance of LiFePO$_4$ batteries at low temperatures will decrease significantly. At the same time, as the discharge rate increases, the impact of temperature on battery performance will become more and more obvious. The curve fitting method by means of *least squares* is employed to identify the non-linear functional relation between OCV and SOC.

For the sake of accurately fitting the experimental result of the lithium-ion battery, the sixth-order polynomial could be performed by means of the function:

$$V_{ocv}(SOC) = K_6 \cdot SOC^6 + K_5 \cdot SOC^5 + K_4 \cdot SOC^4 + K_3 \cdot SOC^3 + K_2 \cdot SOC^2 + K_1 \cdot SOC^1 + K_0 \quad (6)$$

where $K_i$ ($i = 0, 1, 2, \ldots$) is in reference to the specific ambient temperature, it influences the accuracy of the characteristic polynomial, which could be derived through the experimental pulse discharge results. The derived values of $K_i$ under the various ambient temperatures are displayed in Table 2. Based on the sixth-order polynomial, a three-dimensional map of Temperature-SOC-OCV is illustrated in Figure 3 correspondingly.

**Table 2.** Coefficient values under different operating temperatures.

| Coefficient | −25 °C | −20 °C | −10 °C | 0 °C | +25 °C | +40 °C | +55 °C |
|---|---|---|---|---|---|---|---|
| $K_0$ | 2.5802 | 2.7285 | 2.9621 | 3.0937 | 3.2595 | 3.2159 | 3.2143 |
| $K_1$ | 1.2526 | 0.958 | 0.7823 | 0.9278 | 1.0965 | 2.8663 | 3.2648 |
| $K_2$ | −0.7987 | −3.2591 | −7.4156 | −9.5828 | −14.302 | −30.61 | −34.814 |
| $K_3$ | −25.623 | 4.133 | 35.141 | 43.355 | 65.984 | 133 | 151.43 |
| $K_4$ | 99.131 | −9.5266 | −91.694 | −99.723 | −141.04 | −277.11 | −314.99 |
| $K_5$ | −146.39 | 24.44 | 116.65 | 109.91 | 140.22 | 273.9 | 310.24 |
| $K_6$ | 73.74 | −23.711 | −57.662 | −46.669 | −52.579 | −103.23 | −116.37 |

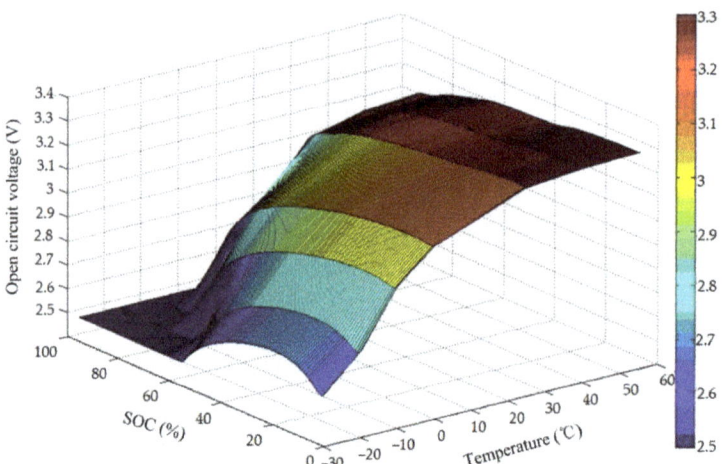

**Figure 3.** Three-dimensional map of Temperature-SOC-OCV.

The functional relation between OCV and SOC is non-linear, and parameter identification in the battery state space model could be achieved through the change characteristics of the terminal voltage during the pulse discharge experiment. At the ambient condition of 25 °C, the power battery is discharged along with the current of 1 C for 5 min, the discharge is stopped for 10 min, and then the battery is discharged for 5 min at the same current. A single pulse is selected to perform parameter identification on the battery model, and the dynamic curve of the battery's terminal voltage is shown in Figure 4.

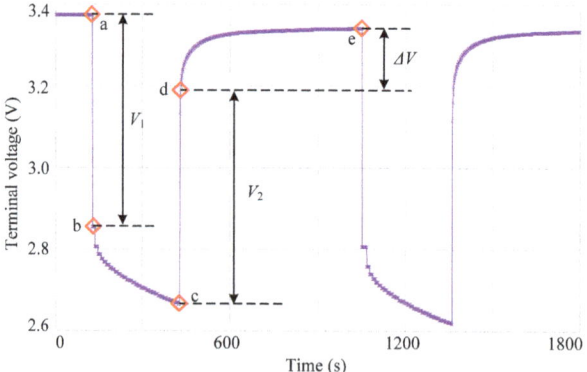

**Figure 4.** Terminal voltage of battery after discharging current pulse.

As to the *Thevenin* model shown in Figure 1, the terminal voltage changes corresponding to the discharging current pulse. Since the voltage of polarization capacitor $C$ is not suddenly changed while starting discharging, the ohmic internal resistance $R_1$ and the polarization resistance $R_2$ are then identified according to the terminal voltage curve. Furthermore, regarding to the transient characteristics of $RC$ topology, the battery's terminal voltage increased to 86.5% $\Delta V$ needs twice constant $\Gamma$ ($\Gamma = R_2 \times C$), which is identified with the terminal voltage curve. Consequently, the model parameters can be performed by means of the following form:

$$\begin{cases} R_1 = \dfrac{|V_T(t_a) - V_T(t_b)| + |V_T(t_d) - V_T(t_c)|}{2|I_{1C}|} \\ R_2 = \dfrac{|V_T(t_e) - V_T(t_d)|}{|I_{1C}|} \end{cases} \quad (7)$$

The dynamic battery parameter identification model based on temperature, voltage, and SOC is established by employing the least squares principle to fit the polarization effect parameters, and the established battery mathematical model could better adapt to the lithium battery performed with rich experimental waveforms. However, it can be seen that at different temperatures, the OCV–SOC curves are different, which indicates that the electrode characteristics of the battery are influenced by temperature, which will affect the SOC estimation. During the parameter identification process for the physical model of a fixed lithium-ion battery, the model parameters are consumed to be constant, which could affect the practicability of the presented estimation method for the sake of the variable temperatures. In order to further improve the estimation accuracy, the ongoing study will be towards the iterative learning-based on-line prediction approach that can precisely identify the parameters associated with the temperature and SOC.

## 3. Estimation Mechanism Based on Optimized MEGM(1,1)

### 3.1. General Principle of GM(1,1)

The traditional GM(1,1) is an approach that employs a small amount of incomplete data samples to establish a grey prediction model, and then describes the development trend in a long-term predictive period. GM(1,1) approach is to take the data sequence of lithium-ion batteries which changes with time as the original data sequence [44]. Through the cumulative calculation of the original data, a new data sequence is obtained. Furthermore, the relevant whitening differential equation is established with the solution derivation. Thereby, the grey estimation model for the lithium-ion battery state can be obtained, which could reflect the real-time features of a battery system. The derivation process of traditional GM(1,1) for SOC estimation is briefly described as follows [45].

(1) Extraction of battery history data: $x_{SOC}$, $x_V$, $x_I$ and $x_T$. Where $x_V$, $x_I$ and $x_T$ are correlation factor series for input parameters, $x_{SOC}$ is target SOC sequence for output parameters.

$$\begin{cases} x_V = (V_1, V_2, \cdots, V_n) \\ x_I = (I_1, I_2, \cdots, I_n) \\ x_T = (T_1, T_2, \cdots, T_n) \\ x_{SOC} = (SOC_1, SOC_2, \cdots, SOC_n) \end{cases} \tag{8}$$

(2) Perform 1st-order Accumulated Generating Operation (1-AGO) for the battery's data.

$$\begin{cases} x_V^1 = (V_1^1, V_2^1, \cdots, V_n^1) \\ x_I^1 = (I_1^1, I_2^1, \cdots, I_n^1) \\ x_T^1 = (T_1^1, T_2^1, \cdots, T_n^1) \\ x_{SOC}^1 = (SOC_1^1, SOC_2^1, \cdots, SOC_n^1) \end{cases} \tag{9}$$

(3) Generate the mean values of consecutive neighbor sequences.

$$\begin{cases} Z_V = (V_1, V_2, \cdots, V_m) \\ Z_I = (I_1, I_2, \cdots, I_m) \\ Z_T = (T_1, T_2, \cdots, T_m) \\ Z_{SOC} = (SOC_1, SOC_2, \cdots, SOC_m) \end{cases} \tag{10}$$

(4) Construct the whitening function for GM(1,1):

$$x_{SOC}(t_k) + a\Delta t_k Z_{SOC}(t_k - \tau) = \begin{array}{l} b_2 x_V^1(t_k - \tau) t_k^\gamma \Delta t_k + \\ b_3 x_I^1(t_k - \tau) t_k^\gamma \Delta t_k + b_4 x_T^1(t_k - \tau) t_k^\gamma \Delta t_k \end{array} \tag{11}$$

where the time interval $\Delta t_k = t_k - t_{k-1}$, $\gamma$ and $\tau$ are the iteration coefficients. In order to make the model as accurate as possible and to minimize the average relative error as the optimization objective of the whitening function, the constructed function related to parameters $\gamma$ and $\tau$ is as:

$$f(\tau,\gamma) = \frac{1}{n-1}\sum_{i=2}^{n}\frac{x_{SOC}^1(i) - x_{SOC}(i)}{x_{SOC}(i)} \quad (12)$$

It could be seen this method could be easily utilized due to the advantage of rapid prediction. However the coefficients $\gamma$ and $\tau$ are fixed for the conventional GM(1,1), it is unsuitable for massive data estimation. With the extension of battery life, the parameters and performance of the battery also change correspondingly, and there're obvious uncertainties in estimating the SOC value with raw data. The interference noise with external fluctuation and various driving conditions should be regarded during practical application. Thus the cumulative error is relatively large and cannot reflect the periodic changes in the data.

### 3.2. Principle of Optimized MEGM(1,1)

While the GM(1,1) model is employed for state evaluation of lithium-ion batteries, the closer the battery data is to the original time, the more accurate the prediction will be. Therefore, the modeling sequence should follow the change in the battery system by removing the oldest data sequence to reflect the updated characteristics. The MEGM(1,1) model is automatically updated and identified with each prediction, so the model has strong adaptability. After obtaining the closest information by the estimation, the original data sequence $x(0)$ is removed from this sequence, and the metabolic data sequence $x(m + 1)$ is introduced as the original sequence to reconstruct MEGM(1,1) model [44,45,47]. Tracking the test data information as input, and using the current information to establish the prediction model until the accurate estimation target is reached, the input sequence and prediction sequence are updated iteratively, as shown in Figure 5.

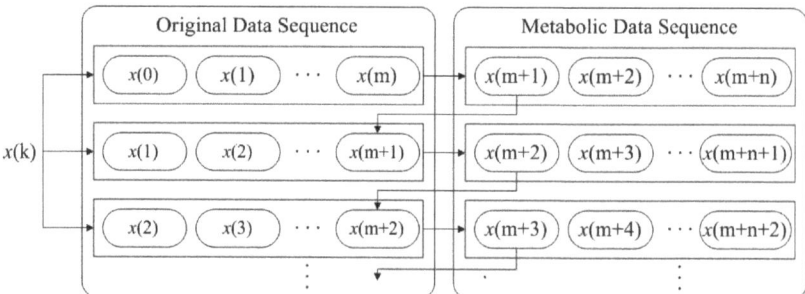

**Figure 5.** Diagram of rolling optimization strategy for MEGM(1,1).

In order to increase the accuracy of the extended data, the original data sequence is divided into $q$ sub-sequences with different sample numbers.

$$\begin{cases} X_1 = \{x_1(0), x_1(1), \cdots, x_1(r)\} \\ X_2 = \{x_2(1), x_2(2), \cdots, x_2(r)\} \\ \cdots \\ X_q = \{x_q(q-1), x_q(q-2), \cdots, x_q(r)\} \end{cases} \quad (13)$$

where

$$x_\alpha(\beta) = x_{\alpha+1}(\beta)|_{\alpha=1,2,\cdots q; \beta=1,2,\cdots r} \quad (14)$$

The MEGM(1,1) dynamic equivalent topology could be represented through the first-order differential function including one single variable, which is fundamental for the grey estimation. Define the feature data sequence in the form:

$$X = \{x(0), x(1), \cdots, x(r)\} \tag{15}$$

The data sequence generated by the means of the first-order accumulation is:

$$\begin{cases} x^{(1)}(m) = \sum_{i=1}^{m} x(i) \\ X^{(1)} = \{x^{(1)}(1), x^{(1)}(2), \cdots, x^{(1)}(r)\} \end{cases} \tag{16}$$

where sequence $x^{(1)}$ is defined as 1-AGO. Furthermore, sequence $W^{(1)}$ is assumed as the information data produced through the average value of consecutive neighbors of $X^{(1)}$, given by:

$$\begin{cases} W^{(1)} = \{w^{(1)}(2), w^{(1)}(3), \cdots, w^{(1)}(n)\} \\ w^{(1)}(\zeta)|_{\zeta=1,2,\cdots,n-1} = \frac{1}{2}\left[x^{(1)}(\zeta+1) + x^{(1)}(\zeta)\right] \end{cases} \tag{17}$$

Then the fundamental form of the MEGM(1,1) estimation model is established as [45]:

$$x(k) + \varepsilon x^{(1)}(k) = \vartheta \tag{18}$$

The whitening differential equation for $X^{(1)}$ is performed as:

$$\frac{dx^{(1)}}{dt} + \varepsilon \cdot x^{(1)} = \vartheta \tag{19}$$

where $\varepsilon$ is the development coefficient, $\vartheta$ is grey input. The following discretized expression can be derived:

$$\begin{cases} dx^{(1)} = x^{(1)}(\zeta+1) - x^{(1)}(\zeta) \\ dt = \zeta + 1 - \zeta = 1 \end{cases} \tag{20}$$

Define the equation $x^{(1)}(\zeta) = \frac{1}{2}(x^{(1)}(\zeta+1)+x^{(1)}(\zeta))$, hence Equation (19) could be expressed as follows:

$$x(\zeta+1) = \varepsilon\left(-\frac{1}{2}\left(x^{(1)}(\zeta+1) + x^{(1)}(\zeta)\right)\right) + \vartheta \tag{21}$$

In order to derive the solving solutions of parameters $\varepsilon$ and $\vartheta$, Equation (21) can be transformed as:

$$\begin{bmatrix} x(2) \\ x(3) \\ \cdots \\ x(l) \end{bmatrix} = \varepsilon \begin{bmatrix} -\frac{1}{2}\left(x^{(1)}(1) + x^{(1)}(2)\right) \\ -\frac{1}{2}\left(x^{(1)}(2) + x^{(1)}(3)\right) \\ \cdots \\ -\frac{1}{2}\left(x^{(1)}(l-1) + x^{(1)}(l)\right) \end{bmatrix} + \vartheta \begin{bmatrix} 1 \\ 1 \\ \cdots \\ 1 \end{bmatrix} \tag{22}$$

In addition, according to Equation (22), the following equations are defined as:

$$\begin{cases} Y_r = \begin{bmatrix} x(2) & x(3) & \cdots & x(r) \end{bmatrix}^T \\ W^{(1)} = \begin{bmatrix} -\frac{1}{2}\left(x^{(1)}(1) + x^{(1)}(2)\right) \\ -\frac{1}{2}\left(x^{(1)}(2) + x^{(1)}(3)\right) \\ \vdots \\ -\frac{1}{2}\left(x^{(1)}(r-1) + x^{(1)}(r)\right) \end{bmatrix} \\ V = \begin{bmatrix} 1 & 1 & \cdots & 1 \end{bmatrix}^T \end{cases} \quad (23)$$

Based on the *least squares* estimation algorithm, the parameter matrix for MEGM(1,1) model could be performed by the form [55]:

$$\hat{a} = \begin{bmatrix} \varepsilon & \vartheta \end{bmatrix}^T = \left(\begin{bmatrix} W^{(1)} & V \end{bmatrix}\begin{bmatrix} W^{(1)} & V \end{bmatrix}^T\right)^{-1}\begin{bmatrix} W^{(1)} & V \end{bmatrix}^T Y_r \quad (24)$$

Thereby, the following expression can be written to compute solutions of parameters $\varepsilon$ and $\vartheta$, considering Equations (23) and (24), as:

$$Y_r = \varepsilon W^{(1)} + \vartheta V = \begin{bmatrix} W^{(1)} & V \end{bmatrix}\begin{bmatrix} \varepsilon \\ \vartheta \end{bmatrix} \quad (25)$$

Define $x^{(1)}(1) = x(1)$, the solution of the whitening differential equation can be calculated as follows:

$$\hat{x}^{(1)}(\zeta + 1) = \left[x(1) - \frac{\vartheta}{\varepsilon}\right] e^{-\varepsilon \zeta} + \frac{\vartheta}{\varepsilon} \quad (26)$$

In order to extend the original data sequence of the lithium-ion battery, the restore sequence is derived by reductive generation, as:

$$\begin{aligned} x(\zeta + 1)|_{\zeta = r+1, r+2, \dots} &= \hat{x}^{(1)}(\zeta + 1) - \hat{x}^{(1)}(\zeta) \\ &= (1 - e^{-\varepsilon})\left[x(1) - \frac{\vartheta}{\varepsilon}\right] e^{-\varepsilon \zeta} \end{aligned} \quad (27)$$

According to Equation (27), the fitting calculation of $q$ sub-sequences in Equation (15) is carried out, and the result of fitting calculation $\hat{X}_i|_{i=1,2,\cdots q}$ is:

$$\begin{cases} \hat{X}_1 = \{\hat{x}_1(0), \hat{x}_1(1), \cdots, \hat{x}_1(r), \hat{x}_1(r+1)\} \\ \hat{X}_2 = \{\hat{x}_2(1), \hat{x}_2(2), \cdots, \hat{x}_2(r), \hat{x}_2(r+1)\} \\ \cdots \\ \hat{X}_q = \{\hat{x}_q(q-1), \hat{x}_q(q-2), \cdots, \hat{x}_q(r), \hat{x}_q(r+1)\} \end{cases} \quad (28)$$

The Grey Relation Function (GRF) $Y(X_\alpha, \hat{X}_\beta)$ is derived by calculating the original sequence and metabolic fitting sequence, as:

$$\begin{cases} Y(X_\alpha, \hat{X}_\alpha) = \frac{1}{r-\alpha} \sum_{\beta=\alpha}^{r} Y_\alpha(x_\alpha(\beta), \hat{x}_\alpha(\beta)) \\ Y_\alpha(x_\alpha(\beta), \hat{x}_\alpha(\beta)) = \frac{MIN + \rho MAX}{|x_\alpha(\beta) - \hat{x}_\alpha(\beta)| + \rho MAX} \\ MIN = \min|x_\alpha(\beta) - \hat{x}_\alpha(\beta)| \\ MAX = \max|x_\alpha(\beta) - \hat{x}_\alpha(\beta)| \end{cases} \quad (29)$$

As a result, the whitening equation with $-N-1$ input parameters and one output parameter is established as follows [47]:

$$\frac{dX_1^{(1)}}{dt} + a_0 X_1^{(1)} = b_1 X_2^{(1)} + b_2 X_3^{(1)} + \cdots + b_{N-1} X_N^{(1)} \tag{30}$$

where $a_0, b_1, b_2, \ldots, b_{N-1}$ are function coefficients, $X_1^{(1)}$ is the output parameter, $X_2^{(1)}, X_3^{(1)}, \ldots, X_{N-1}^{(1)}$ are the input parameters. Furthermore, the parameters are with strong coupling. Accordingly, the estimation result of the output parameter could be expressed as:

$$\overset{\wedge}{X}_1^{(1)}(t) = \left( X_1^{(1)}(1) - \tfrac{b_1}{a_0} X_2^{(1)}(t) - \tfrac{b_2}{a_0} X_3^{(1)}(t) - \cdots - \tfrac{b_{N-1}}{a_0} X_N^{(1)}(t) \right) e^{-a_0 t} + \tfrac{b_1}{a_0} X_2^{(1)}(t) + \tfrac{b_2}{a_0} X_3^{(1)}(t) + \cdots + \tfrac{b_{N-1}}{a_0} X_N^{(1)}(t) \tag{31}$$

In this research, parameters $x_V$, $x_I$ and $x_T$ are input data sequence, $x_{SOC}$ is SOC sequence for output parameters. So the expression can be transformed as:

$$\overset{\wedge}{X}_{SOC}^{(1)}(t) = \left( X_{SOC}^{(1)}(1) - \tfrac{b_1}{a_0} X_V^{(1)}(t) - \tfrac{b_2}{a_0} X_I^{(1)}(t) - \tfrac{b_3}{a_0} X_T^{(1)}(t) \right) e^{-a_0 t} + \tfrac{b_1}{a_0} X_V^{(1)}(t) + \tfrac{b_2}{a_0} X_I^{(1)}(t) + \tfrac{b_3}{a_0} X_T^{(1)}(t) \tag{32}$$

Define:

$$D = \frac{b_1}{a_0} X_V^{(1)}(t) + \frac{b_2}{a_0} X_I^{(1)}(t) + \frac{b_3}{a_0} X_T^{(1)}(t) \tag{33}$$

Correspondingly, the SOC estimation value for lithium-ion batteries could be derived by means of the following equation.

$$\overset{\wedge}{X}_{SOC}^{(1)}(t) = \left\{ \left[ X_{SOC}^{(1)}(1) - D \right] - X_{SOC}^{(1)}(1) \cdot \tfrac{1}{1+e^{-a_0 t}} + 2D \tfrac{1}{1+e^{-a_0 t}} \right\} \left( 1 \mid e^{-a_0 t} \right) \tag{34}$$

On the basis of the original data of lithium-ion batteries, the initial mathematical model and parameter identification are established. Furthermore, the Grey Relation Analysis (GRA) is introduced to figure out the new data by employing the original data, and then a new model is established instead of the original data. With the continuous use of metabolic update-data modeling, the state estimation of lithium-ion batteries is accurately established. Aiming at the requirements of SOC estimation and the non-linear characteristics of the dynamic battery system, an optimized MEGM(1,1) model is proposed to expand battery state data and introduce temperature factors into the prediction process to make SOC estimation much more accurate.

## 4. Experimental Verification and Analysis

In order to verify the effectiveness and accuracy of the presented estimation algorithm, an experimental platform for the battery system is established. The block diagram of the battery system's test bench includes a High-Low temperature incubator (LJGDP-20R-E, LIK Industry Co., Ltd., Dongguan, China), a heavy-duty dual channel cycling station (AV-900, AeroVironment Inc., Simi Valley, CA, USA), a set of the measurement tool and control software, such as CANoe (VN1630A, Vector Co., Ltd., Stuttgart, Germany) and Labview, and a computer, as shown in Figure 6.

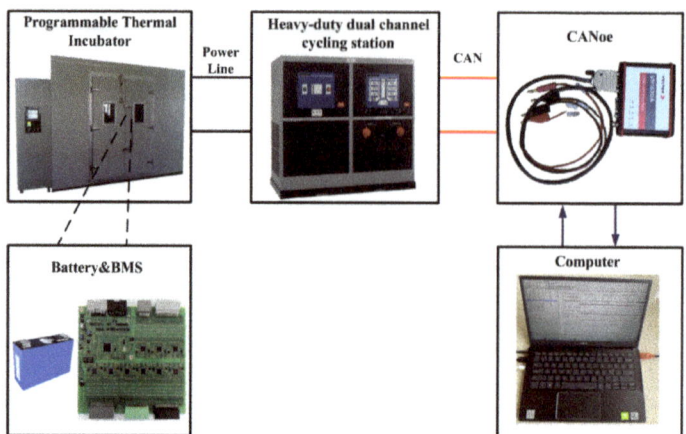

**Figure 6.** Block diagram of battery system test bench.

Under the CLTC driving condition, there is both driving discharge and regenerative charge, which fully considers the complexity of the road situations. For the single static charging of Li-ion battery, the charging station usually adopts constant current for charge and the SOC estimation is much more simplified. With regard to this research, we mainly discuss the SOC estimation performance of CB2P0 LiFePO$_4$ lithium-ion battery cells under CLTC driving conditions. This study focuses on characterizing the battery in dynamic real-time operation during vehicle CLTC driving conditions to optimize EV drivelines and accurate state estimation for EV manufacturers. In the experimental research, the battery cell is placed in an LJGDP-20R-E High-Low temperature incubator and stands for 16 h to achieve thermal balance. The AV-900 cycling station is utilized to simulate the demanded power sequence of CLTC working conditions, the driving cycles are performed, and one cycle period is 1800 s with a total of 36,000 s until the discharge cutoff voltage is achieved. In this research, the time interval of 10 Hz is employed for the battery system's data acquisition.

In the battery experiment system, the adopted CB2P0 cells are from the same batch with good consistency, so that random lithium cell is selected for experimental analysis, and the feasibility and effectiveness of the analysis principle are validated through numerical calculation and measurement results. Verification results under the CLTC driving condition at various temperature conditions of $-10$ °C, 0 °C, 15 °C, 25 °C, and 35 °C are obtained with estimation curves of the GM(1,1) and the optimized MEGM(1,1) approach, respectively. Furthermore, the enlarged view of discharging terminal voltage in one cycle for the time range of 16.8 ks–18.6 ks is also displayed in the following figures.

Figure 7 is the comparison results among the experimentally measured voltage, GM(1,1), and MEGM(1,1) estimation voltage with the enlarged view of discharging terminal cell voltage in one cycle for the time range of 16.8 ks–18.6 ks at the different ambient temperatures. On the basis of the measured big data, the battery's electrical characteristics are temperature-dependent with a three-dimensional map of Temperature-SOC-OCV, as the estimation of the terminal voltage performs much more accurately at 25 °C and 35 °C, compared with the performance at $-10$ °C, 0 °C and 10 °C under the CLTC driving condition. This is mainly because the conductivity of the electrolyte varies at different temperatures, and the migration speed of the lithium ions fluctuates with hysteretic characteristics regarding the OCV–SOC relation, which leads to the accumulation error for the model's parameter estimation. It can be seen that the estimated errors of terminal voltage are within 15 mV with rapid convergence characteristics.

Additionally, the initial SOC value would affect the convergence performance and the estimation accuracy. In order to demonstrate the performance of the proposed SOC estimation approach and the characteristics of convergence, the initial SOC value is set to be 60% in the dynamic experiment. Figure 8 is the SOC comparison among the experimentally measured data, GM(1,1) and MEGM(1,1) estimation values with the enlarged view in one cycle for the time range of 16.8 ks–18.6 ks at the different ambient temperatures. In the research, the quantitative analysis approaches for SOC error evaluation after convergence are Root Mean Square Error (RMSE) and Mean Absolute Error (MAE), as are employed to illustrate the performance of the proposed SOC prediction model shown in Table 3. As can be displayed from the experimental results, regardless of the difference between the initial value and the real value of SOC, as the number of iterations increases, the SOC estimation value approximates the real value rapidly and the steady state error is within 1.00% after convergence by the optimized MEGM(1,1).

**Table 3.** Estimation analysis for GM(1,1) and MEGM(1,1) after convergence under CLTC.

| Temperature | Model | RMSE | MAE |
|---|---|---|---|
| −10 °C | GM(1,1) | 0.0175 | 0.0142 |
| | MEGM(1,1) | 0.0099 | 0.0079 |
| 0 °C | GM(1,1) | 0.0174 | 0.0140 |
| | MEGM(1,1) | 0.0089 | 0.0061 |
| +15 °C | GM(1,1) | 0.0163 | 0.0129 |
| | MEGM(1,1) | 0.0072 | 0.0051 |
| +25 °C | GM(1,1) | 0.0144 | 0.0115 |
| | MEGM(1,1) | 0.0059 | 0.0046 |
| +35 °C | GM(1,1) | 0.0157 | 0.0122 |
| | MEGM(1,1) | 0.0069 | 0.0048 |

Generally, the optimized algorithm presents well robustness against initial SOC deviation and temperature variation, and the initial SOC value only affects the time taken for the SOC estimate to approach the real value, which does not affect the accuracy of the steady state SOC estimation even if the initial SOC has a large deviation. The close agreement between simulation results and experimental data on Li-ion batteries indicates that the presented MEGM(1,1) approach is capable of real-time updating battery model parameters, restraining system variation via self adaption, and achieving accurate SOC prediction with less estimation error under the critical CLTC driving condition at the various ambient temperatures.

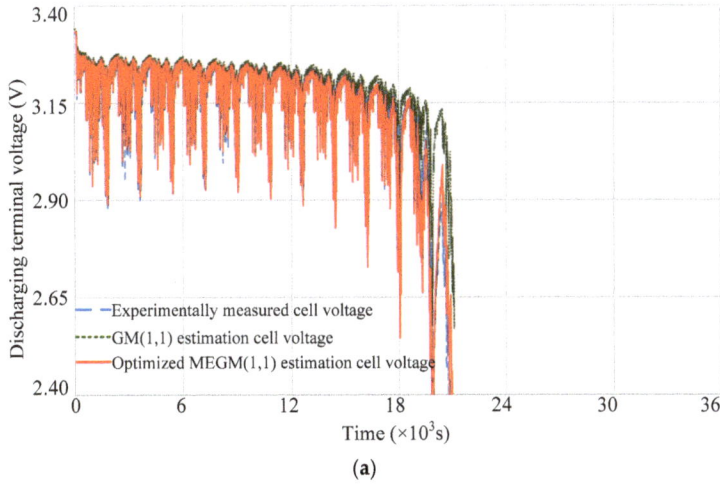

(a)

**Figure 7.** *Cont.*

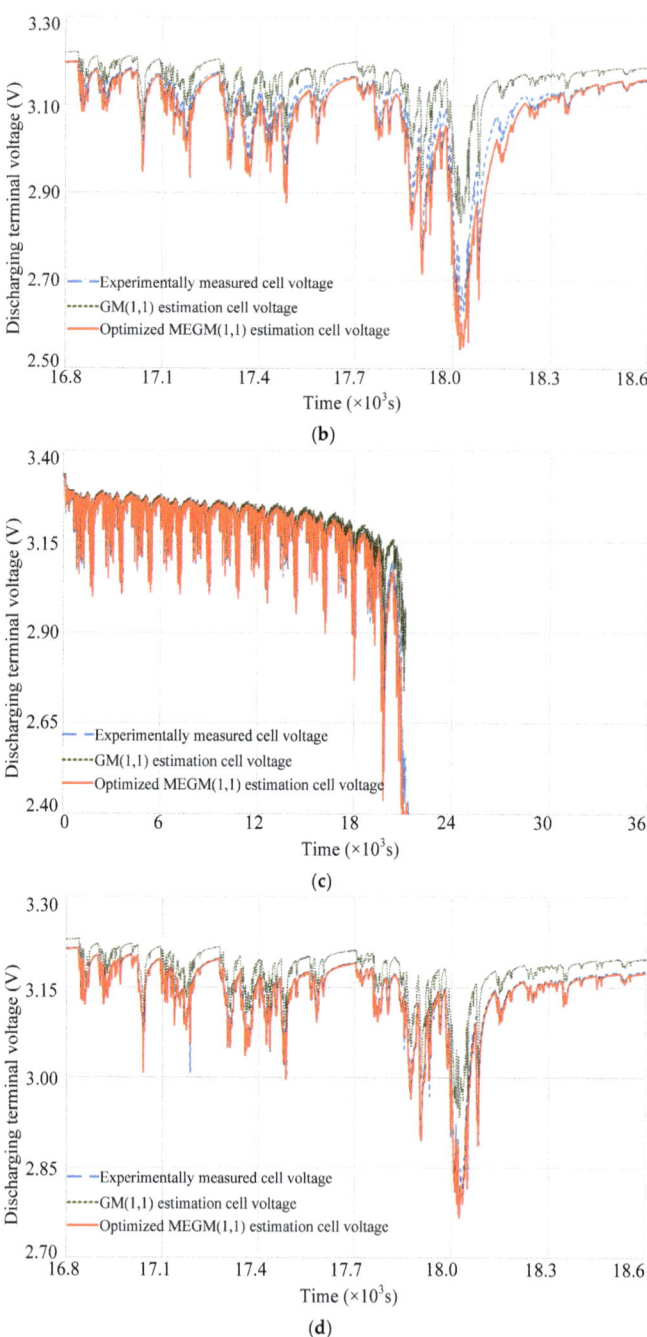

Figure 7. Cont.

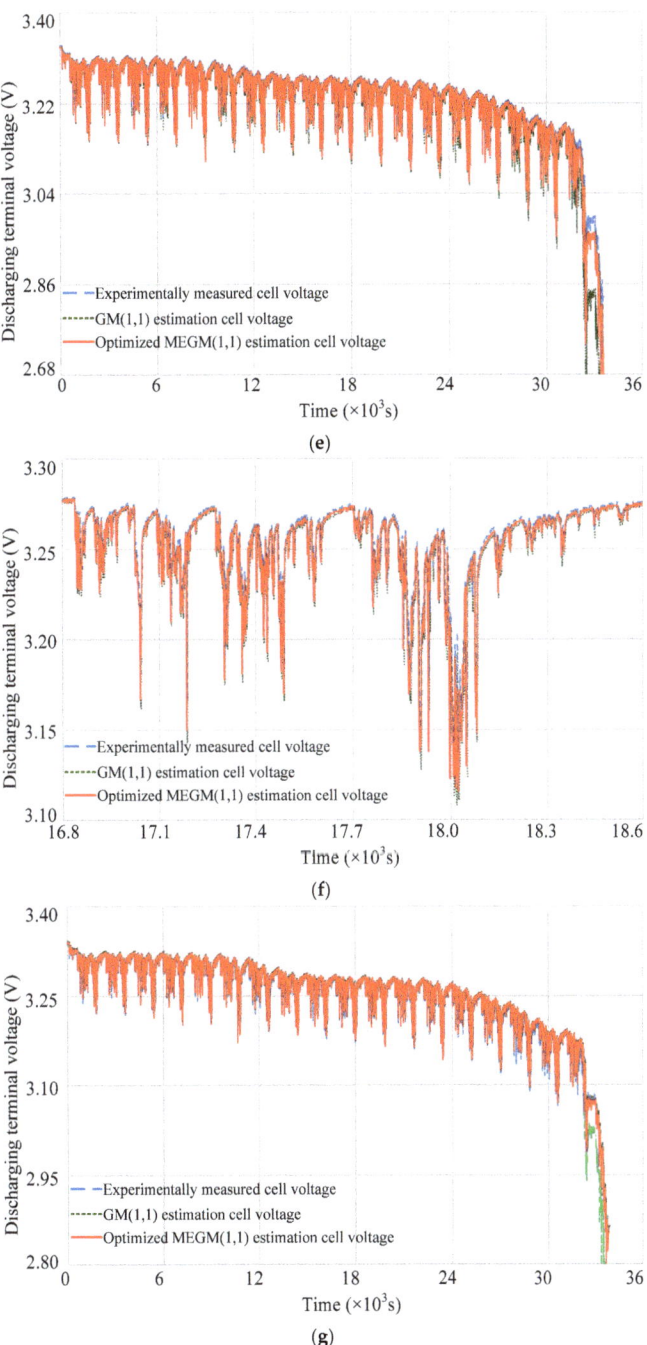

**Figure 7.** *Cont.*

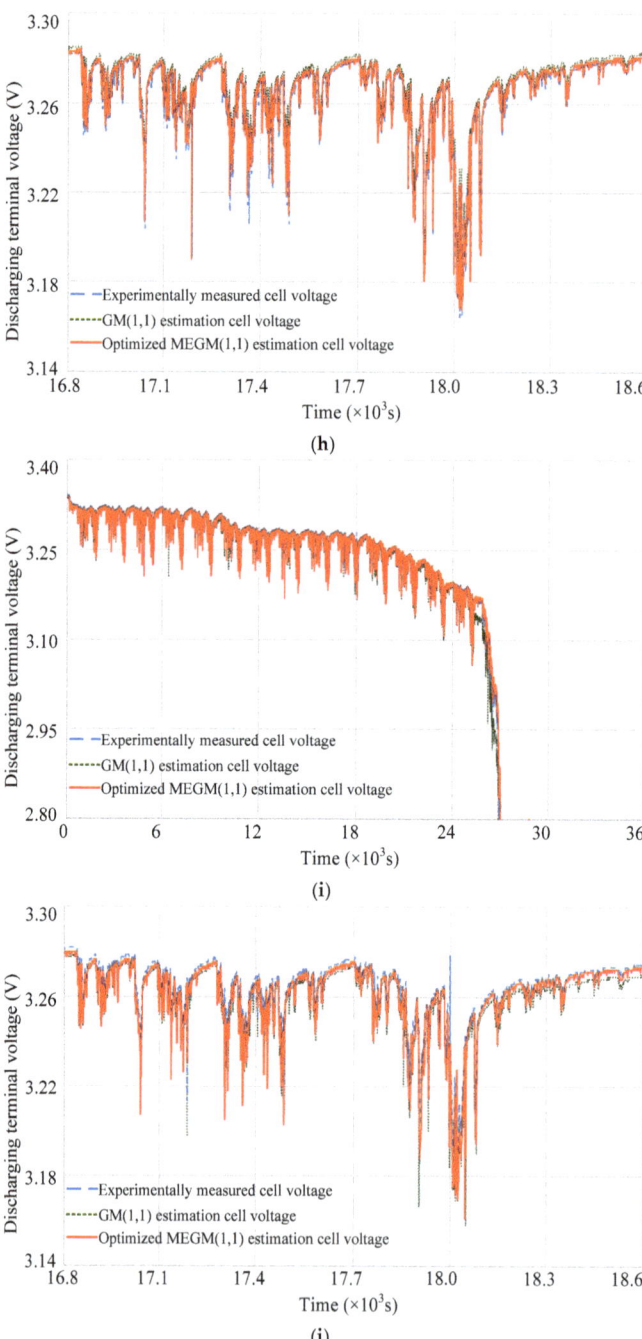

**Figure 7.** Verification results under the CLTC driving condition at the various temperatures: (**a**) voltage comparison among the measured data, GM(1,1) and MEGM(1,1) estimation at −10 °C; (**b**) the enlarged view of discharging terminal voltage in one cycle for the time range of 16.8 ks–18.6 ks

at −10 °C; (**c**) voltage comparison at 0 °C; (**d**) the enlarged view at 0 °C; (**e**) voltage comparison at 15 °C; (**f**) the enlarged view at 15 °C; (**g**) voltage comparison at 25 °C; (**h**) the enlarged view at 25 °C; (**i**) voltage comparison at 35 °C; (**j**) the enlarged view at 35 °C.

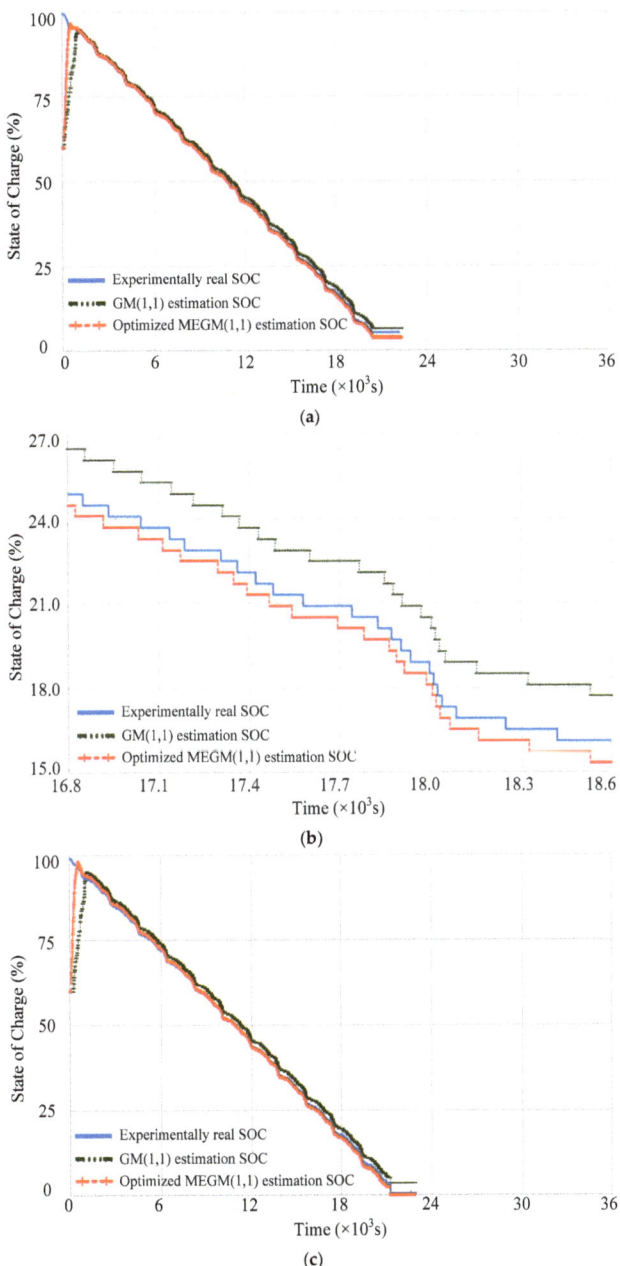

Figure 8. *Cont.*

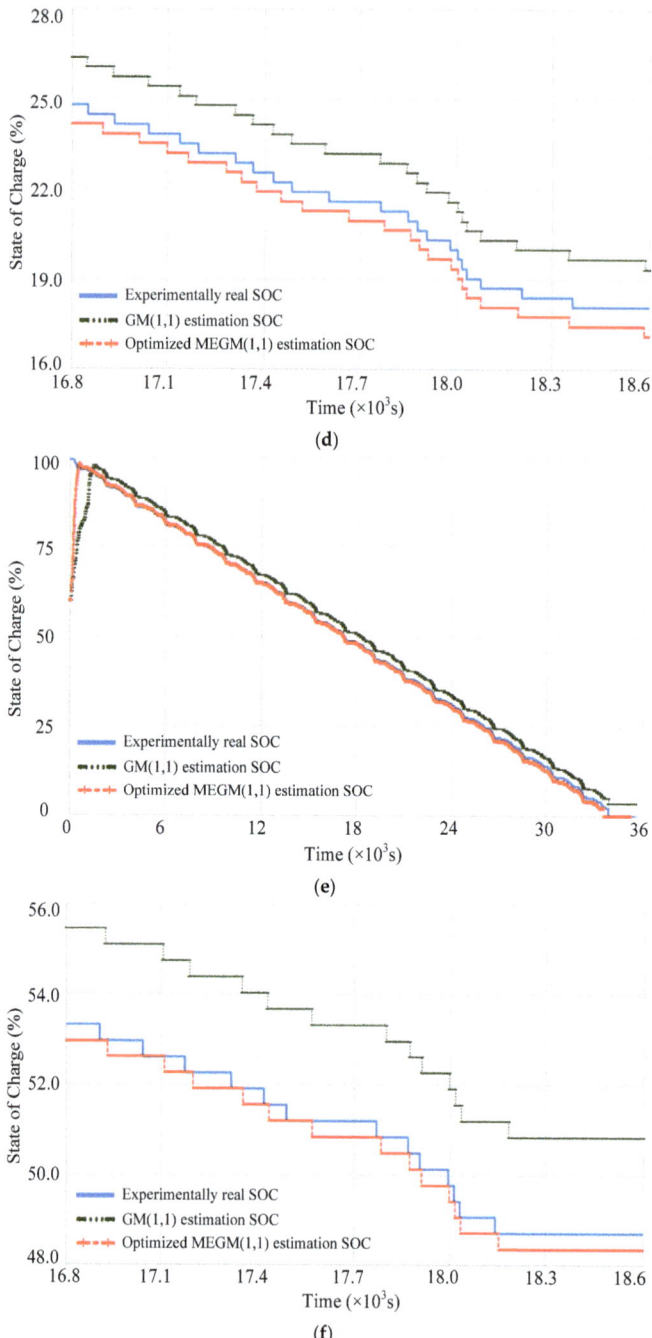

**Figure 8.** *Cont.*

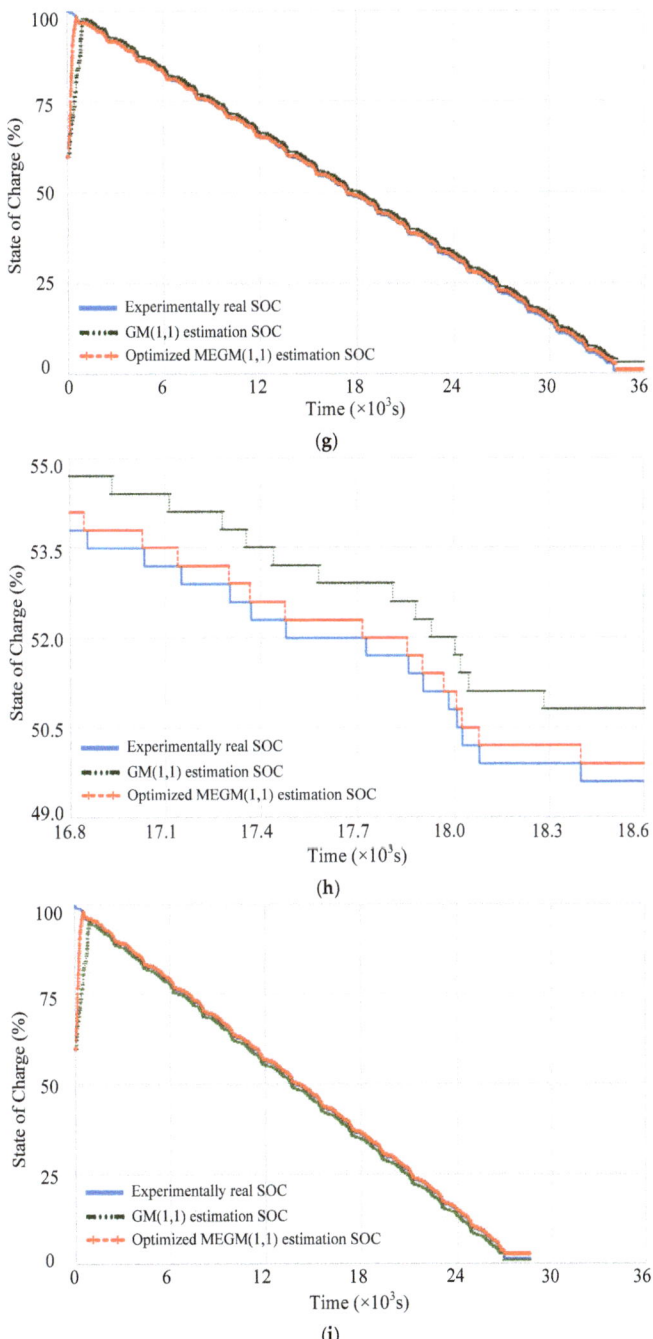

**Figure 8.** *Cont.*

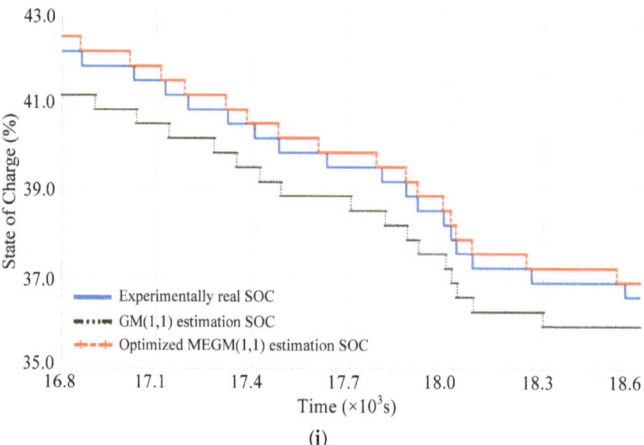

(j)

**Figure 8.** Verification results under the CLTC driving condition with initial SOC errors at the various temperatures: (**a**) SOC comparison among the experimental data, GM(1,1) and MEGM(1,1) estimation at −10 °C; (**b**) the enlarged view of SOC in one cycle for the time range of 16.8 ks–18.6 ks at −10 °C; (**c**) SOC comparison at 0 °C; (**d**) the enlarged view at 0 °C; (**e**) SOC comparison at 15 °C; (**f**) the enlarged view at 15 °C; (**g**) SOC comparison at 25 °C; (**h**) the enlarged view at 25 °C; (**i**) SOC comparison at 35 °C; (**j**) the enlarged view at 35 °C.

## 5. Conclusions

Through the statistical analysis of the experimental data obtained by charging and discharging stations and temperature control boxes to carry out battery test experiments at different temperatures, the battery capacity, ohmic resistance, and open circuit voltage for the parameter model are calibrated. The dynamic battery parameter identification model based on temperature, voltage, and SOC is established by employing the least squares principle to fit the polarization effect parameters, and the established battery mathematical model can better adapt to the lithium battery under the experiment. Aiming at the requirements of battery SOC estimation and the non-linear characteristics of dynamic battery systems, an optimized MEGM(1,1) model is proposed to expand battery state data and introduce temperature factors in the estimation process to make SOC estimation more accurate. The simulation results show that, compared with the traditional GM(1,1) algorithm, the SOC estimation based on the MEGM(1,1) strategy converges faster and the overall error is reduced. Therefore, the proposed optimization algorithm can make the estimated value meet the requirements of fast convergence and small error, which presents well robustness against initial SOC deviation and temperature variation. Experimental results also illustrate that the SOC estimation based on the proposed strategy for power lithium batteries at different temperatures could achieve the goal of an overall error within 1% under CLTC conditions with well robustness and accuracy.

**Author Contributions:** Conceptualization, Q.S. and S.W.; methodology, Q.S. and H.L.; software, S.W. and S.G.; validation, Q.S., S.W. and J.D.; formal analysis, K.W. and J.L.; investigation, S.G. and L.W.; resources, J.D. and L.W.; data curation, Q.S., S.W. and J.L.; writing—original draft preparation, Q.S.; writing—review and editing, Q.S. and H.L.; supervision, Q.S.; project administration, Q.S. and J.L.; funding acquisition, Q.S. and S.G. All authors have read and agreed to the published version of the manuscript.

**Funding:** This research was funded by the Scientific Research Project of Tianjin Municipal Education Commission, grant number 2020KJ093; and in the part by the National Key Research and Development Project of China, grant number 2018YFB0106102.

**Data Availability Statement:** Not applicable.

**Conflicts of Interest:** The authors declare no conflict of interest.

## References

1. Kurzweil, P.; Frenzel, B.; Scheuerpflug, W. A Novel Evaluation Criterion for the Rapid Estimation of the Overcharge and Deep Discharge of Lithium-Ion Batteries Using Differential Capacity. *Batteries* **2022**, *8*, 86. [CrossRef]
2. Li, A.; Yuen, A.C.Y.; Wang, W.; Chen, T.B.Y.; Lai, C.S.; Yang, W.; Wu, W.; Chan, Q.N.; Kook, S.; Yeoh, G.H. Integration of Computational Fluid Dynamics and Artificial Neural Network for Optimization Design of Battery Thermal Management System. *Batteries* **2022**, *8*, 69. [CrossRef]
3. Dubarry, M.; Beck, D. Analysis of Synthetic Voltage vs. Capacity Datasets for Big Data Li-ion Diagnosis and Prognosis. *Energies* **2021**, *14*, 2371. [CrossRef]
4. Zhang, S.; Guo, X.; Dou, X.; Zhang, X. A rapid online calculation method for state of health of lithium-ion battery based on coulomb counting method and differential voltage analysis. *J. Power Sources* **2020**, *479*, 228740. [CrossRef]
5. Solaymani, S. $CO_2$ emissions patterns in 7 top carbon emitter economies: The case of transport sector. *Energy* **2019**, *168*, 989–1001. [CrossRef]
6. Theiler, M.; Schneider, D.; Endisch, C. Kalman Filter Tuning Using Multi-Objective Genetic Algorithm for State and Parameter Estimation of Lithium-Ion Cells. *Batteries* **2022**, *8*, 104. [CrossRef]
7. Sun, Q.; Lv, H.; Wang, S.; Gao, S.; Wei, K. Optimized state of charge estimation of lithium-ion battery in smes/battery hybrid energy storage system for electric vehicles. *IEEE Trans. Appl. Supercond.* **2021**, *31*, 3091119. [CrossRef]
8. Lv, J.; Jiang, B.; Wang, X.; Liu, Y.; Fu, Y. Estimation of the State of Charge of Lithium Batteries Based on Adaptive Unscented Kalman Filter Algorithm. *Electronics* **2020**, *9*, 1425. [CrossRef]
9. Xu, Y.; Hu, M.; Fu, C.; Cao, K.; Su, Z.; Yang, Z. State of Charge Estimation for Lithium-Ion Batteries Based on Temperature-Dependent Second-Order RC Model. *Electronics* **2019**, *8*, 1012. [CrossRef]
10. Hu, M.; Li, Y.; Li, S.; Fu, C.; Qin, D.; Li, Z. Lithium-ion battery modeling and parameter identification based on fractional theory. *Energy* **2018**, *165*, 153–163. [CrossRef]
11. Li, X.; Yuan, C.; Wang, Z. State of health estimation for Li-ion battery via partial incremental capacity analysis based on support vector regression. *Energy* **2020**, *203*, 117852. [CrossRef]
12. Huang, B.; Hu, M.; Chen, L.; Jin, G.; Liao, S.; Fu, C.; Wang, D.; Cao, K. A Novel Electro-Thermal Model of Lithium-Ion Batteries Using Power as the Input. *Electronics* **2021**, *10*, 2753. [CrossRef]
13. Hossain Lipu, M.S.; Hannan, M.A.; Hussain, A.; Ayob, A.; Saad, M.H.M.; Muttaqi, K.M. State of Charge Estimation in Lithium-Ion Batteries: A Neural Network Optimization Approach. *Electronics* **2020**, *9*, 1546. [CrossRef]
14. Li, X.; Huang, Z.; Tian, J.; Tian, Y. State-of-charge estimation tolerant of battery aging based on a physics-based model and an adaptive cubature Kalman filter. *Energy* **2021**, *220*, 119767. [CrossRef]
15. Ye, M.; Guo, H.; Cao, B. A model-based adaptive state of charge estimator for a lithium-ion battery using an improved adaptive particle filter. *Appl. Energy* **2017**, *190*, 740–748. [CrossRef]
16. Lian, B.; Sims, A.; Yu, D.; Wang, C.; Dunn, R.W. Optimizing LiFePO4 battery energy storage systems for frequency response in the UK system. *IEEE Trans. Sustain. Energy* **2016**, *8*, 385–394. [CrossRef]
17. Jiang, K.; Gu, P.; Huang, P.; Zhang, Y.; Duan, B.; Zhang, C. The Hazards Analysis of Nickel-Rich Lithium-Ion Battery Thermal Runaway under Different States of Charge. *Electronics* **2021**, *10*, 2376. [CrossRef]
18. Yuan, W.; Jeong, S.; Sean, W.; Chiang, Y. Development of Enhancing Battery Management for Reusing Automotive Lithium-Ion Battery. *Energies* **2020**, *13*, 3306. [CrossRef]
19. Wu, L.; Pang, H.; Geng, Y.; Liu, X.; Liu, J.; Liu, K. Low-complexity state of charge and anode potential prediction for lithium-ion batteries using a simplified electrochemical model-based observer under variable load condition. *Int. J. Energy Res.* **2022**, *46*, 11834–11848. [CrossRef]
20. Armand, M.; Axmann, P.; Bresser, D.; Copley, M.; Edström, K.; Ekberg, C.; Guyomard, D.; Lestriez, B.; Novák, P.; Petranikova, M. Lithium-ion batteries–Current state of the art and anticipated developments. *J. Power Sources* **2020**, *479*, 228708. [CrossRef]
21. Li, Y.; Vilathgamuwa, M.; Xiong, B.; Tang, J.; Su, Y.; Wang, Y. Design of minimum cost degradation-conscious lithium-ion battery energy storage system to achieve renewable power dispatchability. *Appl. Energy* **2020**, *260*, 114282. [CrossRef]
22. Chen, T.; Jin, Y.; Lv, H.; Yang, A.; Liu, M.; Chen, B.; Xie, Y.; Chen, Q. Applications of lithium-ion batteries in grid-scale energy storage systems. *Trans. Tianjin Univ.* **2020**, *26*, 208–217. [CrossRef]
23. Wang, Q.; Shen, J.; He, Y.; Ma, Z. Design and management of lithium-ion batteries: A perspective from modeling, simulation, and optimization. *Chin. Phys. B* **2020**, *29*, 068201. [CrossRef]
24. Li, J.; Ye, M.; Jiao, S.; Meng, W.; Xu, X. A Novel State Estimation Approach Based on Adaptive Unscented Kalman Filter for Electric Vehicles. *IEEE Access* **2020**, *8*, 185629–185637. [CrossRef]
25. Guo, F.; Hu, G.; Xiang, S.; Zhou, P.; Hong, R.; Xiong, N. A multi-scale parameter adaptive method for state of charge and parameter estimation of lithium-ion batteries using dual Kalman filters. *Energy* **2019**, *178*, 79–88. [CrossRef]
26. Wang, Z.; Gladwin, D.T.; Smith, M.J.; Haass, S. Practical state estimation using Kalman filter methods for large-scale battery systems. *Appl. Energy* **2021**, *294*, 117022. [CrossRef]
27. Jiang, Z.; Li, H.; Qu, Z.; Zhang, J. Recent progress in lithium-ion battery thermal management for a wide range of temperature and abuse conditions. *Int. J. Hydrogen Energy* **2022**, *47*, 9428–9459. [CrossRef]
28. Feng, F.; Teng, S.; Liu, K.; Xie, J.; Xie, Y.; Liu, B.; Li, K. Co-estimation of lithium-ion battery state of charge and state of temperature based on a hybrid electrochemical-thermal-neural-network model. *J. Power Sources* **2020**, *455*, 227935. [CrossRef]

29. Shi, Y.; Ahmad, S.; Liu, H.; Lau, K.T.; Zhao, J. Optimization of air-cooling technology for LiFePO$_4$ battery pack based on deep learning. *J. Power Sources* **2021**, *497*, 229894. [CrossRef]
30. Yetik, O.; Karakoc, T.H. Estimation of thermal effect of different busbars materials on prismatic Li-ion batteries based on artificial neural networks. *J. Energy Storage* **2021**, *38*, 102543. [CrossRef]
31. Zhang, C.; Jiang, J.; Gao, Y.; Zhang, W.; Liu, Q.; Hu, X. Charging optimization in lithium-ion batteries based on temperature rise and charge time. *Appl. Energy* **2017**, *194*, 569–577. [CrossRef]
32. Corno, M.; Pozzato, G. Active adaptive battery aging management for electric vehicles. *IEEE Trans. Veh. Technol.* **2019**, *69*, 258–269. [CrossRef]
33. Castaings, A.; Lhomme, W.; Trigui, R.; Bouscayrol, A. Energy management of a multi-source vehicle by λ-control. *Appl. Sci.* **2020**, *10*, 6541. [CrossRef]
34. Eckert, J.J.; Silva, L.C.; Dedini, F.G.; Correa, F.C. Electric Vehicle Powertrain and Fuzzy Control Multi-objective Optimization, considering Dual Hybrid Energy Storage Systems. *IEEE Trans. Veh. Technol.* **2020**, *69*, 3773–3782. [CrossRef]
35. Vidal, C.; Malysz, P.; Kollmeyer, P.; Emadi, A. Machine learning applied to electrified vehicle battery state of charge and state of health estimation: State-of-the-art. *IEEE Access* **2020**, *8*, 52796–52814. [CrossRef]
36. Stroe, D.I.; Schaltz, E. Lithium-Ion Battery State-of-Health Estimation Using the Incremental Capacity Analysis Technique. *IEEE Trans. Ind. Appl.* **2020**, *56*, 678–685. [CrossRef]
37. Fotouhi, A.; Auger, D.J.; Propp, K.; Longo, S. Lithium–Sulfur Battery State-of-Charge Observability Analysis and Estimation. *IEEE Trans. Power Electron.* **2018**, *33*, 5847–5859. [CrossRef]
38. Benveniste, G.; Rallo, H.; Canals, L.; Merino, A.; Amante, B. Comparison of the state of lithium-sulphur and lithium-ion batteries applied to electromobility. *J. Environ. Manag.* **2018**, *226*, 1–12. [CrossRef] [PubMed]
39. Eckert, J.J.; Barbosa, T.P.; da Silva, S.F.; Silva, F.L.; Silva, L.C.; Dedini, F.G. Electric hydraulic hybrid vehicle powertrain design and optimization-based power distribution control to extend driving range and battery life cycle. *Energy Convers. Manag.* **2022**, *252*, 115094. [CrossRef]
40. Castanho, D.; Guerreiro, M.; Silva, L.; Eckert, J.; Antonini Alves, T.; Tadano, Y.d.S.; Stevan, S.L., Jr.; Siqueira, H.V.; Corrêa, F.C. Method for SoC Estimation in Lithium-Ion Batteries Based on Multiple Linear Regression and Particle Swarm Optimization. *Energies* **2022**, *15*, 6881. [CrossRef]
41. Zerrahn, A.; Schill, W.-P.; Kemfert, C. On the economics of electrical storage for variable renewable energy sources. *Eur. Econ. Rev.* **2018**, *108*, 259–279. [CrossRef]
42. Omariba, Z.B.; Zhang, L.; Kang, H.; Sun, D. Parameter Identification and State Estimation of Lithium-Ion Batteries for Electric Vehicles with Vibration and Temperature Dynamics. *World Electr. Veh. J.* **2020**, *11*, 50. [CrossRef]
43. Chen, Q.; Jiang, J.; Ruan, H.; Zhang, C. Simply designed and universal sliding mode observer for the SOC estimation of lithium-ion batteries. *IET Power Electron.* **2017**, *10*, 697–705. [CrossRef]
44. Li, S.; Li, G.; Ma, X. Grey prediction of lithium battery lifetime based on Markov rolling optimization. *J. Hefei Univ. Technol. Nat. Sci.* **2019**, *42*, 763–769.
45. Liu, S.; Dang, Y.; Fang, Z.; Xie, N. *Grey System Theory and Its Application*, 5th ed.; Science Press: Beijing, China, 2010.
46. Duan, H.; Wang, D.; Pang, X.; Liu, Y.; Zeng, S. A novel forecasting approach based on Multi-Kernel Nonlinear Multivariable Grey Model: A case report. *J. Clean. Prod.* **2020**, *260*, 120929. [CrossRef]
47. Wei, H.; Chen, X.; Lv, Z.; Wang, Z.; Pan, H.; Chen, L. Online Estimation of Lithium-Ion Battery State of Health Using Grey Neural Network. *Power Syst. Technol.* **2017**, *41*, 4038–4044.
48. Wu, W.; Ma, X.; Wang, Y.; Cai, W.; Zeng, B. Predicting China's energy consumption using a novel grey Riccati model. *Appl. Soft Comput.* **2020**, *95*, 106555. [CrossRef]
49. Liu, Y.; Wu, Z.X.; Zhou, H.; Zheng, H.; Yu, N.; An, X.P.; Li, J.Y.; Li, M.L. Development of China Light-Duty Vehicle Test Cycle. *Int. J. Automot. Technol.* **2020**, *21*, 1233–1246. [CrossRef]
50. Liu, Y.; Zhou, H.; Xu, Y.; Qin, K.; Yu, H. *Feasibility Study of Using WLTC for Fuel Consumption Certification of Chinese Light-Duty Vehicles*; SAE International: Warrendale, PA, USA, 2018.
51. Tucki, K. A Computer Tool for Modelling CO$_2$ Emissions in Driving Cycles for Spark Ignition Engines Powered by Biofuels. *Energies* **2021**, *14*, 1400. [CrossRef]
52. Deng, Z.; Hu, X.; Lin, X.; Che, Y.; Xu, L.; Guo, W. Data-driven state of charge estimation for lithium-ion battery packs based on Gaussian process regression. *Energy* **2020**, *205*, 118000. [CrossRef]
53. Plett, G.L. Extended Kalman filtering for battery management systems of LiPB-based HEV battery packs: Part 3. State and parameter estimation. *J. Power Sources* **2004**, *134*, 277–292. [CrossRef]
54. Deng, Z.; Yang, L.; Cai, Y.; Deng, H.; Sun, L. Online available capacity prediction and state of charge estimation based on advanced data-driven algorithms for lithium iron phosphate battery. *Energy* **2016**, *112*, 469–480. [CrossRef]
55. Wang, L. Research on Reliability Predication and Life Cycle Cost Assessment of Low-voltage Switchgear. Ph.D. Thesis, Hebei University of Technology, Tianjin, China, 2017.

Article

# State of Health Estimation of Lithium-Ion Batteries Using a Multi-Feature-Extraction Strategy and PSO-NARXNN

Zhong Ren [1,2,3], Changqing Du [1,2,3,*] and Weiqun Ren [4]

[1] Hubei Key Laboratory of Advanced Technology for Automotive Components, Wuhan University of Technology, Wuhan 430070, China
[2] Foshan Xianhu Laboratory of the Advanced Energy Science and Technology Guangdong Laboratory, Foshan 528200, China
[3] Hubei Research Center for New Energy & Intelligent Connected Vehicle, Wuhan University of Technology, Wuhan 430070, China
[4] Dongfeng Commercial Vehicle Technical Center of DFCV, Wuhan 430056, China
* Correspondence: cq_du@whut.edu.cn

**Abstract:** The lithium-ion battery state of health (SOH) estimation is critical for maintaining reliable and safe working conditions for electric vehicles (EVs). However, accurate and robust SOH estimation remains a significant challenge. This paper proposes a multi-feature extraction strategy and particle swarm optimization-nonlinear autoregressive with exogenous input neural network (PSO-NARXNN) for accurate and robust SOH estimation. First, eight health features (HFs) are extracted from partial voltage, capacity, differential temperature (DT), and incremental capacity (IC) curves. Then, qualitative and quantitative analyses are used to evaluate the selected HFs. Second, the PSO algorithm is adopted to optimize the hyperparameters of NARXNN, including input delays, feedback delays, and the number of hidden neurons. Third, to verify the effectiveness of the multi-feature extraction strategy, the SOH estimators based on a single feature and fusion feature are comprehensively compared. To verify the effectiveness of the proposed PSO-NARXNN, a simple three-layer backpropagation neural network (BPNN) and a conventional NARXNN are built for comparison based on the Oxford aging dataset. The experimental results demonstrate that the proposed method has higher accuracy and stronger robustness for SOH estimation, where the average mean absolute error (MAE) and root mean square error (RMSE) are 0.47% and 0.56%, respectively.

**Keywords:** state of health; lithium-ion battery; machine learning; battery management system

**Citation:** Ren, Z.; Du, C.; Ren, W. State of Health Estimation of Lithium-Ion Batteries Using a Multi-Feature-Extraction Strategy and PSO-NARXNN. *Batteries* **2023**, *9*, 7. https://doi.org/10.3390/batteries9010007

Academic Editor: Carlos Ziebert

Received: 16 November 2022
Revised: 12 December 2022
Accepted: 21 December 2022
Published: 23 December 2022

**Copyright:** © 2022 by the authors. Licensee MDPI, Basel, Switzerland. This article is an open access article distributed under the terms and conditions of the Creative Commons Attribution (CC BY) license (https://creativecommons.org/licenses/by/4.0/).

## 1. Introduction

Vehicle electrification has been proven to be one of the most promising directions to reduce carbon dioxide emissions and solve the energy crisis. With the advantages of high power and energy density, high energy efficiency, and relatively long cycles of life, Lithium-ion batteries (LiBs) have become the primary power source of electric vehicles (EVs) [1]. However, during long-term cycling or storage, it is inevitable for LiBs to degrade, resulting in performance degradation or safety problems. Therefore, an accurate estimation of the state of health (SOH) is essential for the energy management system to maintain safe and high-efficient working conditions for EVs. Generally, the degradation of LiBs is an integrated consequence of internal and external factors. The internal factors mainly include the loss of active material (LAM), the loss of lithium inventory (LLI), resistance increase (RI), and solid electrolyte interface (SEI) growth [2,3]. Moreover, the external factors include operating temperature, charge and discharge rate, discharge depth, and cut-off voltage [4]. From the perspective of onboard applications, the loss of capacity and the increase of internal resistance are two widely used indicators to reflect the battery SOH, expressed as follows:

$$SOH_c = \frac{C_t}{C_0} \times 100\%,$$
$$SOH_r = \left|\frac{R_{EOL} - R_{act}}{R_{EOL} - R_{NEW}}\right| \times 100\% \quad (1)$$

where $SOH_c$ and $SOH_r$ represent the capacity-based SOH and resistance-based SOH, respectively; $C_t$ and $C_0$ are the actual and nominal capacity, respectively; $R_{act}$ denotes the current resistance, and $R_{EOL}$ and $R_{NEW}$ are resistances of the end-of-life (EOL) and new battery, respectively. Compared to resistance-based SOH estimation methods, capacity-based SOH estimation methods draw more attention because capacity directly decides the driving range for EVs [5].

Generally, the existing SOH estimation methods can be divided into two categories, as shown in Figure 1 [6]: the direct measurement methods, like the capacity measurement method, internal resistance measurement test, and impedance measurement method, are suitable for laboratory condition but not practical in actual operations. Indirect analytical methods mainly include model-based, data-driven, and hybrid methods. According to the modeling mechanism, the model-based method can be divided into equivalent circuit model (ECM)-based method and electrochemical model (EM)-based method. The ECM employs lumped components, such as resistors, capacitors, and voltage sources, to describe the battery's dynamic behavior [7]. It is one of the most promising approaches for online battery parameter identification and state estimation, owing to its ease of implementation and acceptable accuracy for EV applications. Based on the principle of ECM, a state equation and an observe equation are established. Then, filter-based methods, such as extended Kalman filter (EKF) [8] and unscented Kalam filter (UKF) [9], are used for online SOH prediction. The EM aims to describe the thermodynamics, Li-ion diffusion process, SEI film thickness, and side reaction kinetics inside the battery to realize the most accurate battery modeling and state estimation theoretically. However, with many parameters to be identified and partial differential equations, the onboard application of the EM-based SOH estimation remains a significant challenge. The trade-off between model complexity and accuracy still needs further research.

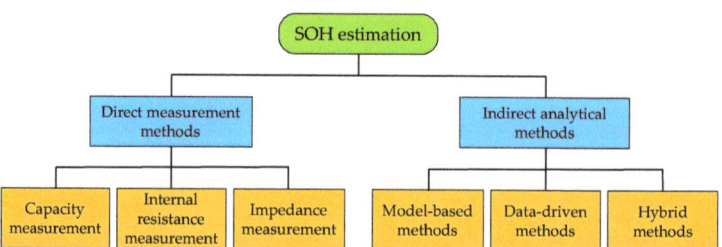

**Figure 1.** Classification of SOH estimation methods.

With the development of artificial intelligence, the machine learning (ML)-based data-driven method has gradually become the most popular method for SOH estimation [10]. The data-driven method can estimate battery states based on measured data. It does not require pre-knowledge about the battery aging mechanism or the battery models mentioned above, making it suitable and easily implementable for different LiBs. Typical procedures for developing data-driven SOH estimation methods are shown in Figure 2. In short, the first step is to conduct battery aging tests and collect raw data, such as voltage, current, and temperature. Since the raw data cannot provide sufficient information to reflect SOH, it cannot guarantee an accurate and robust SOH estimation either. The second step is hence to extract high-related health features (HFs) from the raw data using different techniques, such as model-based analysis, incremental curve analysis (ICA), and differential voltage analysis (DVA). In addition, correlation analysis is applied to analyze the correlation of the HFs.

After that, the HFs and the reference SOH constitute the training dataset. Subsequently, different ML methods are used to learn and validate the nonlinear relationship between the input features and output based on the training dataset. Finally, the established ML algorithms can be used to estimate SOH for new data.

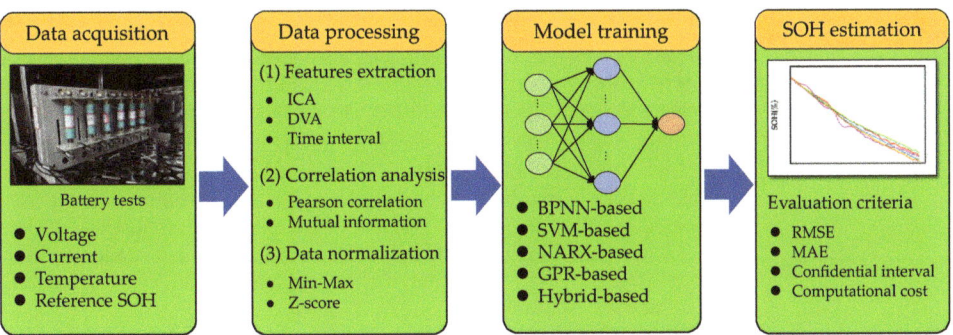

**Figure 2.** Procedures for developing data-driven SOH estimation algorithms.

In summary, two key processes for building an accurate data-driven SOH estimation method are health feature extraction and ML algorithm implementation. How to extract high-related and easy-obtained features is the basis for developing a data-driven SOH estimation method and has become a current research hotspot. The current selection of aging features can be divided into three categories:

(1) Features extracted from voltage and temperature curves during the charging and discharging process, especially the constant–current constant–voltage (CCCV) charging and constant–current (CC) discharging processes. For example, Cui et al. [11] extracted eight HFs from the voltage and temperature curves during CC discharging process and built a SOH estimation method. Liu et al. [12] used the discharging voltage difference of equal time intervals as an HF. However, the CC discharging mode rarely occurs in practical applications, making these HFs unusable for EV operations. Cao et al. [13] first analyzed the CC charging and constant–voltage (CV) charging phases, respectively, and then extracted seventeen HFs. The results of Grey relational analysis concluded that the HFs extracted from the CV phase were less closely related to battery degradation. According to the geometrical analysis of the complete CCCV charging profile, Yang et al. [14] extracted four HFs, such as the time of CC mode, the time of CV mode, the slope of the curve at the end of CC charging mode, and the vertical slope at the corner of the CC charging curve. Undeniably, the HFs extracted from the complete CCCV charging profile can reflect the battery degradation, but for the actual charging condition of EVs, the initial charging SOC is not necessarily 0%, and the terminal charging SOC is not 100%.

(2) Features extracted from constructed curves, such as incremental capacity (IC) curve [15], differential voltage (DV) curve [16], and differential temperature (DT) curve [17]. Take the IC curve as an example. Because the IC curve has prominent peaks, many studies have selected relevant features as the HFs to build data-driven SOH estimation methods. For example, Li et al. [18] extracted eleven HFs in the voltage range from 3.8 V to 4.1 V at the voltage interval of 30 mV. Zhao et al. [19] selected the peak and valley values as HFs to construct the SOH prediction method. Moreover, other geometrical characteristics, such as the width of the peak [20], the area under the peak [21], and the slope of the peak [22] are considered HFs. Although valuable features can be extracted from these constructed curves, these curves are easily disturbed by noise in actual operations. Additionally, an appropriate filtering algorithm is required to smooth the original curve, and then accurate HFs can be identified.

(3) Features obtained from electrochemical impedance, or parameters of the ECM, such as polarization capacitance, polarization resistance, and ohmic resistance. For example, Lyu et al. [23] utilized the recursive least squares (RLS) method to identify the parameters of the Thevenin model. The identified ohmic and polarization resistance were used as HFs to train a linear regression model. Similarly, Yang et al. [24] chose ohmic resistance, polarization resistance, polarization capacitance, and state of charge (SOC) as the inputs of the particle swam optimization-least square support vector regression (PSO-LSSVR) method to estimate SOH. Generally, these features need to be identified using additional algorithms, which increases its difficulty in practical applications.

In addition, other HFs, such as sample entropy [25], and Kullback–Leibler distance [26], are employed to reflect the aging states of LiBs. Based on the above review, there are some principles to bear in mind when choosing HFs [27]: (1) suitability for practical working conditions; (2) easy access to acquire; (3) strong adaptability and robustness; (4) considering thermal factor; and (4) high relevant degree. Therefore, this paper employs a multi-feature extraction strategy to extract HFs from partial charging voltage, capacity, and temperature curves to match these principles.

After selecting HFs and constituting the training dataset, different data-driven methods are used to train the SOH estimation model. Widely used ML methods include:

(1) Shallow neural networks (NNs), such as backpropagation neural network (BPNN) [28], extreme learning machine (ELM) [29], radial basis function neural network (RBFNN) [30], are employed owing to their simple implementation.
(2) Deep learning (DL) methods, such as long-short term memory (LSTM) [31], gated recurrent unit (GRU) [32], and convolutional neural network (CNN) [33], are utilized owing to their superior accuracy, adaptation ability, and good generalization.
(3) Probabilistic-based methods, such as Gaussian process regression (GPR) [34], and deep brief network (DBN) [13], are applied owing to their capability to provide the uncertainty of the estimated value.
(4) Ensemble learning methods, such as random forest (RF) [17], AdaBoost [35], and gradient boosting decision tree (GBDT) [36], are used because they do not easily fall into over-fitting.
(5) Support vector machine (SVM)-based methods [37,38] are utilized owing to their simple implementation and high accuracy.

Nonlinear autoregressive with exogenous input neural network (NARXNN) is a subclass of the recurrent neural network (RNN), which is suitable for predicting complex and nonlinear systems [39]. Compared to other RNNs, such as the LSTM and GRU, the NARXNN has a more straightforward structure, fewer parameters, and reasonable accuracy. Although many researchers employed the NARXNN for state of charge (SOC) estimation [40–42], only a few researchers applied it for SOH estimation. For example, Khaleghi et al. [43] utilized the open-mode NARXNN to capture the dependency between the HFs and battery SOH. In another work, Cui et al. [11] built the closed-mode NARXNN to estimate the battery SOH. Moreover, the existing methods have used a time-consuming trial-and-error approach for finding the appropriate hyperparameters, which is inefficient. Recently, an effective strategy for hyperparameter tuning has been to combine data-driven algorithms and heuristic optimization techniques. For example, Ren et al. [44] utilized the particle swarm optimization (PSO) algorithm to optimize the number of hidden neurons, the learning rate, and the maximum epochs of the LSTM. Zhang et al. [45] employed the PSO algorithm to optimize the kernel parameters ($w$ and $\sigma$) of the RBFNN. Hossain et al. [46] used the gravitational search algorithm (GSA) to find the best number of hidden neurons of the ELM. Compared with other heuristic optimization techniques, the PSO algorithm has the advantages of easy implementation, strong robustness, and global exploration. Therefore, this paper attempts to use the PSO algorithm to find the best values of input delays, feedback delays, and the number of hidden neurons of the NARXNN. Then the NARXNN is employed to build a multi-feature-based SOH estimation model. In summary, the main contributions of this paper are as follows:

- To comprehensively describe the battery aging characteristics, a multi-feature extraction strategy is employed to extract HFs from partial voltage, capacity, and temperature curves. Qualitative and quantitative analysis is used to evaluate the selected HFs.
- The performance of the NARXNN is highly dependent on the number of input delays, feedback delays, and neurons in the hidden layer. Hence, the PSO algorithm is applied to improve the training efficiency of NARXNN by searching for the optimal values of input delays, feedback delays, and the number of hidden neurons.
- The SOH estimators based on a single feature and fusion feature are comprehensively compared to verify the validity of the muti-feature extraction strategy. Moreover, to verify the effectiveness of the proposed PSO-NARXNN, a simple three-layer BPNN and a conventional NARXNN are built for comparison.

The remainder of the paper is organized as follows: Section 2 gives data analysis and feature extraction. The related algorithms are introduced in Section 3. Results and discussion are given in Section 4. Finally, Section 5 summarizes the conclusions.

## 2. Data Analysis and Feature Extraction

### 2.1. Oxford Battery Degradation Dataset

In this paper, a public dataset from the University of Oxford is utilized for LiBs aging analysis and SOH estimation algorithm development. As introduced in Ref. [47], the Oxford aging dataset contains measurements of battery aging data from eight Kokam pouch batteries with a nominal capacity of 740 mAh, noted as cell 1 to cell 8. The negative electrode material is graphite, and the positive electrode material is $LiMO_2$ (where M means a combination of Ni, Co, and Mn, commercially known as NMC) [48]. The cells were all tested in a thermal chamber at 40 °C. The cells were exposed to a CCCV charging profile, followed by a drive cycle discharging profile obtained from the urban Artemis profile. Characterization measurements were taken every 100 cycles. The whole test procedure is summarized in Table 1. The voltage, current, and temperature data is recorded at a sampling interval of 1 s. More details can be found in Ref. [47]. The typical EOL threshold for LiBs is when the SOH decreases to 80%, and the LiBs are suggested to be retired. Hence, only the data with SOH higher than 80% are selected for LiBs aging analysis and SOH estimation algorithm development, as depicted in Figure 3. It is worth noting that even though these 8 LiBs with the same cathode material use the same aging experimental settings, the aging paths are significantly different. One possible reason for this phenomenon is internal variations in material properties from cell manufacturing. Another reason could be the effects of non-uniform environmental temperatures in the thermal chamber.

**Table 1.** Test schedule of the Oxford dataset [49].

| Step 1: Characterization test | |
|---|---|
| (1) | 1 C cycles: Charge and discharge the battery with 1 C (740 mA) current. |
| (2) | Pseudo-OCV: Charge and discharge the battery with C/18.5 (40 mA) current. |

| Step 2: Drive cycle test (repeat 100 times) | |
|---|---|
| (1) | Charge the battery with 2 C (1480 mA) current |
| (2) | Discharge the battery with Artemis Urban driving profile (average current = 1.36 A). |

| Step 3: EOL judgment | |
|---|---|
| (1) | Repeat steps 1 and 2 until the cell loses at least 20% of its rated capacity. |

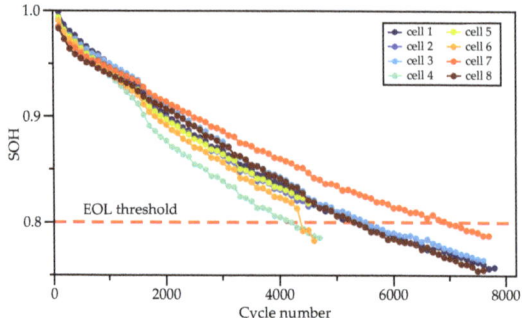

**Figure 3.** SOH curves of cell 1 to cell 8.

*2.2. Health Feature Extraction*

The proposed multi-feature extraction strategy will be explained in detail in this section.

*2.2.1. Voltage Feature Extraction*

The terminal voltage and capacity curves of cell 1 under different aging states are shown in Figure 4. Note that only the CC charging phase is recorded in the Oxford dataset. Additionally, as concluded in Ref. [13], the HFs extracted from CC charging phase are more related to the battery SOH. Therefore, we only extract features from CC charging phase.

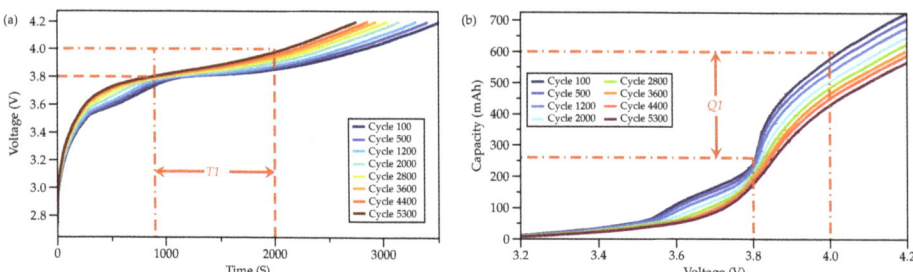

**Figure 4.** (**a**) Terminal voltage curves of cell 1; (**b**) Charging capacity curves of cell 1.

As shown in Figure 4a, the time for LiBs to reach 4.2 V gradually decreases as the number of cycles increases, which directly reflects the reduction in usable capacity. This phenomenon can also be demonstrated in Figure 4b, where the charged capacity gradually decreases with the battery SOH decreases. Thus, the charged time and charged capacity of the CC phase can be selected as HFs. However, considering the practical operations of EVs, LiBs are rarely charged from 0% to 100% SOC but in a specific SOC range (e.g., 40% to 80%). Therefore, the charged time and charged capacity from partial CC curves of voltage varying from 3.8 V to 4.0 V (about 35% to 80% SOC) are selected as HFs to reflect the battery degradation, denoted as T1 and Q1, respectively.

*2.2.2. Temperature Feature Extraction*

As shown in Figure 5a, the raw temperature curves under different aging states are vulnerable to the impact of temperature sensor noise, resulting in difficulty in extracting temperature-related features. Therefore, a finite difference method is utilized in this paper to pre-process the raw temperature curves and then obtain the differential temperature (DT) curves [17]. The expression is as follows:

$$DT(t) \approx \frac{T(t+\Delta) - T(t)}{\Delta} \qquad (2)$$

where ∆ is the pre-determined sample interval. Generally, when ∆ chooses a larger value, the DT curves cannot present subtle temperature changes. However, if ∆ takes a smaller value, the finite difference method cannot eliminate the influence of temperature sensor noise and produce errors. After parameter tuning, the interval sampling ∆ = 40 s is selected in this paper. In addition, a Gaussian filter is used to smooth the original DT curves to eliminate the impact of noise further.

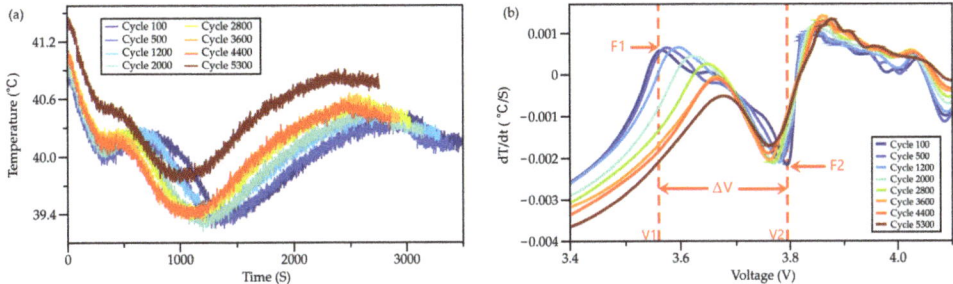

**Figure 5.** (a) Raw temperature curves of cell 1; (b) Smoothed DT curves of cell 1.

Figure 5b shows the DT curve of cell 1 under different aging states. Overall, the DT curve can be divided into three parts: (1) The DT curve first undergoes a period of rapid ascent, reaching its first peak value (denoted as F1); (2) After that, it undergoes a sharp decline and reaches its first valley value (denoted as F2); (3) Then, it experiences a rapid increase. Note that the DT curve represents the temperature change rate during the charging process. Thus, the entire DT curve shows a particular trend with the SOH decreases. Specifically, the first peak value, F1, gradually decreases, and the voltage corresponding to F1 (denoted as V1) increases with the SOH decreases. Moreover, the voltage corresponding to the valley value F2 (denoted as V2) gradually declines with the SOH decreases, but the valley value F2 does not show a clear upward or downward trend. In addition, the voltage difference between F1 and F2 (denoted as ∆V) shows a decreasing trend. As for the third part, although the DT curve shows an overall increasing trend and then declines, health features are not noticeable. Overall, the first and second parts of the DT curve present an obvious change with the SOH decreases. Therefore, the peak value F1, the voltage corresponding to F1, the voltage corresponding to F2, and the voltage difference ∆V between them, are chosen as HFs to describe the battery degradation.

### 2.2.3. IC Feature Extraction

The ICA is a widely used method to analyze the aging mechanism of LiBs from the electrochemical level. The IC curve is an effective tool for analyzing capacity loss and extracting HFs. The most important function of ICA is translating the flat capacity curve into the IC curve with clearly identifiable peaks, which can reflect the phase change characteristics of LiBs during active material insertion and delamination. Usually, the IC curve is obtained from the charging process under the CC charging phase by using a differential equation:

$$IC = \frac{dQ}{dV} = I \cdot \frac{dt}{dV} \qquad (3)$$

where $Q$ represents the capacity, $V$ denotes the voltage, and $t$ is the sampling time.

As shown in Figure 6b, there are two peaks in the IC curve's middle range (e.g., 3.4–4.0 V), and each peak can reflect the phase change inside the LiBs during the charging process with a 1-C charging rate. It can be found that the area under the peaks decreases with the cycle increases, indicating LAM and LLI [21]. In addition, the peak values decrease with a clear trend. Note that compared with the first peak (between 3.5 V and 3.7 V), the second peak (between 3.7 V and 4.0 V) shows a more noticeable trend during the aging process. Therefore, the value of the second peak (denoted as P1) and the area under the second peak between 3.7 V and 4.0 V (denoted as A1) are chosen as HFs to describe the degradation of battery capacity.

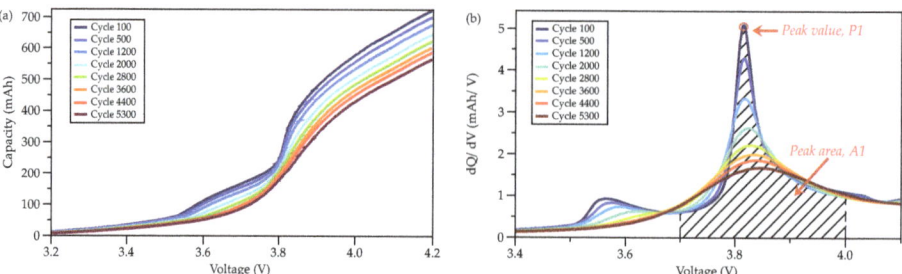

**Figure 6.** (**a**) Charged capacity curves of cell 1; (**b**) Smoothed IC curves of cell 1.

### 2.2.4. Correlation Analysis

In summary, eight HFs are extracted from partial voltage, capacity, DT, and IC curves, respectively, which can match the feature-selecting principles mentioned in the Introduction. First, the voltage-related features are extracted from partial charging curves, which are suitable for practical conditions and easy to obtain. Second, the temperature-related features consider the thermodynamic factor. Third, the multi-feature extraction strategy can improve adaptability and robustness. Figure 7 shows the tendencies of the eight HFs of cell 1 with the increase of cycle numbers, which can reflect the qualitative relationship between HFs and battery SOH. It can be seen that only the V1 shows an upward trend, while other HFs all show a downward trend with the SOH decreases. In addition, HFs of other cells show a similar change trend as in Figure 7.

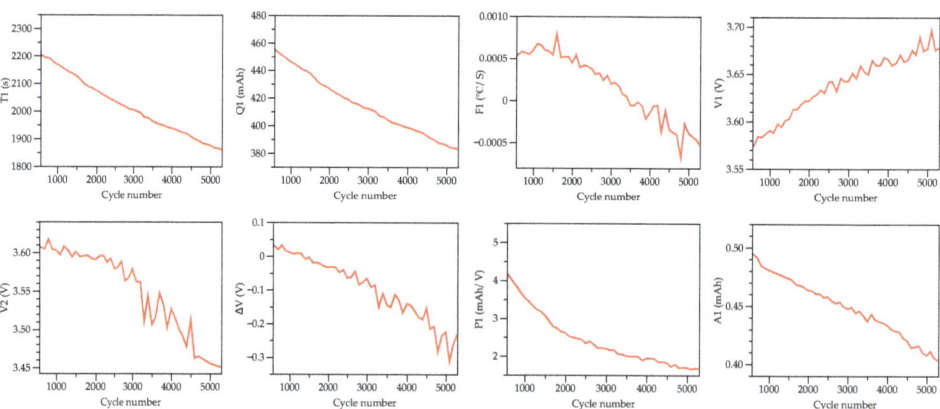

**Figure 7.** Relationship between HFs and battery degradation.

To further evaluate the correlation of the selected HFs quantitatively, the Pearson correlation analysis is employed to calculate the correlation coefficient between the HFs and battery SOH. The equation is as follows:

$$\rho = \frac{\sum_{i=1}^{n}(F_i - \overline{F})(C_i - \overline{C})}{\sqrt{\sum_{i=1}^{n}(F_i - \overline{F})^2 \sum_{i=1}^{n}(C_i - \overline{C})^2}} \quad (4)$$

where $F_i$ is the sequence of HF, $C_i$ is the sequence of battery SOH, $\overline{F}$ and $\overline{C}$ are their average values.

Table 2 summarizes the Pearson correlation coefficients between the HFs and battery SOH of eight cells. Generally, the absolute value of the correlation coefficient is closer to 1, indicating that the relational degree is greater. According to the correlation analysis results in Table 2, the absolute Pearson correlation coefficients of the HFs of eight cells are all greater than 0.9, indicating a high relational grade with the battery SOH. Therefore, using these HFs to build a data-driven method for SOH estimation is reasonable.

**Table 2.** Correlation analysis results.

|  | Cell 1 | Cell 2 | Cell 3 | Cell 4 | Cell 5 | Cell 6 | Cell 7 | Cell8 | Average |
|---|---|---|---|---|---|---|---|---|---|
| T1 | 0.9994 | 0.9977 | 0.9993 | 0.9978 | 0.9985 | 0.9973 | 0.9990 | 0.9979 | 0.9984 |
| Q1 | 0.9992 | 0.9976 | 0.9994 | 0.9975 | 0.9985 | 0.9973 | 0.9990 | 0.09978 | 0.9983 |
| F1 | 0.9080 | 0.9169 | 0.8520 | 0.8919 | 0.8476 | 0.8836 | 0.9084 | 0.9120 | 0.9010 |
| V1 | −0.9916 | −0.9796 | −0.9597 | −0.9795 | −0.9609 | −0.9779 | −0.9671 | −0.9828 | −0.9749 |
| V2 | 0.9147 | 0.9185 | 0.8702 | 0.9149 | 0.9324 | 0.9162 | 0.9211 | 0.9447 | 0.9166 |
| ΔV | 0.9538 | 0.9546 | 0.9334 | 0.9557 | 0.9538 | 0.9571 | 0.9695 | 0.9686 | 0.9558 |
| P1 | 0.9647 | 0.9742 | 0.9682 | 0.9669 | 0.9719 | 0.9715 | 0.9658 | 0.9684 | 0.9690 |
| A1 | 0.9890 | 0.9923 | 0.9894 | 0.9932 | 0.9926 | 0.9929 | 0.9908 | 0.9910 | 0.9914 |

## 3. Related Algorithms

### 3.1. Nonlinear Autoregressive with Exogenous Input Neural Network

NARXNN is a sort of RNN that can learn to predict one time series by means of giving past values of the same time series and another time series called the external or exogenous time series. The structure of NARXNN is depicted in Figure 8, where TDL represents the time–delay line. According to the feedback mechanism of NARXNN, it can be regarded as a variant of the Jordan NN [50]. The expression of NARXNN is as follows:

$$y(n) = f_o[b_o + \sum_{h=1}^{l} w_{ho}f_h(b_h + \sum_{i=0}^{d_u} w_{ih}x(n-i) + \sum_{j=0}^{d_y} w_{jh}y(n-j)))] \quad (5)$$

where $f_o(\cdot)$ and $f_h(\cdot)$ are activation functions of the output layer and hidden layer respectively, $[w_{ih}, w_{jh}, w_{ho}]$ and $[b_h, b_o]$ are weights and biases between the corresponding layers, $d_u$ and $d_y$ represent the input and feedback delays, respectively, and $l$ is the number of hidden neurons.

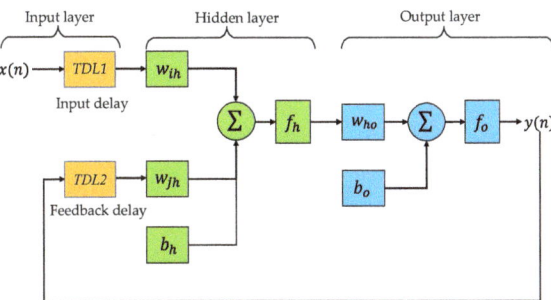

**Figure 8.** Structure of NARXNN [43].

The most important hyperparameters of NARXNN are the input delays, feedback delays, number of hidden neurons, feedback mode, and training methods. In this paper, the input delays, feedback delays, and the number of hidden neurons are optimized using the

PSO algorithm (introduced in the next section). Regarding the training methods, Levenberg–Marquardt (LM) and Bayesian regularization (BR) methods are the most convenient and common functions. Referring to [11], BR is chosen as the training function in this study. Regarding the feedback mode, there are two types: close and open. The former feeds back the predicted output to the input, while the latter feeds back the target output to the input. Although the open-mode NARXNN has a higher estimation theoretically, the close-mode NARXNN is adopted in this work because of the unavailability of the actual target in practical operations. In addition, the sigmoid transfer function and a linear transfer function are used at the hidden layer and output layer, respectively [51].

### 3.2. Particle Swarm Optimization

PSO was first proposed by Kennedy and Eberhart in 1995 [52]. Owing to its advantages of easy implementation and strong robustness, PSO has been employed in numerous applications. The basic idea of PSO is to search for the best results of particles with optimal values through an iterative process. Two locations are used for searching for the best solutions in PSO. The first is obtained through the current iteration, denoted as local best, *pbest*. The second is achieved in earlier iterations, denoted as global best, *gbest*. By calculating the objective function of each particle, the best *pbest* can be found in every iteration. Moreover, the best *gbest* can be found through a continuous update of particles. The position and velocity of every particle are updated as follows:

$$v_i^{k+1} = wv_i^k + c_1 r_1 \left( pbest_i^k - x_i^k \right) + c_2 r_2 \left( gbest^k - x_i^k \right) \tag{6}$$

$$x_i^{k+1} = x_i^k + v_i^{k+1} \tag{7}$$

where $v_i^k$ and $x_i^k$ represent the velocity and position of *i*th particle at *k*th iteration, respectively, $pbest_i^k$ is the optimal solution of *i*th particle at *k*th iteration, $gbest^k$ is the global optimal solution of all particles until *k*th iteration, $w$ represents weigh factor, $r_1$ and $r_2$ are random numbers between 0 and 1, and $c_1$ and $c_2$ are positive learning factor.

### 3.3. Flowchart of the PSO-NARXNN

It is well-known that building a high-performance neural network requires appropriate hyperparameter settings. As for the NARXNN, input delays, feedback delays, and the number of hidden neurons are the three most essential hyperparameters determining its overall performance. Hence, in this paper, the PSO algorithm is applied to improve the performance of NARXNN by searching for the optimal value of input delays, feedback delays, and the number of hidden neurons. The flowchart of the PSO-NARXNN is depicted in Figure 9, and the specific steps are as follows:

Step 1: Data processing. Feature extraction and data normalization are the basis for model training. In this paper, the eight HFs mentioned in Section 2.2 are first extracted, then the Z-score normalization method is utilized to transform the original data to no-dimensional forms.

Step 2: PSO algorithm is used to optimize the hyperparameters of NARXNN.

1. The main parameters of the PSO algorithm are assigned as follows: the particle dimension $D$ is 3, population size $N$ is 10, maximum iteration $M$ is 100, the boundary limit of input and feedback delays is set between '1' and '5', and the boundary limit of hidden neurons is set between '1' and '20'. Then, the initial position is generated randomly within the boundary.
2. According to the initial position, which contains the values of input delays, feedback delays, and the number of hidden neurons, the NARXNN is trained based on BR algorithm. The mean square error (MSE) is taken as the objective function to calculate the fitness value, and the lowest value is considered *'gbest'*.

3. The particle velocity and position are updated according to Equations (6) and (7), and then the fitness value is calculated to update the '*pbest*' and '*gbest*'. In addition, the position of particles is verified by whether they are situated in the boundary.
4. If the termination conditions are met, the algorithm ends and outputs the optimization results; otherwise, return to 3 in Step 2.

Step 3: The optimized hyperparameters are used to train the NARXNN. Then, the trained NARXNN is tested based on the testing datasets. Moreover, several statistical metrics are used to evaluate the model error. The experimental results will be discussed in the next section.

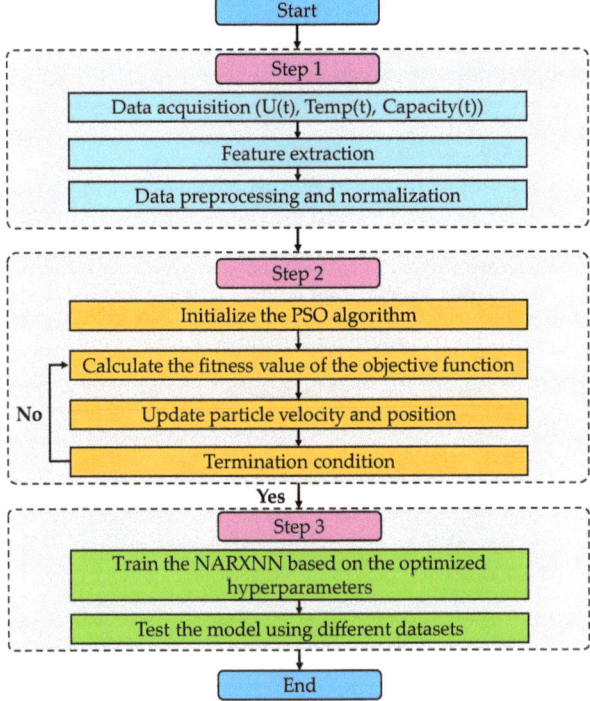

**Figure 9.** Flowchart of the PSO-NARXNN.

Note that MATLAB version 2022b (MathWorks, Natick, MA, USA) is used to develop all related algorithms proposed in this paper.

## 4. Results and Discussion

This section discusses the experimental results of the proposed muti-feature extraction strategy and PSO-NARXNN for SOH estimation. It should be noted that the reference SOH is calculated based on the CC charging characteristic test, as summarized in Table 1. Several statistical metrics, such as the mean absolute error (*MAE*), the root mean square error (*RMSE*), and the maximum error (MaxE), are employed to evaluate the estimation results quantitatively. The *MAE* can measure the average error size, while the *RMSE* describes the dispersion and convergence performance. The expressions are as follows:

$$MAE = \frac{1}{N}\sum_{i=1}^{N}\left|\widehat{SOH}_i - SOH_i\right| \qquad (8)$$

$$RMSE = \sqrt{\frac{1}{N}\sum_{i=1}^{N}\left(\widehat{SOH_i} - SOH_i\right)^2} \quad (9)$$

where $\widehat{SOH_i}$ represents the predicted value, $SOH_i$ represents the reference value, and $N$ is the number of samples.

To fully use the Oxford aging dataset and verify the generalization of the proposed SOH estimation method, the aging datasets of eight cells are constructed into eight groups for experiments. For example, experimental group 1 represents that the aging dataset of cell 1 is used as a training dataset for model training. Then, the aging datasets of the other seven cells are used as testing datasets. In this way, the feasibility of the selected HFs and the generalization of the proposed SOH method can be evaluated comprehensively. Sections 4.1.1–4.1.3 explain the results of group 1, and Section 4.1.4 gives the results of the other seven experimental groups.

*4.1. Results*

4.1.1. Optimal Parameters

Firstly, the optimal values of input delays, output delays, and the number of hidden neurons are optimized by the PSO algorithm. The convergence curve is shown in Figure 10. It can be seen that the minimum value of the objective function is achieved after 32 iterations when the relative change in the objective function value over the last iteration is less than the tolerance. Moreover, the optimal values of input delays, output delays, and the number of hidden neurons are attained as 1, 2, and 15, respectively. Then, the NARXNN is trained using the optimal values and compared with other algorithms in the following sections.

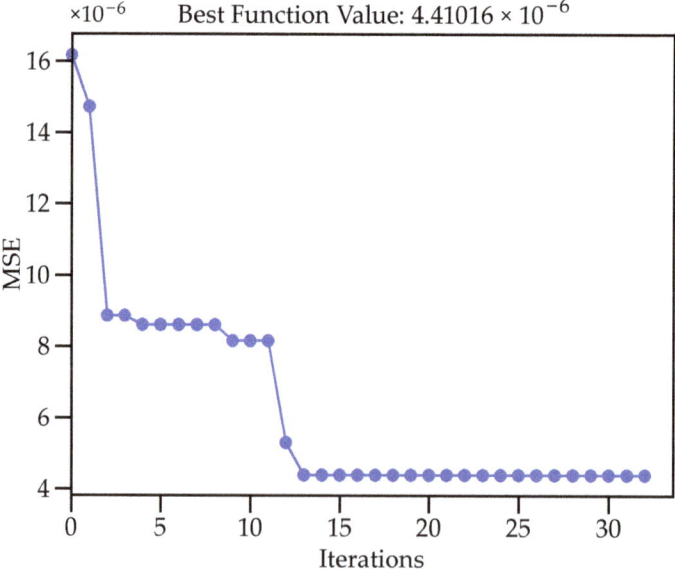

**Figure 10.** Convergence curve of PSO algorithm.

4.1.2. Comparison with Different Feature Extraction Strategies

As mentioned in the Introduction, health feature extraction is essential for building a high-performance SOH estimation model. To verify the effectiveness of the proposed multi-feature extraction strategy, the voltage, temperature, IC, and fusion features are separately used to train the PSO-NARXNN and compared in this section. The SOH estimation results based on different feature extraction strategies are shown in Figure 11, and the statistical metrics are given in Table 3.

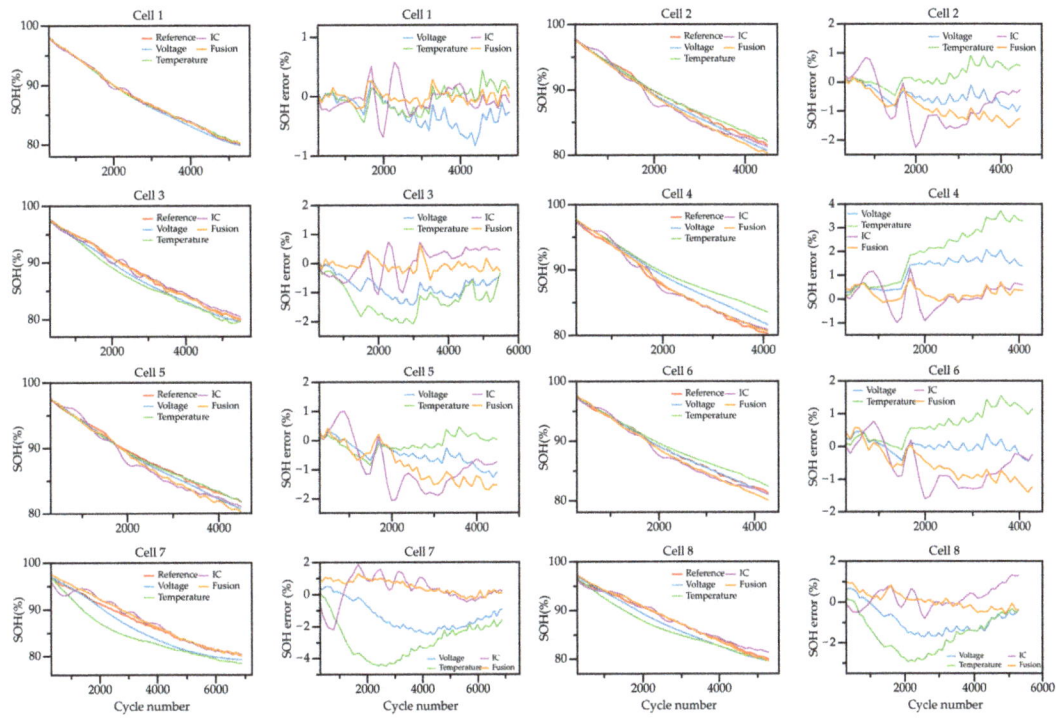

**Figure 11.** SOH estimation results of different feature extraction strategies.

**Table 3.** Summary of MAE and RMSE of different feature extraction strategies.

| Cell | Voltage | | Temperature | | IC | | Fusion | |
|---|---|---|---|---|---|---|---|---|
| | MAE | RMSE | MAE | RMSE | MAE | RMSE | MAE | RMSE |
| 1 | 0.29 | 0.36 | 0.16 | 0.20 | 0.18 | 0.26 | 0.09 | 0.116 |
| 2 | 0.51 | 0.57 | 0.29 | 0.39 | 0.90 | 1.06 | 0.84 | 0.95 |
| 3 | 0.79 | 0.87 | 1.24 | 1.35 | 0.43 | 0.49 | 0.17 | 0.22 |
| 4 | 1.21 | 1.34 | 1.96 | 2.27 | 0.47 | 0.59 | 0.28 | 0.36 |
| 5 | 0.51 | 0.59 | 0.24 | 0.31 | 1.03 | 1.16 | 0.85 | 1.01 |
| 6 | 0.17 | 0.22 | 0.67 | 0.81 | 0.71 | 0.83 | 0.65 | 0.74 |
| 7 | 1.40 | 1.60 | 2.82 | 3.04 | 0.73 | 0.94 | 0.55 | 0.66 |
| 8 | 0.98 | 1.10 | 1.56 | 1.80 | 0.43 | 0.56 | 0.32 | 0.41 |
| Average | 0.73 | 0.83 | 1.12 | 1.27 | 0.61 | 0.73 | 0.47 | 0.56 |

Overall, the fusion feature-based method can obtain accurate and robust estimation results under all testing datasets. In contrast, single feature-based methods can only achieve acceptable estimation results on specific testing datasets. For example, the temperature feature-based method can obtain great estimation accuracy for cell 2 and cell 5 with the MAEs less than 0.3% and RMSEs less than 0.4%. However, the estimation results of other cells are the worst, where the MAEs of cell 3, cell 4, cell 7, and cell 8 are 1.24%, 1.96%, 2.82%, and 1.56%, respectively, while the RMSEs are 1.35%, 2.27%, 3.04%, and 1.80%, respectively. Moreover, the MaxE of cell 7 exceeds 4%, which is unacceptable. The DT curve can reflect the thermodynamic characteristics of LiBs during the degradation process. However, the differential operation may magnify the noise in measurement, and the filtering algorithm may influence the feature extraction process. Thus, only using the temperature feature cannot obtain accurate and robust estimation results for all testing datasets. Regarding the voltage feature, the overall estimation performance is better than the temperature feature, according to the error curves in Figure 11. However, only using the voltage feature cannot guarantee the estimation robustness under different testing datasets. It can be seen that the MAE of cell 6 achieves the lowest value, while the MAEs of cell 4 and cell 7 exceed 1%. The MaxEs of cell 4 and cell 7 exceed 2%. Regarding the IC feature, the IC curve can describe the electrochemical characteristics of LiBs during the aging process. It can be seen that the overall SOH estimation results based on the IC feature are better than the temperature and voltage features, especially for cell 3, cell 4, cell 7, and cell 8. However, compared with the fusion feature-based method, the error curves of the IC feature-based method show more fluctuations, resulting in larger RMSEs and MaxEs, where the MaxEs of cell 2, cell 5, and cell 7 exceed 2%. Finally, the multi-feature extraction strategy can fully use the advantages of different kinds of features and avoid their disadvantages, resulting in more accurate and robust estimation results. It can be seen from Figure 12 that although the MAEs of the fusion feature-based method for cell 2 and cell 5 are not the lowest, it remains within 1%. Moreover, the fusion feature-based method can obtain the best estimation performance for other cells. The average MAE of the fusion feature-based method is 0.47%, which is 57.79%, 35.63%, and 22.73% lower than the temperature, voltage, and IC feature-based methods, respectively. As such, the effectiveness of the proposed multi-feature extraction strategy is verified based on the above analysis.

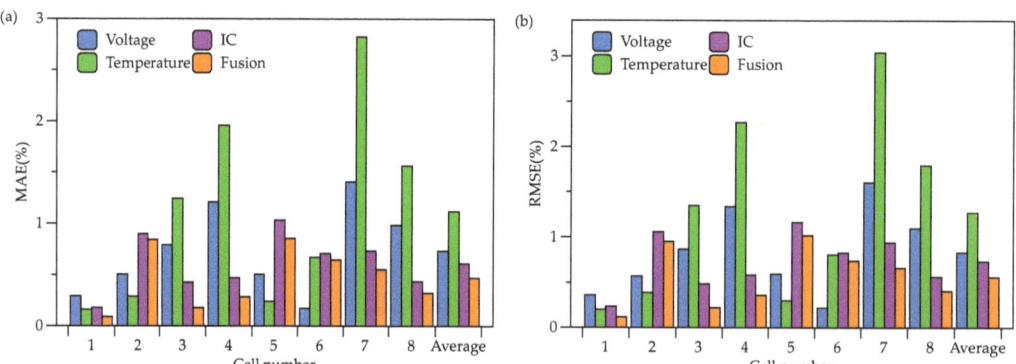

**Figure 12.** (**a**) MAE of different feature extraction strategies; (**b**) RMSE of different feature extraction strategies.

4.1.3. Comparison with Different Algorithms

As concluded in Section 4.1.1, the optimal values of input delays, feedback delays, and the number of hidden neurons are optimized by the PSO algorithm. To verify the effectiveness of the optimal hyperparameters, a conventional NARXNN whose input delays, feedback delays, and the number of hidden neurons are randomly assigned is built for comparison. Additionally, to further verify the validity of the selected HFs, a simple three-layer BPNN is trained for comparison, too. For a fair comparison, the hyperparameters and training settings of the above two methods, including the number of hidden neurons, activation function, and training method, are entirely consistent with the PSO-NARXNN method. Figure 13 shows the SOH estimation results, and Figure 14 visually compares the MAE and RMSE of the above three methods. The statistical metrics are given in Table 4.

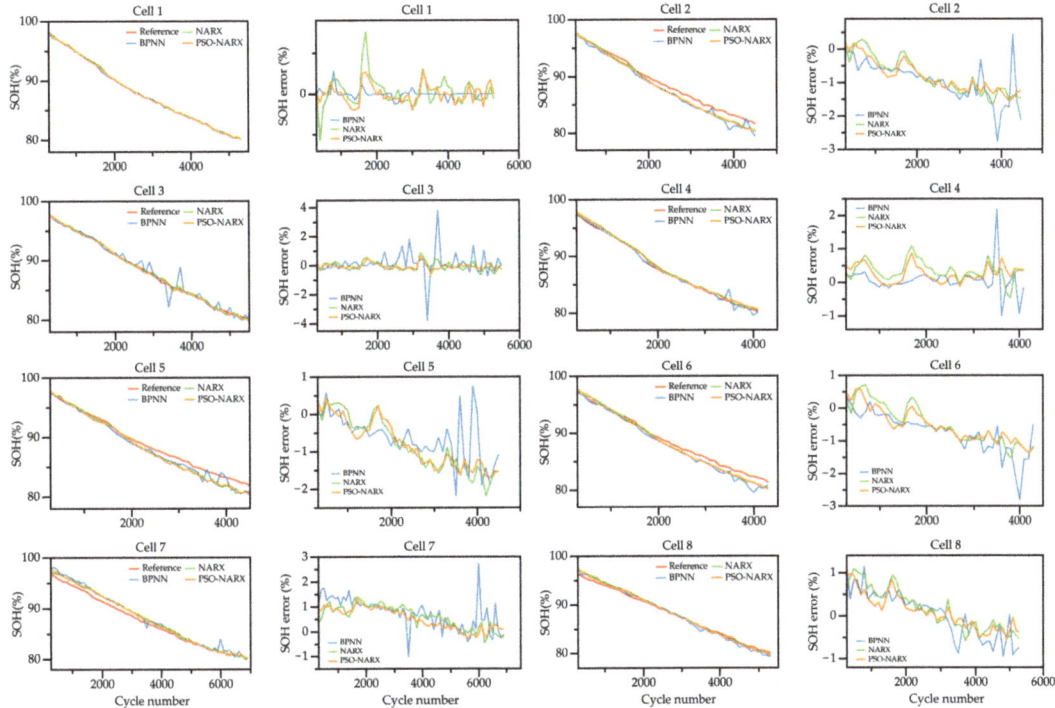

**Figure 13.** SOH estimation results of different algorithms.

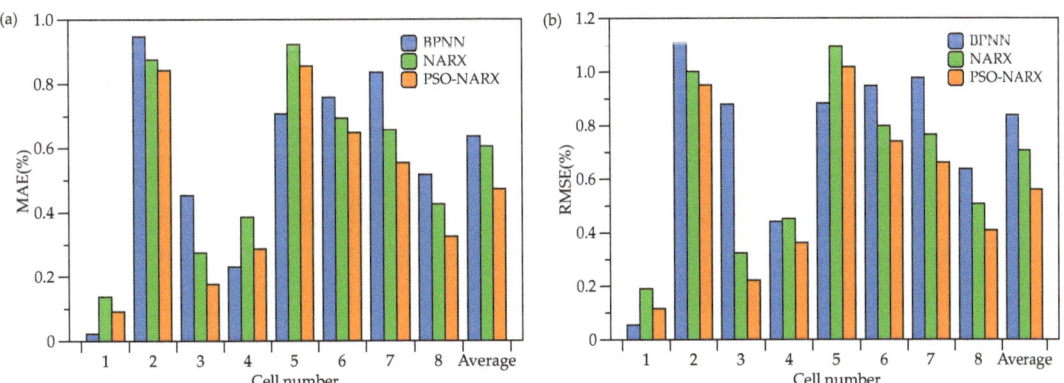

**Figure 14.** (**a**) MAE of different algorithms; (**b**) RMSE of different algorithms.

According to the SOH estimation curves shown in Figure 13, the estimated SOH of the three methods can generally follow the aging path. However, the SOH curves estimated based on the PSO-NARXNN method have better consistency and smoothness with the real SOH trajectory. In contrast, the estimated curves based on the BPNN and conventional NARXNN methods show different degrees of fluctuations. Specifically, regarding the BPNN method, the MAEs of all cells are less than 1%, and only the RMSE of cell 2 exceeds 1%, indicating 1.10%. Moreover, the MAEs of cell 4 and cell 5 achieve the lowest values among the three methods. Therefore, the effectiveness of the proposed multi-feature extraction strategy is further verified because a simple BPNN can obtain a relatively satisfactory estimation performance. However, the MaxEs of the BPNN method are all larger than 2%, and the estimated SOH curves show significant fluctuations, especially during the EOL of LiBs, according to the error curves in Figure 13. As a result of the fluctuation, the BPNN method has an average MAE and RMSE of 0.64% and 0.84%, which are larger than the other two methods. Regarding the conventional NARXNN method, it can achieve an overall better performance in comparison to the BPNN method. Especially, the estimation errors of cell 3 and cell 7 are reduced to a larger extent. The average MAE is 0.60%, 6.7% lower than the BPNN method. Moreover, the average RMSE is 0.70%, which is 16.7% lower than the BPNN method. In addition, according to the error curves in Figure 13, the MaxEs of the conventional NARXNN method are all less than those of the BPNN method, showing a more robust estimation result. This is because the feedback mechanism of the NARXNN can learn information from the past values of output and previous values of exogenous input data. Regarding the PSO-NARXNN, it can be seen from Figure 14 that all MAEs and RMSEs are reduced in comparison to the conventional NARXNN. The error curves in Figure 13 also demonstrate that the PSO-NARXNN has an overall better estimation performance than the conventional NARXNN. The MAE of cell 3 reaches the lowest value, 0.17%. The average MAE and RMSE are 0.47% and 0.56%, which are 21.67% and 20% lower than the conventional NARXNN method and 26.56% and 33.33% lower than the BPNN method. Therefore, the effectiveness of the PSO algorithm is verified.

**Table 4.** Summary of MAE and RMSE of different algorithms.

| Cell | BPNN | | NARXNN | | PSO-NARXNN | |
|---|---|---|---|---|---|---|
| | MAE | RMSE | MAE | RMSE | MAE | RMSE |
| 1 | 0.06 | 0.33 | 0.14 | 0.19 | 0.09 | 0.116 |
| 2 | 0.95 | 1.10 | 0.88 | 1.00 | 0.84 | 0.95 |
| 3 | 0.45 | 0.88 | 0.27 | 0.32 | 0.17 | 0.22 |
| 4 | 0.23 | 0.44 | 0.39 | 0.45 | 0.28 | 0.36 |
| 5 | 0.71 | 0.88 | 0.93 | 1.09 | 0.85 | 1.01 |
| 6 | 0.76 | 0.95 | 0.69 | 0.80 | 0.65 | 0.74 |
| 7 | 0.84 | 0.98 | 0.66 | 0.76 | 0.55 | 0.66 |
| 8 | 0.52 | 0.64 | 0.43 | 0.51 | 0.32 | 0.41 |
| Average | 0.64 | 0.84 | 0.60 | 0.70 | 0.47 | 0.56 |

In summary, by comparing the BPNN and conventional NARXNN methods, the effectiveness of the multi-feature extraction strategy is further demonstrated. Moreover, by comparing the conventional NARXNN and PSO-NARXNN methods, the effectiveness of the optimal values by the PSO algorithm is verified.

4.1.4. Results of Other Experimental Groups

As explained at the beginning of Section 4, there are eight experimental groups to comprehensively evaluate the effectiveness of the proposed multi-feature extraction strategy and PSO-NARXNN model. Section 4.1.2 and 4.1.3 has discussed the experimental results of group 1, which uses the aging dataset of cell 1 to train the model, and then the aging datasets of the other seven cells to test it. In this section, the results of other experimental groups are explained. Owing to space limitation, the average *MAE* and *RMSE* of 7 experimental groups are given in Table 5, and Figure 15 intuitively compares them.

Table 5. Summary of *MAE* and *RMSE* of other experimental groups.

|  | Group 1 | Group 2 | Group 3 | Group 4 | Group 5 | Group 6 | Group 7 | Group 8 | Average |
|---|---|---|---|---|---|---|---|---|---|
| *MAE* | 0.47 | 0.71 | 0.49 | 0.55 | 0.69 | 0.57 | 0.74 | 0.48 | 0.59 |
| *RMSE* | 0.56 | 0.77 | 0.58 | 0.65 | 0.79 | 0.63 | 0.80 | 0.54 | 0.66 |

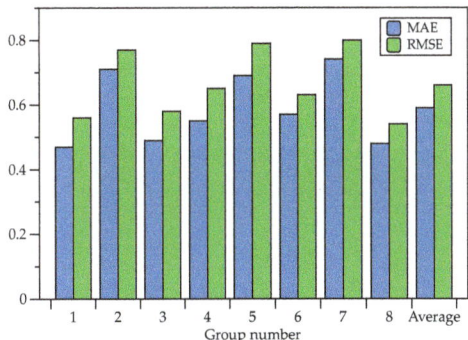

**Figure 15.** Comparison of MAE and RMSE of other experimental groups.

As shown in Figure 15, it can be seen that average *MAE* and *RMSE* are all less than 1%, which means that no matter which cell aging dataset is used to train the PSO-NARXNN SOH estimation model, it can achieve excellent estimation accuracy for other aging datasets. This result demonstrates the generalization of the proposed PSO-NARXNN SOH estimation method and proves that the multi-feature extraction strategy is valid for different cells. This conclusion also coincides with the observation in the above two sections. To further validate the superior performance of the proposed PSO-NARXNN method, similar methods, especially those with SOH estimation works based on the Oxford dataset, are compared in Table 6. It is evident from Table 6 that the proposed method has higher or at least comparable accuracy than those of other existing approaches. In addition, as explained in the Introduction, the NARXNN has a more straightforward structure and fewer parameters than the GRU, CNN, and RF regressor.

**Table 6.** Comparison with other similar works.

| Method | Ref. | MAE | RMSE | MaxE |
|---|---|---|---|---|
| GRU-CNN | [53] | 0.62 | - | 1.62 |
| GRU | [49] | 0.73 | 0.86 | 1.75 |
| GPR | [54] | 0.83 | 1.13 | - |
| RF regressor | [17] | 0.64 | 0.70 | - |
| SRU-decoder3 | [55] | 0.46 | 0.51 | 1.02 |
| PSO-NARXNN | - | 0.47 | 0.56 | 1.48 |

Overall, the analysis above comprehensively verifies the effectiveness of the multi-feature extraction strategy and the proposed PSO-NARXNN SOH estimation method on the Oxford aging dataset.

*4.2. Discussion*

The above results demonstrate the validity of the multi-feature extraction strategy and the performance of the PSO-NARXNN for SOH estimation, but there are still some limitations and shortcomings. First, the selected HFs and the proposed method are verified on one type of LiB. The feasibility on other types of LiBs needs to be further validated. Second, it can be concluded from Section 4.1.3 that a simple three-layer BPNN can obtain a relatively satisfactory estimation accuracy when the HFs are reasonable. Therefore, how to further optimize the HFs and then achieve high-precision estimation using only a simple algorithm deserves further investigation. Third, this research only focuses on SOH estimation methods for cell, while SOH estimation methods for the battery pack are not covered. More studies need to determine whether the selected HFs are suitable for battery pack SOH estimation. Additionally, the computation requirements would become larger when applying the cell estimation method to pack estimation. How to maintain a trade-off between model accuracy and complexity deserves further investigation. Therefore, our future work will employ other aging datasets, such as the NASA dataset [56] and CALCE (Center Advanced Life Cycle Engineering) dataset [57], to validate the feasibility of the proposed multi-feature extraction strategy as well as the PSO-NARXNN. In addition, SOH estimation methods for the battery pack will be investigated based on the research in this paper.

## 5. Conclusions

To improve the SOH estimation performance, this paper proposes a multi-feature extraction strategy and PSO-NARXNN for accurate SOH estimation of LiBs. Firstly, eight HFs are extracted from partial voltage, capacity, DT, and IC curves to reflect the battery aging process comprehensively. Then, qualitative and quantitative analyses are used to evaluate the effectiveness of the selected HFs. Second, owing to the advantages of simple structure, easy implementation, and high estimation accuracy, the NARXNN is adopted to build an accurate SOH estimation model. To improve the training efficiency, the PSO algorithm is applied to optimize the hyperparameters of NARXNN, including input delays, feedback delays, and the number of hidden neurons. Finally, the proposed multi-feature extraction strategy and PSO-NARXNN are systematically validated using the Oxford aging dataset. The results show that in comparison to a simple three-layer BPNN and a conventional NARXNN, the proposed PSO-NARXNN can achieve higher accuracy and stronger robustness, where the average MAE and RMSE of eight experimental groups are 0.59% and 0.66%, respectively.

Our future work will use other types of LiBs to validate the proposed multi-feature extraction strategy and PSO-NARXNN method. Moreover, the multi-feature extraction strategy will be further optimized, and a simpler algorithm will be used for accurate SOH estimation. Then, the optimized algorithm will be used for battery pack SOH estimation.

**Author Contributions:** Conceptualization, Z.R. and C.D.; Funding acquisition, C.D. and W.R.; Methodology, Z.R.; Software, Z.R.; Supervision, C.D. and W.R.; Validation, Z.R.; Visualization, Z.R.; Writing—original draft, Z.R.; Writing—review and editing, Z.R., C.D. and W.R. All authors have read and agreed to the published version of the manuscript.

**Funding:** This research was funded by the Key R&D project of Hubei Province China, grant number 2021AAA006; Foshan Xianhu Laboratory of the Advanced Energy Science and Technology Guangdong Laboratory, grant number XHD2020-003.

**Data Availability Statement:** The data of this paper are available on the following website: https://ora.ox.ac.uk/objects/uuid:03ba4b01-cfed-46d3-9b1a-7d4a7bdf6fac (accessed on 2 November 2022).

**Conflicts of Interest:** The authors declare no conflict of interest.

## References

1. Balasingam, B.; Ahmed, M.; Pattipati, K. Battery Management Systems-Challenges and Some Solutions. *Energies* **2020**, *13*, 2825. [CrossRef]
2. Pastor-Fernández, C.; Yu, T.F.; Widanage, W.D.; Marco, J. Critical Review of Non-Invasive Diagnosis Techniques for Quantification of Degradation Modes in Lithium-Ion Batteries. *Renew. Sustain. Energy Rev.* **2019**, *109*, 138–159. [CrossRef]

3. Waldmann, T.; Iturrondobeitia, A.; Kasper, M.; Ghanbari, N.; Aguesse, F.; Bekaert, E.; Daniel, L.; Genies, S.; Gordon, I.J.; Löble, M.W.; et al. Review—Post-Mortem Analysis of Aged Lithium-Ion Batteries: Disassembly Methodology and Physico-Chemical Analysis Techniques. *J. Electrochem. Soc.* **2016**, *163*, A2149–A2164. [CrossRef]
4. Han, X.; Lu, L.; Zheng, Y.; Feng, X.; Li, Z.; Li, J.; Ouyang, M. A Review on the Key Issues of the Lithium Ion Battery Degradation among the Whole Life Cycle. *eTransportation* **2019**, *1*, 100005. [CrossRef]
5. Che, Y.; Deng, Z.; Li, P.; Tang, X.; Khosravinia, K.; Lin, X.; Hu, X. State of Health Prognostics for Series Battery Packs: A Universal Deep Learning Method. *Energy* **2022**, *238*, 121857. [CrossRef]
6. Xiong, R.; Li, L.; Tian, J. Towards a Smarter Battery Management System: A Critical Review on Battery State of Health Monitoring Methods. *J. Power Sources* **2018**, *405*, 18–29. [CrossRef]
7. Ren, Z.; Du, C.; Wu, Z.; Shao, J.; Deng, W. A Comparative Study of the Influence of Different Open Circuit Voltage Tests on Model-Based State of Charge Estimation for Lithium-Ion Batteries. *Int. J. Energy Res.* **2021**, *45*, 13692–13711. [CrossRef]
8. Vennam, G.; Sahoo, A.; Ahmed, S. A Novel Coupled Electro-Thermal-Aging Model for Simultaneous SOC, SOH, and Parameter Estimation of Lithium-Ion Batteries. In Proceedings of the 2022 American Control Conference (ACC), Atlanta, GA, USA, 8–10 June 2022; pp. 5259–5264.
9. Zeng, M.; Zhang, P.; Yang, Y.; Xie, C.; Shi, Y. SOC and SOH Joint Estimation of the Power Batteries Based on Fuzzy Unscented Kalman Filtering Algorithm. *Energies* **2019**, *12*, 3122. [CrossRef]
10. Sui, X.; He, S.; Vilsen, S.B.; Meng, J.; Teodorescu, R.; Stroe, D.I. A Review of Non-Probabilistic Machine Learning-Based State of Health Estimation Techniques for Lithium-Ion Battery. *Appl. Energy* **2021**, *300*, 117346. [CrossRef]
11. Cui, Z.; Wang, C.; Gao, X.; Tian, S. State of Health Estimation for Lithium-Ion Battery Based on the Coupling-Loop Nonlinear Autoregressive with Exogenous Inputs Neural Network. *Electrochim. Acta* **2021**, *393*, 139047. [CrossRef]
12. Liu, D.; Zhou, J.; Liao, H.; Peng, Y.; Peng, X. A Health Indicator Extraction and Optimization Framework for Lithium-Ion Battery Degradation Modeling and Prognostics. *IEEE Trans. Syst. Man, Cybern. Syst.* **2015**, *45*, 915–928. [CrossRef]
13. Cao, M.; Zhang, T.; Wang, J.; Liu, Y. A Deep Belief Network Approach to Remaining Capacity Estimation for Lithium-Ion Batteries Based on Charging Process Features. *J. Energy Storage* **2022**, *48*, 103825. [CrossRef]
14. Yang, D.; Zhang, X.; Pan, R.; Wang, Y.; Chen, Z. A Novel Gaussian Process Regression Model for State-of-Health Estimation of Lithium-Ion Battery Using Charging Curve. *J. Power Sources* **2018**, *384*, 387–395. [CrossRef]
15. Li, X.; Wang, Z.; Yan, J. Prognostic Health Condition for Lithium Battery Using the Partial Incremental Capacity and Gaussian Process Regression. *J. Power Sources* **2019**, *421*, 56–67. [CrossRef]
16. Wang, Z.; Yuan, C.; Li, X. Lithium Battery State-of-Health Estimation via Differential Thermal Voltammetry with Gaussian Process Regression. *IEEE Trans. Transp. Electrif.* **2021**, *7*, 16–25. [CrossRef]
17. Lin, M.; Wu, D.; Meng, J.; Wu, J.; Wu, H. A Multi-Feature-Based Multi-Model Fusion Method for State of Health Estimation of Lithium-Ion Batteries. *J. Power Sources* **2022**, *518*, 230774. [CrossRef]
18. Li, X.; Yuan, C.; Li, X.; Wang, Z. State of Health Estimation for Li-Ion Battery Using Incremental Capacity Analysis and Gaussian Process Regression. *Energy* **2020**, *190*, 116467. [CrossRef]
19. Zhao, Q.; Jiang, H.; Chen, B.; Wang, C.; Chang, L. Research on the SOH Prediction Based on the Feature Points of Incremental Capacity Curve. *J. Electrochem. Soc.* **2021**, *168*, 110554. [CrossRef]
20. Yang, S.; Luo, B.; Wang, J.; Kang, J.; Zhu, G. State of Health Estimation for Lithium-Ion Batteries Based on Peak Region Feature Parameters of Incremental Capacity Curve. *Diangong Jishu Xuebao/Transactions China Electrotech. Soc.* **2021**, *36*, 2277–2287. [CrossRef]
21. Zhou, R.; Zhu, R.; Huang, C.G.; Peng, W. State of Health Estimation for Fast-Charging Lithium-Ion Battery Based on Incremental Capacity Analysis. *J. Energy Storage* **2022**, *51*, 104560. [CrossRef]
22. Zhang, S.; Zhai, B.; Guo, X.; Wang, K.; Peng, N.; Zhang, X. Synchronous Estimation of State of Health and Remaining Useful Lifetime for Lithium-Ion Battery Using the Incremental Capacity and Artificial Neural Networks. *J. Energy Storage* **2019**, *26*, 100951. [CrossRef]
23. Lyu, Z.; Wang, G.; Tan, C. A Novel Bayesian Multivariate Linear Regression Model for Online State-of-Health Estimation of Lithium-Ion Battery Using Multiple Health Indicators. *Microelectron. Reliab.* **2022**, *131*, 114500. [CrossRef]
24. Yang, D.; Wang, Y.; Pan, R.; Chen, R.; Chen, Z. State-of-Health Estimation for the Lithium-Ion Battery Based on Support Vector Regression. *Appl. Energy* **2018**, *227*, 273–283. [CrossRef]
25. Cao, M.; Zhang, T.; Yu, B.; Liu, Y. A Method for Interval Prediction of Satellite Battery State of Health Based on Sample Entropy. *IEEE Access* **2019**, *7*, 141549–141561. [CrossRef]
26. Lin, M.; Zeng, X.; Wu, J. State of Health Estimation of Lithium-Ion Battery Based on an Adaptive Tunable Hybrid Radial Basis Function Network. *J. Power Sources* **2021**, *504*, 230063. [CrossRef]
27. Fan, L.; Wang, P.; Cheng, Z. A Remaining Capacity Estimation Approach of Lithium-Ion Batteries Based on Partial Charging Curve and Health Feature Fusion. *J. Energy Storage* **2021**, *43*, 103115. [CrossRef]
28. Kashkooli, A.G.; Fathiannasab, H.; Mao, Z.; Chen, Z. Application of Artificial Intelligence to State-of-Charge and State-of-Health Estimation of Calendar-Aged Lithium-Ion Pouch Cells. *J. Electrochem. Soc.* **2019**, *166*, A605–A615. [CrossRef]
29. Ren, Z.; Du, C. State of Charge Estimation for Lithium-Ion Batteries Using Extreme Learning Machine and Extended Kalman Filter. *IFAC Pap.* **2022**, *55*, 197–202. [CrossRef]

30. Mao, L.; Hu, H.; Chen, J.; Zhao, J.; Qu, K.; Jiang, L. Online State of Health Estimation Method for Lithium-Ion Battery Based on CEEMDAN for Feature Analysis and RBF Neural Network. *IEEE J. Emerg. Sel. Top. Power Electron.* **2021**, *6777*. [CrossRef]
31. Kim, S.; Choi, Y.Y.; Kim, K.J.; Choi, J. Il Forecasting State-of-Health of Lithium-Ion Batteries Using Variational Long Short-Term Memory with Transfer Learning. *J. Energy Storage* **2021**, *41*, 102893. [CrossRef]
32. Rouhi Ardeshiri, R.; Ma, C. Multivariate Gated Recurrent Unit for Battery Remaining Useful Life Prediction: A Deep Learning Approach. *Int. J. Energy Res.* **2021**, *45*, 16633–16648. [CrossRef]
33. Yang, Y. A Machine-Learning Prediction Method of Lithium-Ion Battery Life Based on Charge Process for Different Applications. *Appl. Energy* **2021**, *292*, 116897. [CrossRef]
34. Liu, K.; Li, Y.; Hu, X.; Lucu, M.; Widanage, W.D. Gaussian Process Regression with Automatic Relevance Determination Kernel for Calendar Aging Prediction of Lithium-Ion Batteries. *IEEE Trans. Ind. Inform.* **2020**, *16*, 3767–3777. [CrossRef]
35. Li, R.; Li, W.; Zhang, H. State of Health and Charge Estimation Based on Adaptive Boosting Integrated with Particle Swarm Optimization/Support Vector Machine (AdaBoost-PSO-SVM) Model for Lithium-Ion Batteries. *Int. J. Electrochem. Sci.* **2022**, *17*, 1–17. [CrossRef]
36. Qin, P.; Zhao, L.; Liu, Z. State of Health Prediction for Lithium-Ion Battery Using a Gradient Boosting-Based Data-Driven Method. *J. Energy Storage* **2022**, *47*, 103644. [CrossRef]
37. Li, X.; Yuan, C.; Wang, Z. State of Health Estimation for Li-Ion Battery via Partial Incremental Capacity Analysis Based on Support Vector Regression. *Energy* **2020**, *203*, 117852. [CrossRef]
38. Guo, Y.F.; Huang, K.; Hu, X.Y. A State-of-Health Estimation Method of Lithium-Ion Batteries Based on Multi-Feature Extracted from Constant Current Charging Curve. *J. Energy Storage* **2021**, *36*, 102372. [CrossRef]
39. Sun, W.; Qiu, Y.; Sun, L.; Hua, Q. Neural Network-Based Learning and Estimation of Battery State-of-Charge: A Comparison Study between Direct and Indirect Methodology. *Int. J. Energy Res.* **2020**, *44*, 10307–10319. [CrossRef]
40. Hannan, M.A.; Lipu, M.S.H.; Hussain, A.; Ker, P.J.; Mahlia, T.M.I.; Mansor, M.; Ayob, A.; Saad, M.H.; Dong, Z.Y. Toward Enhanced State of Charge Estimation of Lithium-Ion Batteries Using Optimized Machine Learning Techniques. *Sci. Rep.* **2020**, *10*, 4687. [CrossRef]
41. Hossain Lipu, M.S.; Hussain, A.; Saad, M.H.M.; Ayob, A.; Hannan, M.A. Improved Recurrent NARX Neural Network Model for State of Charge Estimation of Lithium-Ion Battery Using Pso Algorithm. In Proceedings of the 2018 IEEE Symposium on Computer Applications & Industrial Electronics (ISCAIE), Penang, Malaysia, 28–29 April 2018; pp. 354–359. [CrossRef]
42. Wang, Q.; Gu, H.; Ye, M.; Wei, M.; Xu, X. State of Charge Estimation for Lithium-Ion Battery Based on NARX Recurrent Neural Network and Moving Window Method. *IEEE Access* **2021**, *9*, 83364–83375. [CrossRef]
43. Khaleghi, S.; Karimi, D.; Beheshti, S.H.; Hosen, M.S.; Behi, H.; Berecibar, M.; Van Mierlo, J. Online Health Diagnosis of Lithium-Ion Batteries Based on Nonlinear Autoregressive Neural Network. *Appl. Energy* **2021**, *282*, 116159. [CrossRef]
44. Ren, X.; Liu, S.; Yu, X.; Dong, X. A Method for State-of-Charge Estimation of Lithium-Ion Batteries Based on PSO-LSTM. *Energy* **2021**, *234*, 121236. [CrossRef]
45. Zhang, L.; Zheng, M.; Du, D.; Li, Y.; Fei, M.; Guo, Y.; Li, K. State-of-Charge Estimation of Lithium-Ion Battery Pack Based on Improved RBF Neural Networks. *Complexity* **2020**, *2020*, 8840240. [CrossRef]
46. Hossain Lipu, M.S.; Hannan, M.A.; Hussain, A.; Saad, M.H.; Ayob, A.; Uddin, M.N. Extreme Learning Machine Model for State-of-Charge Estimation of Lithium-Ion Battery Using Gravitational Search Algorithm. *IEEE Trans. Ind. Appl.* **2019**, *55*, 4225–4234. [CrossRef]
47. Christoph, R.B. Diagnosis and Prognosis of Degradation in Lithium-Ion Batteries. Ph.D. Thesis, Department of Engineering Science, University of Oxford, Oxford, UK, 2017.
48. Birkl, C.R.; McTurk, E.; Roberts, M.R.; Bruce, P.G.; Howey, D.A. A Parametric Open Circuit Voltage Model for Lithium Ion Batteries. *J. Electrochem. Soc.* **2015**, *162*, A2271–A2280. [CrossRef]
49. Chen, Z.; Zhao, H.; Zhang, Y.; Shen, S.; Shen, J.; Liu, Y. State of Health Estimation for Lithium-Ion Batteries Based on Temperature Prediction and Gated Recurrent Unit Neural Network. *J. Power Sources* **2022**, *521*, 230892. [CrossRef]
50. Jordan, M.I. *Serial Order: A Parallel Distributed Processing Approach*; Ies Report 8604; Institute for Cognitive Science University of California: San Diego, CA, USA, 1986.
51. Lipu, M.S.H.; Hannan, M.A.; Hussain, A.; Saad, M.H.M.; Ayob, A.; Blaabjerg, F. State of Charge Estimation for Lithium-Ion Battery Using Recurrent NARX Neural Network Model Based Lighting Search Algorithm. *IEEE Access* **2018**, *6*, 28150–28161. [CrossRef]
52. Kennedy, J.; Eberhart, R. Particle Swarm Optimization. In Proceedings of the ICNN'95–International Conference on Neural Networks, Perth, WA, Australia, 27 November–01 December 1995; Volume 4, pp. 1942–1948.
53. Fan, Y.; Xiao, F.; Li, C.; Yang, G.; Tang, X. A Novel Deep Learning Framework for State of Health Estimation of Lithium-Ion Battery. *J. Energy Storage* **2020**, *32*, 101741. [CrossRef]
54. Goh, H.H.; Lan, Z.; Zhang, D.; Dai, W.; Kurniawan, T.A.; Goh, K.C. Estimation of the State of Health (SOH) of Batteries Using Discrete Curvature Feature Extraction. *J. Energy Storage* **2022**, *50*, 104646. [CrossRef]
55. Gong, Q.; Wang, P.; Cheng, Z. An Encoder-Decoder Model Based on Deep Learning for State of Health Estimation of Lithium-Ion Battery. *J. Energy Storage* **2022**, *46*, 103804. [CrossRef]

56. Saha, B.; Goebel, K. *Battery Data Set*; NASA Ames Prognostics Data Repository; NASA Ames: Moffett Field, CA, USA, 2007. Available online: http://ti.arc.nasa.gov/project/prognostic-data-repository (accessed on 5 November 2022).
57. University of Maryland Battery Data | Center for Advanced Life Cycle Engineering (CALCE). Available online: https://calce.umd.edu/battery-data (accessed on 21 September 2021).

**Disclaimer/Publisher's Note:** The statements, opinions and data contained in all publications are solely those of the individual author(s) and contributor(s) and not of MDPI and/or the editor(s). MDPI and/or the editor(s) disclaim responsibility for any injury to people or property resulting from any ideas, methods, instructions or products referred to in the content.

*Article*

# Online State of Health Estimation of Lithium-Ion Batteries Based on Charging Process and Long Short-Term Memory Recurrent Neural Network

Kang Liu [1,*], Longyun Kang [1] and Di Xie [1,2]

1. School of Electric Power, South China University of Technology, Guangzhou 510641, China
2. Guangdong Hynn Technology Co., Ltd., Dongguan 518109, China
* Correspondence: epkangliu@mail.scut.edu.cn; Tel.: +86-134-1816-5071

**Abstract:** Accurate state of health (SOH) estimation is critical to the operation, maintenance, and replacement of lithium-ion batteries (LIBs), which have penetrated almost every aspect of our life. This paper introduces a new approach to accurately estimate the SOH for rechargeable lithium-ion batteries based on the corresponding charging process and long short-term memory recurrent neural network (LSTM-RNN). In order to learn the mapping function without employing battery models and filtering techniques, the LSTM-RNN is initially fed into the health indicators (HIs) extracted from the charging process and trained to encode the dependencies of the related data sequence. Subsequently, the trained LSTM-RNN can properly estimate online SOHs of LIBs using extracted HIs. We experiment on two public datasets for model construction, validation, and comparison. Conclusively, the trained LSTM-RNN achieves an overall root mean square error (RMSE) lower than 1% on the cases with the same discharging current rate and an RMSE of 1.1198% above 80% SOH on another testing case that underwent a different discharging current rate.

**Keywords:** state of health (SOH); lithium-ion batteries (LIBs); long short-term memory recurrent neural network (LSTM-RNN); health indicators (HIs)

**Citation:** Liu, K.; Kang, L.; Xie, D. Online State of Health Estimation of Lithium-Ion Batteries Based on Charging Process and Long Short-Term Memory Recurrent Neural Network. *Batteries* **2023**, *9*, 94. https://doi.org/10.3390/batteries9020094

Academic Editors: Carlos Ziebert and Pascal Venet

Received: 30 November 2022
Revised: 12 January 2023
Accepted: 24 January 2023
Published: 30 January 2023

**Copyright:** © 2023 by the authors. Licensee MDPI, Basel, Switzerland. This article is an open access article distributed under the terms and conditions of the Creative Commons Attribution (CC BY) license (https:// creativecommons.org/licenses/by/ 4.0/).

## 1. Introduction

Lithium-ion batteries are deployed in many fields, such as consumer electronics, electric vehicles (EV), and aerospace technologies, due to their high energy density, long lifetime, environmental friendliness, and low self-discharge rate [1–3]. However, as the charge/discharge cycle increases, the material inside the lithium-ion battery is irreversibly consumed [4,5], which means that the cell ages, manifesting in decreasing capacity, declining power, and thermal instability. When the battery ages beyond a certain point, the battery performance consequently becomes unreliable and is prone to failure. State of health (SOH) is proposed to represent the battery aging degree and can reflect the total capacity reduction and resistance increment. Most companies set 80% SOH as the criterion for the decommission of used batteries [6]. Additionally, a reliable SOH estimation method is crucial for secure and reliable battery operation and is the backbone of the battery management system (BMS) [7].

This paper utilizes the capacity to indicate the battery state of health (SOH), which can be defined in terms of battery capacity [8,9], internal resistance [10–12], and peak power [13]. Technically, battery SOH estimation methods mainly fall into two categories. In the first category, SOH estimation is based on battery models, including the equivalent circuit models [14,15], electrochemical models [16–18], and empirical models [19,20], in combination with advanced filter techniques such as the particle filter (PF) and Kalman filter (KF) techniques. Bi et al. [15] proposed a second-order equivalent circuit model of an RC circuit for battery packs and subsequently developed a genetic resampling particle filter (GPF) technique to cope with the inaccuracy of the equivalent circuit model. In addition,

Li et al. [17] developed an SP-based degradation model including solid electrolyte interface (SEI) layer formation. This model can quickly estimate capacity fade and voltage profile changes with high accuracy. Additionally, Guha et al. [19] obtained a degradation model to monitor the SOH of a battery by fusing the capacity degradation model and an empirical model for internal resistance growth. In [21], Ossai et al. estimated the SOH of batteries using the Weibull distribution function based on a nonlinear mixed effect degradation model framework. Since lithium-ion batteries have intricate electrochemical properties and complex aging mechanisms, the model-based approach requires computationally intensive algorithms to accurately estimate SOH [9]. Moreover, the identified parameters always need to be modified according to the different working conditions and battery types, which obstructs the promotion of related model-based approaches [22,23].

The second category of SOH estimating methods is built on data-driven (machine learning) methodologies. In contrast to model-based approaches, data-driven methods are dependent on the offline charging/discharging data to learn the mapping function to estimate the SOH, ignoring the complex aging mechanisms and intricate internal electrochemical properties of lithium-ion batteries. Due to significant strides made in the field of machine learning, these data-driven approaches, mainly including the support vector machine (SVM) [24,25], Gaussian process regression (GPR) [26–28], and the neural network (NN) [9,29–32], typically produce more accurate SOH estimates. In [24], Feng et al. built a predictive diagnosis model based on a support vector machine whose coefficients for cells are identified by determining the support vectors. Richardson et al. [27] proposed Gaussian process (GP) regression for estimating the SOH of batteries and highlighted the advantages of GPs over other SOH forecasting approaches. To obtain more accurate SOH estimates, the effective health indicators fed into the algorithms are initially extracted from the charge/discharge data. Zhao et al. [25] employed a relevance vector machine (RVM) to fit the mapping function between the five types of extracted health indicators and SOH. Wang et al. [28] initially proposed an accurate SOH prediction model using multi-output Gaussian process regression (MOGPR) and subsequently performed the SOH estimation based on extracted health indicators (HIs). Li et al. [9] utilized the convolutional neural network for SOH estimation based on the battery's charging current, voltage, and temperature. Based on an incremental capacity curve, Lin et al. [29] utilized a back-propagation neural network to develop an SOH hybrid estimation method. In [31], Shahriari et al. obtained SOH estimates based on the relationship between health indicators (HIs) from the state of charge (SOC) and the battery open-circuit voltage. Additionally, Liu et al. [32] extracted partial segments of charging and discharging data as health indicators for SOH estimation.

By establishing a non-linear mapping function between the input vectors and SOH, data-driven methods can accurately estimate battery SOH. Since the cyclic charge/discharge data can be viewed as a series of time series data, the sequence prediction problem of SOH estimation can be tackled using RNNs with a "memory" property [33]. To address the gradient vanishing problem of convolutional recurrent neural networks (RNN) [34], the LSTM-RNN is designed to learn long-term dependencies by remembering information over long periods [35]. Due to the suitability of this characteristic for addressing the time series predicting problem, this paper employs a vanilla LSTM-RNN with one single hidden layer to estimate the online SOH with the stable and monolithic charging process of lithium-ion batteries. Specifically, the main advantage of this method is that it can adequately exploit time series characteristics to learn the long-term dependencies from the historical cyclic data. With only one single hidden layer on the LSTM-RNN, this method can be achieved with less model complexity and fewer parameter sets compared with other data-driven approaches. Another advantage is that it can directly map battery measurement signals such as voltage and current to the online SOH, avoiding the inference algorithms and intensively computational filter techniques used in model-based SOH estimators.

After a brief introduction, the second section will introduce the battery cyclic datasets and the health indicator extraction process. The third section will elaborate on the detailed LSTM-RNN for online SOH estimation. The fourth section will describe the procedure for the LSTM-RNN application and show the online SOH estimation results on the introduced battery datasets, and this will be followed by the conclusions in Section 5. All abbreviations are explained in Table A1 of Appendix A.

## 2. Battery Datasets and Health Indicator Extraction

This paper utilizes two public cyclic aging datasets, one from the data repository of the NASA Ames Prognostics Center of Excellence (PCoE) [36] and another provided by the Center for Advanced Life Cycle Engineering (CALCE) [37,38] at the University of Maryland, to verify the effectiveness and performance of the proposed SOH estimator. The SOH is defined as the ratio of the present capacity to the nominal capacity and can be expressed as follows:

$$\text{SOH} = \frac{C_P}{C_N} \times 100\% \quad (1)$$

where $C_N$ denotes the nominal capacity and $C_P$ is the present capacity of the battery.

### 2.1. Description of NASA Battery Dataset

The NASA battery datasets, regarding several commercially available lithium-ion 18 650-sized rechargeable batteries, was collected from a battery prognostics test bed. The test bed setup included a power supply, a programmable load, a voltmeter, a thermocouple sensor, and an environmental chamber to regulate and stabilize the temperature [26]. Among the six battery datasets, the first dataset includes three batteries (labeled B0005, B0006, and B0007) considered suitable for this battery state of health estimation study. This set of cells was run through three operational profiles (charge, discharge, and impedance) at room temperature. As is shown in Figure 1a, the charge was carried out in a constant current–constant voltage (CC–CV) mode where the constant current was 1.5 A, the voltage was 4.2 V, and the constant voltage (CV) mode continued until the charging current dropped to 20 mA. Discharge was carried out at a constant current (CC) level of 2 A until the voltage fell to 2.7 V, 2.5 V, and 2.2 V for batteries B0005, B0006, and B0007, respectively. The charge/discharge cycle was repeated to accelerate the aging until the batteries reached their end-of-life (EOF) criterion of a 30% reduction in rated capacity (from 2 Ah to 1.4 Ah), and all the SOH degradation curves are shown in Figure 1b.

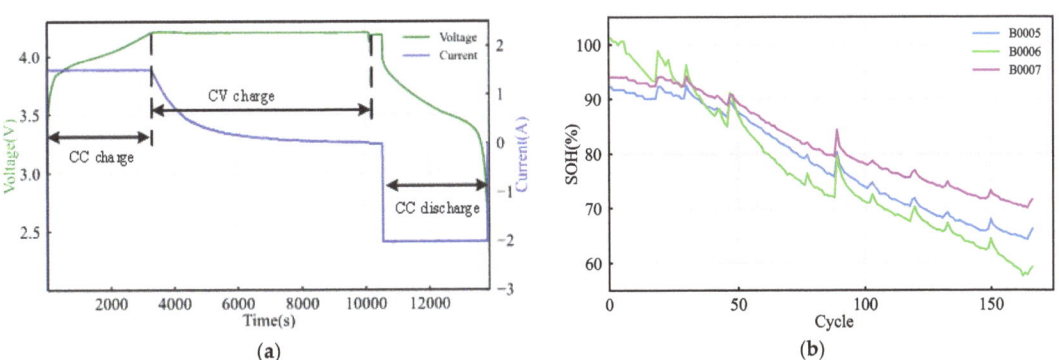

**Figure 1.** (a) Battery terminal voltage and charge/discharge current during one cycle; (b) the curves of degradation capacity.

### 2.2. Description of CALCE Battery Dataset

The second battery dataset, from the Center for Advanced Life Cycle Engineering (CALCE) at the University of Maryland, was obtained from the implementation of cyclic battery testing on the Arbin200 battery testing system at room temperature. Specifically, we employ a set of five CS2 cells labeled CS2-33, CS2-35, CS2-36, CS2-37, and CS2-38 to verify the effectiveness of the LSTM-RNN estimating model. Each cell underwent the same

CC–CV charging mode in which the constant current rate was 0.5 C (this indicates that the charging current was 0.55 A), and the constant voltage was sustained at 4.2 V until the charge current fell below 0.05 A. As for the discharge process, CS2-33 was cycled with a constant discharge current of 0.55 A, which indicates that the discharge current rate was 0.5 C. The rest of the cells (CS2-35, CS2-36, CS2-37, and CS2-38) were discharged with a constant current rate of 1 C. The materials of the cells consisted of graphite on the anode and $LiCoO_2$ on the cathode. Figure 2 represents the SOH degradation curves of all the batteries whose nominal capacity is 1.1 Ah.

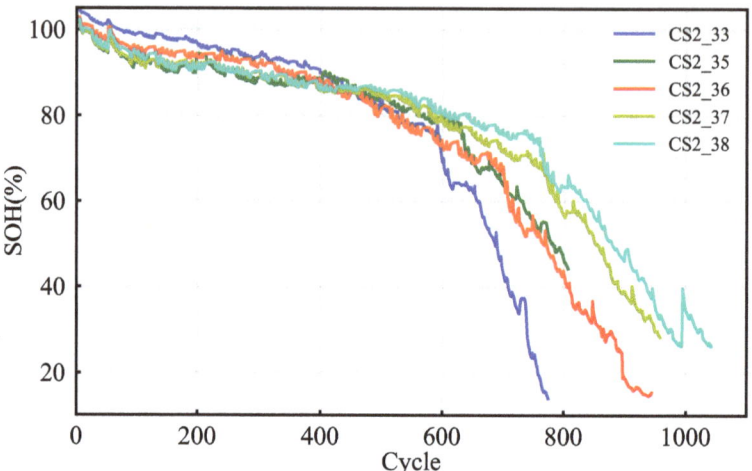

**Figure 2.** The SOH degradation curves during the whole life cycle.

*2.3. Health Indicator Extraction*

In some articles, health indicators are extracted from the discharging process [39], but this is not practical due to the complex and inconsistent discharging scenario, especially for electric vehicles (EVs) [40]. Therefore, in this paper we extract the health indicators from the controllable and monolithic charging process to estimate the current online SOH.

Figure 3 represents the related terminal voltage and current curves for the NASA batteries during the CC–CV charging process. Figure 3a shows the terminal voltage responses extracted from a partial voltage segment of a single cell during different cycles. The terminal voltage gets higher and reaches the cut-off voltage earlier as the cycle number increases. Therefore, the voltage integration from the 3.8–4.2 V terminal voltage varies regularly with the number of cycles. Consequently, this partial voltage integration can indicate the cell's health status, which is denoted as $HI_v$ and calculated as follows:

$$HI_v = \int_{t=t_0}^{t=t_1} v(t)dt \qquad (2)$$

where $v(t)$ represents the terminal voltage function versus time, and $t_0$ and $t_1$ denote the moments at which the voltage equals 3.8 V and 4.2 V, respectively. The plots of $HI_v$ values versus the cycles for different cells from the NASA repository are represented in Figure 4a.

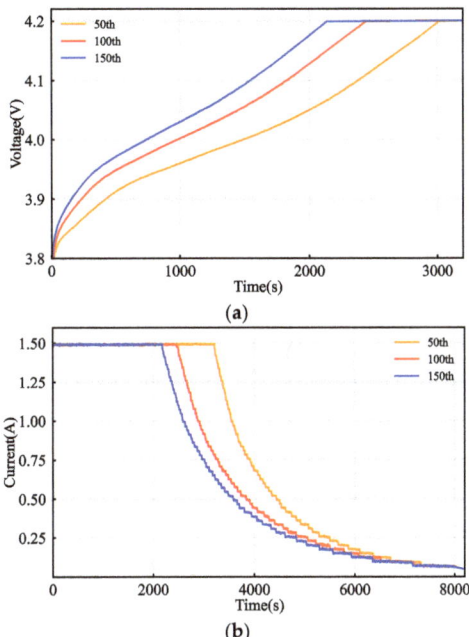

**Figure 3.** (**a**) Charge voltage curves from 3.8 V to 4.2 V during different cycles; (**b**) current curves during the whole charging process during different cycles.

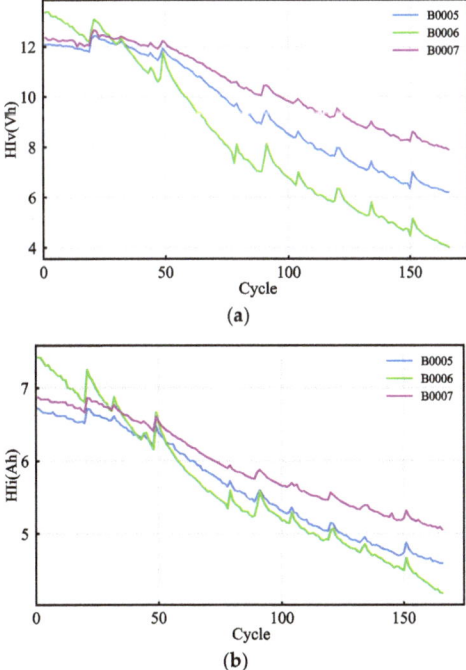

**Figure 4.** (**a**) The values of the $HI_v$ versus the cycle number; (**b**) the values of the $HI_i$ versus the cycle number.

Besides the charge terminal voltage, Figure 3b compares the whole current curves during different cycles and shows that the corresponding current decreases as the number of cycles increases. Consequently, the integration, calculated from the charging process, can be viewed as a health indicator and denoted as $HI_i$. This current integration is the charging capacity and can be calculated as follows:

$$HI_i = \int_{t=t_0}^{t=t_1} i(t)dt \tag{3}$$

where $i(t)$ represents the corresponding charge current, and $t_0$ and $t_1$ denote the charge start and end moments, respectively. The $HI_i$ curves for different cells from the NASA repository are shown in Figure 4b.

As the partial charge voltage segment is more easily accessible than the whole charging current, we only employ the $HI_v$ as the health indicator for the CALCE batteries to avoid current sampling interference. Figure 5a illustrates the corresponding terminal voltage curves during different cycles of battery CS2-35, and the corresponding $HI_v$ values versus the cycle number are represented in Figure 5b. As can be seen in Figure 5b, the $HI_v$ values degrade with more noise than the corresponding SOH values compared with those in Figure 2.

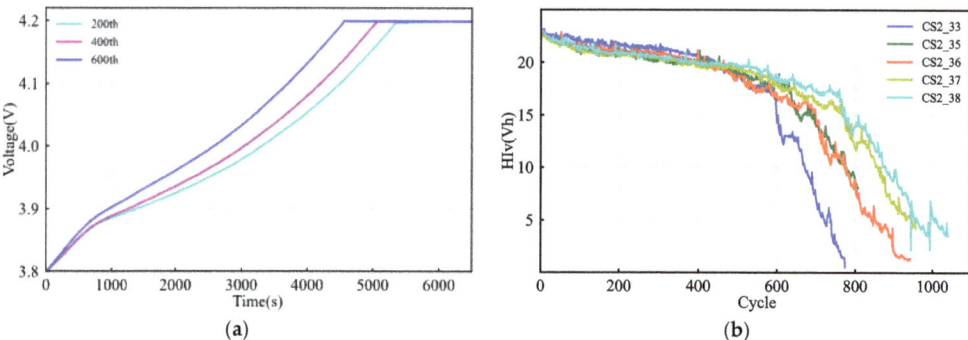

**Figure 5.** (a) Charge voltage curves from 3.8 V to 4.2 V during three different cycles for the CS2-35 battery; (b) the curves of the values of the health indicator ($HI_v$) versus the cycle number for the CALCE battery dataset.

## 3. LSTM-RNN Algorithm

Due to their "memory" characteristics, RNNs are suitable for this type of time-series data estimation [33]. However, conventional RNNs have issues in learning long-term dependencies where the gradients may either vanish or explode during backpropagation [34]. To address this problem, an LSTM-RNN was proposed to capture long-term dependencies within the sequence by remembering information over long periods [35]. Therefore, this paper utilizes an LSTM-RNN to estimate the SOH of lithium-ion batteries.

Figure 6 schematically illustrates the network architecture of LSTM-RNN, which can work as a nonlinear dynamic system by mapping input vectors to output sequences. The LSTM cell is equipped with a memory cell $c_k$, which is the key part of the structure and stores the long-term dependencies. In addition, the input, output, and forget gates are distinctive features of the LSTM-RNN and can regulate the information flow. Each gate is a sigmoid unit ($\sigma$) that is activated from the hidden layer at the last time step $h_{k-1}$ and from the present input layer $\psi_k$. The constructed LSTM-RNN can be represented by the composite function below:

$$\begin{aligned}
i_k &= \sigma(W_{\psi i}\psi_k + W_{hi}h_{k-1} + b_i) \\
f_k &= \sigma(W_{\psi f}\psi_k + W_{hf}h_{k-1} + b_f) \\
c_k &= f_k c_{k-1} + i_k \tanh(W_{\psi c}\psi_k + W_{hc}h_{k-1} + b_c) \\
o_k &= \sigma(W_{\psi o}\psi_k + W_{ho}h_{k-1} + b_o) \\
h_k &= o_k \tanh(c_k)
\end{aligned} \tag{4}$$

where $h_0$ is the initial hidden state, $\sigma$ denotes the sigmoid function, and $i, f, c$, and $o$ are the input gate, forget gate, memory cell, and output gate, respectively. These gates can inhibit the flow of information by setting the value of the sigmoid unit as 0 from the current input layer $\psi_k$ and the hidden layer at the last time step $h_{k-1}$. The $W$ and $b$ values are the network weights and biases, respectively. The subscripts of $W$ describe the interaction occurring between the two corresponding components, e.g., $W_{\psi i}$ denotes the input–input gate matrix and $W_{hi}$ denotes the hidden–input gate matrix. At each gate, a bias $b$ is added to the matrix multiplication to increase the computing flexibility.

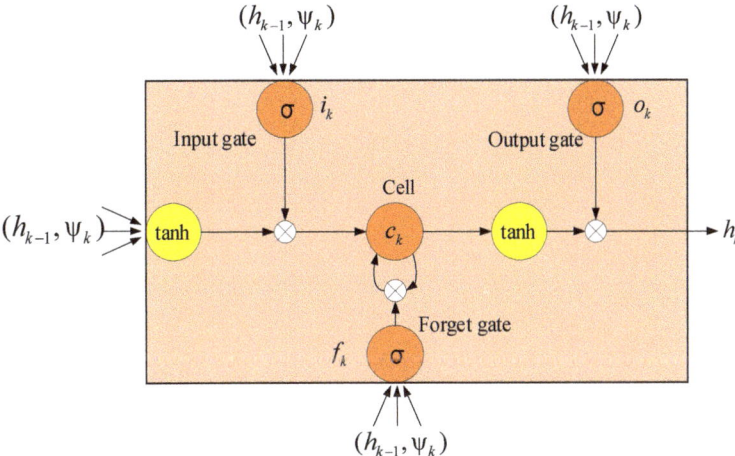

**Figure 6.** Network architecture of the LSTM-RNN.

When we exploit the LSTM-RNN to estimate online SOH, the network initially needs to be trained with a series of sequences from the dataset. In order to take full advantage of the LSTM's ability to capture long-term dependencies, this work utilizes n-step (n equals 10) input vectors to estimate one SOH value. Therefore, a typical dataset used to train the network is given by D = (($\psi_1, \psi_2, \ldots, \psi_n$, SOH$_n$*), ($\psi_2, \psi_3, \ldots, \psi_{n+1}$, SOH$_{n+1}$*), $\ldots$ ($\psi_{k-n+1}, \psi_{k-n+2}, \ldots, \psi_k$, SOH$_k$*), $\ldots$, ($\psi_{N-n+1}, \psi_{N-n+2}, \ldots, \psi_N$, SOH$_N$*)), where SOH$_k$* denotes the ground-true value at cycle $k$ and $\psi_k$ is the vector of inputs at cycle $k$. After the LSTM-RNN, a fully connected layer linearly transforms the hidden state tensor $h_k$ to obtain a single estimated SOH value. The fully connected layer achieving the regression is given by:

$$SOH_k = V_{out}h_k + b_f \tag{5}$$

where $V_{out}$ and $b_f$ are the weight matrix and biases of the fully connected layer, respectively. The whole training and SOH estimating process structure is shown in Figure 7. Consequently, we can estimate the online SOH using the trained LSTM-RNN with the testing input vectors.

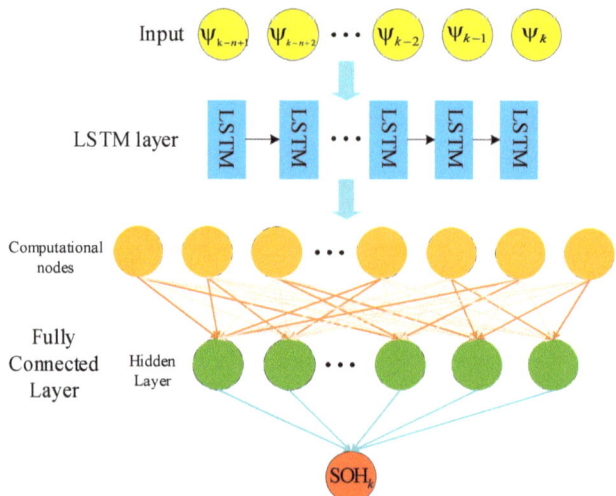

**Figure 7.** SOH estimating process structure.

## 4. SOH Estimation Results and Analysis

This section implements the proposed LSTM-RNN to estimate the online SOH of batteries as they age using the two public datasets introduced above. Considering the stable and monolithic charging scenario, we can extract the health indicators from the partial charging voltage segment and the charging current, respectively, to estimate the corresponding SOH. The $k$th vector of inputs, consisting of the health indicators fed into LSTM-RNN, is described as $\psi_k$, where $k$ denotes the $k$th charging cycle. The following two subsections investigate and verify the performance of the proposed SOH estimation technique using the NASA and CALCE battery datasets, respectively.

### 4.1. SOH Estimation Based on the NASA Battery Dataset

Section 2 introduced the cyclic experiment with the battery datasets. Considering the varying information in the whole life cycle of the batteries, we train the LSTM-RNN using the entire battery samples rather than partial data from the whole cycling life, since this is more compatible with the intended practical application. Specifically, the initial training uses 80% of the battery B0005 and B0006 datasets, whereas cross-validation uses the remaining 20%. The dataset from cell B0007 is for testing. The validation serves the hyper-parameters adjustment, and the testing achieves the performance assessment. After completing the parameter configuration, we retrain the model with the datasets from batteries B0005 and B0006. As mentioned above, the vector of inputs $\psi_k$ fed into the LSTM in the experiment using the NASA dataset is defined as $\psi_k = (HI_v, HI_i)$, where $HI_v$ denotes the voltage integration from the partial voltage segment and $HI_i$ represents the charging capacity in the charging process. All the experiments were implemented using Python 3.9.12 on a laptop equipped with an Intel Core i5-8300H processor.

To perform the online SOH estimation, we explore a vanilla LSTM with one single hidden layer to reduce the model complexity and parameter settings. In the training process, the LSTM network can reach higher predicting accuracy when its layer owns more computational nodes. However, the network is prone to overfitting when the layer is equipped with large computational nodes. Combining with the cross-validation, we set 128 computational nodes in the layer and select the mean square error (MSE) as the optimization goal, which can be calculated as:

$$\text{MSE} = \frac{1}{m}\sum_{i=1}^{m}(\text{SOH}_i - \text{SOH}_i^*) \qquad (6)$$

where $m$ is the number of examples in one batch, $\text{SOH}_i^*$ is the predicted SOH for the $i$th example, and $\text{SOH}_i$ is the $i$th actual value.

For the evaluation of the performance of the algorithms in the testing stage, the error metrics are given in terms of the root mean square error (RMSE) and mean absolute error (MAE) in the following equations:

$$\text{RMSE} = \sqrt{\frac{1}{m}\sum_{i=1}^{m}(\text{SOH}_i - \text{SOH}_i^*)^2} \qquad (7)$$

$$\text{MAE} = \frac{1}{m}\sum_{i=1}^{m}|\text{SOH}_i - \text{SOH}_i^*| \qquad (8)$$

where the root mean square error (RMSE) is sensitive to the large errors, and the mean absolute error reflects how close the estimated SOH is to the real SOH, regardless of the sign.

Having features on a similar scale helps the gradient descent converge more smoothly and quickly toward the minima. We standardize the features by removing the mean and scaling to unit variance to transform the health indicators into a similar scale. The stand score of a sample x is calculated as:

$$z = \frac{(x-u)}{s} \qquad (9)$$

where $u$ is the mean of the training samples and $s$ is the deviation of the training samples. The testing samples are transformed based on the mean and deviation of the training samples.

To minimize the MSE loss, the network weights W and biases b are updated through every training epoch $\epsilon$, which includes one forward and one backward pass. Based on the gradient of the MSE loss function, the Adam optimization algorithm [41] was utilized to update the parameters and is given by the following:

$$\begin{aligned} m_\epsilon &= \beta_1 m_{\epsilon-1} \nabla L(W_{\epsilon-1}) \\ r_\epsilon &= \beta_2 r_{\epsilon-1} \nabla L(W_{\epsilon-1})^2 \\ \widetilde{m}_\epsilon &= m_\epsilon/(1-\beta_1^\epsilon) \\ \widetilde{r}_\epsilon &= r_\epsilon/(1-\beta_2^\epsilon) \\ W_\epsilon &= W_{\epsilon-1} - \alpha \frac{\widetilde{m}_\epsilon}{\widetilde{r}_\epsilon - k} \end{aligned} \qquad (10)$$

where $\beta_1$ and $\beta_2$ are decay rates set to 0.9 and 0.999, respectively, L denotes the loss function, $\alpha$ is the learning rate, and constant term $k$ is set to $10^{-8}$. $W\epsilon$ represents the matrix of network parameters at the current training epoch. More details about the Adam optimization method can be found in [41].

We routinely use cross-validation to evaluate model performance to obtain adequate hyper-parameters. In combination with the cross-validation technique, we set the training epochs as 15,000. Figure 8 displays a plot of the RMSE as a function versus the training epoch. As can be observed, the RMSE drops fast in the first 5000 epochs and almost reaches 0 after 12,000 epochs. Due to the decreasing tendency after 12,000 epochs, 15,000 is considered a reasonable number of training epochs. We choose a batch size of 64 to train the network as a compromise between the large and small batches. All the parameters of the LSTM-RNN are listed in Table 1.

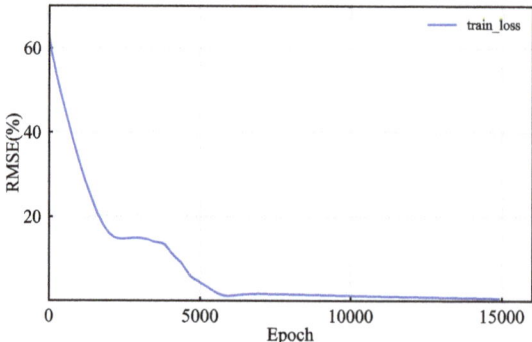

**Figure 8.** RMSEs of the training performance versus the training epochs.

**Table 1.** Parameter settings for the LSTM-RNN.

| Parameter | Value Setting |
|---|---|
| Optimizer | Adam |
| Loss function | MSE |
| Activation function | RELU |
| Computational nodes in one layer | 128 |
| Batch size | 64 |
| Learning rate | 0.00005 |
| Epochs | 15,000 |

In the online SOH estimation, we compare the LSTM-RNN with the gated recurrent unit recurrent neural network (GRU-RNN) and simple recurrent neural network (Sim-RNN) to demonstrate its effectiveness and performance. In order to carry out a fair comparison, the RNN and GRU algorithms are provided with the same structure and parameters as the LSTM, except for the main RNN and GRU working layers. The Sim-RNN and GRU-RNN are given 15,000 and 25,000 epochs, respectively, to achieve converging loss in the training process. The computational time required to train the LSTM-RNN, Sim-RNN, and GRU-RNN is around 18.1 min, 13.8 min, and 30.6 min, respectively. Consequently, more training epochs for the GRU-RNN incur a longer training time.

Figure 9 shows the online SOH estimation results for cell B0007 using the three networks. As can be seen, the SOH predictions of the three networks generally follow the actual SOH values. The estimation errors of the LSTM-RNN, Sim-RNN, and GRU-RNN are plotted in Figure 9d, and all the estimation errors of the three different networks stay approximately within 2%. The test results from the LSTM-RNN, Sim-RNN, and GRU-RNN algorithms are listed in Table 2. The overall RMSE and MAE of the LSTM are 0.5623% and 0.5746%, respectively (slighter smaller than those of the GRU and RNN). Therefore, the proposed SOH estimator can accurately estimate the online SOH with the extracted HIs based on the charging process in this experiment.

**Table 2.** The RMSE and MAE results from the testing of battery B0007.

|  | LSTM (%) | GRU (%) | RNN (%) |
|---|---|---|---|
| RMSE | 0.5623 | 0.6421 | 0.6345 |
| MAE | 0.5746 | 0.7494 | 0.6400 |

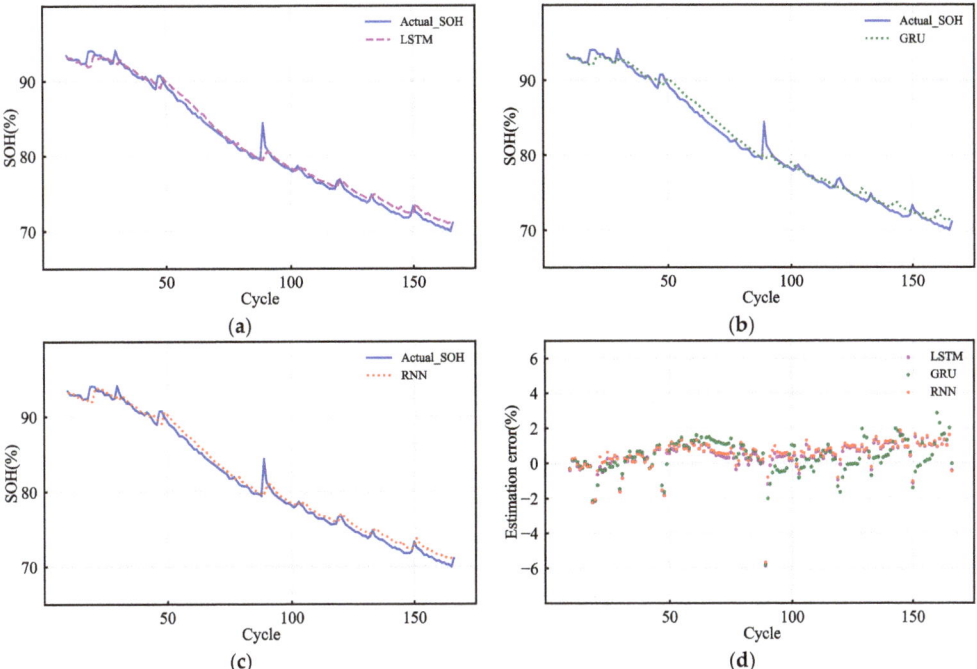

**Figure 9.** (**a**) SOH estimation results of the LSTM-RNN; (**b**) SOH estimation results of the GRU-RNN; (**c**) SOH estimation results of the Sim-RNN; (**d**) estimation errors of the LSTM-RNN, Sim-RNN, and GRU-RNN.

### 4.2. SOH Estimation Based on the CALCE Battery Dataset

To validate the generality of the proposed LSTM-RNN, we further employ the CALCE battery dataset to establish its effectiveness. As mentioned above, the input vector, fed into the LSTM network, only contains one health indicator $HI_v$ in the CALCE battery dataset experiment. Excluding the $HI_i$ optimizes the online SOH estimation process by avoiding interference from the electric current sampling process.

Similarly, the training and cross-validation are performed on the cyclic data of batteries CS2-36 and CS2-38, while the testing is achieved using the remaining battery datasets. The parameters of this LSTM-RNN are virtually the same as those in Table 1, except that the batch size is 128. Additionally, we continue to train the model using 15,000 training epochs. The computation time for training the LSTM-RNN based on cells CS2-36 and CS2-38 is 32.0 min.

Figure 10 shows the estimation performance on batteries CS2-35 and CS2-37, which underwent the 1 C discharging rate. Figure 10a,b represent the corresponding SOH estimation results for batteries CS2-35 and CS2-37. The overall RMSE achieved on the two batteries is 0.9311% and 0.8288%, respectively. The overall RMSE and MAE performance metrics for batteries CS2-35 and CS2-37 are listed in Table 3. Moreover, the estimation errors of each cycle for the two testing cases are plotted in Figure 10c,d, and the errors mainly stay within 3%. Specifically, the estimation root mean square errors above 80% SOH (safe battery operating range) generally fall within 2% and are smaller than those estimated in the final phase. Moreover, the SOH estimation results from the LSTM-RNN eliminate the impact of the input noise from the $HI_v$ extraction process shown in Figure 5b.

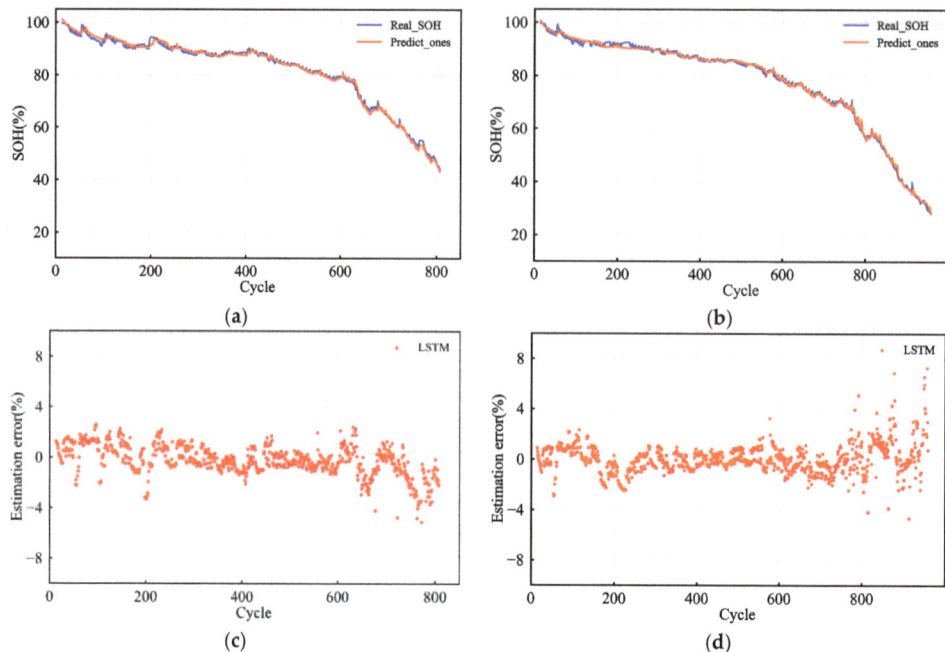

**Figure 10.** SOH estimation results based on the batteries CS2-35 and CS2-37. (**a**) SOH estimation for CS2-35; (**b**) SOH estimation for CS2-37; (**c**) SOH estimation errors of each cycle for battery CS2-35; (**d**) SOH estimation errors of each cycle for battery CS2-37.

**Table 3.** The RMSE and MAE results for the batteries under different charging rates.

| Testing Battery | Discharging Rate | RMSE (%) | MAE (%) |
|---|---|---|---|
| CS2-33 | 0.5 C | 2.038 | 1.4952 |
| CS2-35 | 1 C | 0.9311 | 0.7437 |
| CS2-37 | 1 C | 0.8288 | 0.6373 |

To evaluate the robustness of the trained LSTM-RNN model against different discharging current rates, battery CS2-33, which underwent a 0.5 C discharging rate, was used to validate the estimator. Figure 11a,b show the estimation results and the estimation errors of each cycle, respectively. The SOH estimations above 80% SOH are close to the actual values at different cycles, and the corresponding estimation errors are almost within 3%. However, the SOH estimation errors of the LSTM-RNN model are relatively large for the cycles in which the SOH is below 80%. The overall RMSE of the estimation result for battery CS2-33 is 2.0384%, and the MAE is 1.495%, as listed in Table 3. Since lithium-ion batteries generally operate within a certain range for a safety, evaluating the estimator performance above 80% SOH is more practical and critical. Table 4 shows that the overall RMSE of the estimation results above 80% SOH for the battery CS2-33 is 1.1198%, and the MAE 0.9454%, a result approaching the estimation performance of the developed LSTM-RNN for batteries CS2-35 and CS2-37, which means that the trained LSTM network possesses a certain degree of robustness against the different discharging current rates.

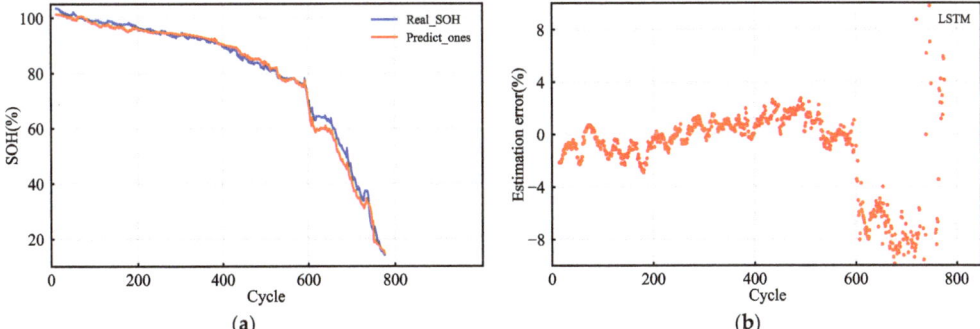

**Figure 11.** SOH estimation results based on the battery CS2-33; (**a**) SOH estimation for CS2-33; (**b**) SOH estimation error for CS2-33.

**Table 4.** The RMSE and MAE results for the different testing cells above 80% SOH.

| Testing Battery | Discharging Rate | RMSE (%) (SOH > 80%) | MAE (%) (SOH > 80%) |
|---|---|---|---|
| CS2-33 | 0.5 C | 1.1198 | 0.9454 |
| CS2-35 | 1 C | 0.9062 | 0.7260 |
| CS2-37 | 1 C | 0.8317 | 0.6370 |

From the preceding experiments and analysis, we can conclude that the proposed LSTM-RNN-based estimator can accurately estimate online SOH with robustness against different discharging current rates for different battery types.

## 5. Conclusions

State of health illustrates the aging of lithium-ion batteries and accurate SOH estimation provides a basis for lithium-ion battery maintenance and replacement. As the characteristics of the LSTM RNN enable it to address the time series predicting problem, this paper mainly uses an LSTM-RNN model to estimate the online SOH of lithium-ion batteries with a stable and monolithic charging process.

In the final analysis, the primary benefit of this approach is that it accurately estimates SOH by effectively leveraging the properties of time series to infer long dependencies from offline battery data. Due to the one hidden layer on the LSTM-RNN, this method can be accomplished with less model complexity and fewer parameter sets than other data-driven approaches. Additionally, the proposed LSTM-RNN can directly map battery measurement signals such as voltage and current to the online SOH, avoiding the inference algorithms and intensively computational filter techniques used in model-based SOH estimators.

Experimental studies using two different battery datasets illustrate the performance and adaptability of this type of data-driven algorithm in the SOH estimation of lithium-ion batteries. In summary, the proposed LSTM-RNN achieves good performance in the NASA battery dataset with an overall RMSE of 0.5623% and shows robustness against the battery discharging rate when applied to the CALCE battery dataset. In addition, the experiment results show that the LSTM-RNN produces more accurate lithium-ion battery SOH estimates than the gated recurrent unit recurrent neural network (GRU-RNN) and simple recurrent neural network (Sim-RNN). Since the SOH estimator is developed based on the charging process rather than the complex discharging scenario, we can use this estimator in many practical applications for online SOH estimation.

**Author Contributions:** Conceptualization, K.L.; methodology, K.L.; software, K.L.; validation, D.X. and L.K.; formal analysis, K.L.; resources, K.L.; data curation, K.L. and D.X.; writing—original draft preparation, K.L.; writing—review and editing, K.L.; supervision, L.K. and D.X.; project administration, L.K.; funding acquisition, L.K. All authors have read and agreed to the published version of the manuscript.

**Funding:** This research was funded by the Guangdong Basic and Applied Basic Research Foundation, grant number No. 2022A1515140009.

**Data Availability Statement:** Not applicable.

**Acknowledgments:** The authors gratefully thank the NASA Ames Centers of Excellence Diagnostic Center and the Center for Advanced Life Cycle Engineering (CALCE) of the University of Maryland for providing the experimental data.

**Conflicts of Interest:** The authors declare no conflict of interest.

## Appendix A

**Table A1.** Abbreviation comparison table.

| Abbreviation | Explanation |
|---|---|
| SOH | State of health |
| LIB | Lithium-ion battery |
| LSTM | Long short-term memory |
| HI | Health indicator |
| SOC | State of charge |
| RMSE | Root mean square error |
| MAE | Mean absolute error |
| EV | Electirc vehicle |
| BMS | Battery manangement system |
| PF | Particle filter |
| KF | Kalma filter |
| SEI | Solid electrolyte interface |
| SVM | Support vector machine |
| GPR | Gaussian process regression |
| NN | Neural network |
| RVM | Relevance vector machine |
| NASA | National Aeronautics and Space Administration |
| PCoE | Prognostics Center of Excellence |
| CALCE | Center for Advanced Life Cycle Engineering |
| CC–CV | Constant current–constant voltage |
| GRU | Gated recurrent unit |
| Sim-RNN | Simple recurrent neural network |

## References

1. Jaguemont, J.; Boulon, L.; Dubé, Y. A comprehensive review of lithium-ion batteries used in hybrid and electric vehicles at cold temperatures. *Appl. Energy* **2016**, *164*, 99–114. [CrossRef]
2. Choi, S.; Wang, G. Advanced lithium-ion batteries for practical applications: Technology, development, and future perspectives. *Adv. Mater. Technol.* **2018**, *3*, 1700376. [CrossRef]
3. Koorata, P.K.; Panchal, S.; Fraser, R.; Fowler, M. Combined influence of concentration-dependent properties, local deformation and boundary confinement on the migration of Li-ions in low-expansion electrode particle during lithiation. *J. Energy Storage* **2022**, *52*, 104908. [CrossRef]
4. Han, X.; Ouyang, M.; Lu, L.; Li, J.; Zheng, Y.; Li, Z. A comparative study of commercial lithium ion battery cycle life in electrical vehicle: Aging mechanism identification. *J. Power Sources* **2014**, *251*, 38–54. [CrossRef]
5. Xie, Y.; Li, W.; Hu, X.; Tran, M.-K.; Panchal, S.; Fowler, M.; Liu, K. Co-estimation of SOC and three-dimensional SOT for lithium-ion batteries based on distributed spatial-temporal online correction. *IEEE Trans. Ind. Electron.* **2022**, 1–10. [CrossRef]
6. Zhang, J.; Lee, J. A review on prognostics and health monitoring of Li-ion battery. *J. Power Sources* **2011**, *196*, 6007–6014. [CrossRef]
7. Yalçın, S.; Panchal, S.; Herdem, M.S. A CNN-ABC model for estimation and optimization of heat generation rate and voltage distributions of lithium-ion batteries for electric vehicles. *Int. J. Heat Mass Transf.* **2022**, *199*, 123486. [CrossRef]

8. Ng, K.S.; Moo, C.-S.; Chen, Y.-P.; Hsieh, Y.-C. Enhanced coulomb counting method for estimating state-of-charge and state-of-health of lithium-ion batteries. *Appl. Energy* **2009**, *86*, 1506–1511. [CrossRef]
9. Chaoran, L.; Fei, X.; Yaxiang, F. An approach to lithium-ion battery SOH estimation based on convolutional neural network. *Trans. China Electrotech. Soc.* **2020**, *35*, 4106–4119. [CrossRef]
10. Remmlinger, J.; Buchholz, M.; Meiler, M.; Bernreuter, P.; Dietmayer, K. State-of-health monitoring of lithium-ion batteries in electric vehicles by on-board internal resistance estimation. *J. Power Sources* **2011**, *196*, 5357–5363. [CrossRef]
11. Iurilli, P.; Brivio, C.; Carrillo, R.E.; Wood, V. Physics-Based SoH Estimation for Li-Ion Cells. *Batteries* **2022**, *8*, 204. [CrossRef]
12. Olarte, J.; Martinez de Ilarduya, J.; Zulueta, E.; Ferret, R.; Garcia-Ortega, J.; Lopez-Guede, J.M. Online Identification of VLRA Battery Model Parameters Using Electrochemical Impedance Spectroscopy. *Batteries* **2022**, *8*, 238. [CrossRef]
13. Rahimi-Eichi, H.; Ojha, U.; Baronti, F.; Chow, M.-Y. Battery management system: An overview of its application in the smart grid and electric vehicles. *IEEE Ind. Electron. Mag.* **2013**, *7*, 4–16. [CrossRef]
14. Xiong, R.; Tian, J.; Mu, H.; Wang, C. A systematic model-based degradation behavior recognition and health monitoring method for lithium-ion batteries. *Appl. Energy* **2017**, *207*, 372–383. [CrossRef]
15. Bi, J.; Zhang, T.; Yu, H.; Kang, Y. State-of-health estimation of lithium-ion battery packs in electric vehicles based on genetic resampling particle filter. *Appl. Energy* **2016**, *182*, 558–568. [CrossRef]
16. Prada, E.; Di Domenico, D.; Creff, Y.; Bernard, J.; Sauvant-Moynot, V.; Huet, F. A simplified electrochemical and thermal aging model of LiFePO4-graphite Li-ion batteries: Power and capacity fade simulations. *J. Electrochem. Soc.* **2013**, *160*, A616. [CrossRef]
17. Li, J.; Adewuyi, K.; Lotfi, N.; Landers, R.G.; Park, J. A single particle model with chemical/mechanical degradation physics for lithium ion battery State of Health (SOH) estimation. *Appl. Energy* **2018**, *212*, 1178–1190. [CrossRef]
18. Bartlett, A.; Marcicki, J.; Onori, S.; Rizzoni, G.; Yang, X.G.; Miller, T. Electrochemical model-based state of charge and capacity estimation for a composite electrode lithium-ion battery. *IEEE Trans. Control. Syst. Technol.* **2015**, *24*, 384–399. [CrossRef]
19. Guha, A.; Patra, A. State of health estimation of lithium-ion batteries using capacity fade and internal resistance growth models. *IEEE Trans. Transp. Electrif.* **2017**, *4*, 135–146. [CrossRef]
20. Tang, X.; Zou, C.; Yao, K.; Chen, G.; Liu, B.; He, Z.; Gao, F. A fast estimation algorithm for lithium-ion battery state of health. *J. Power Sources* **2018**, *396*, 453–458. [CrossRef]
21. Ossai, C.I.; Raghavan, N. Statistical characterization of the state-of-health of lithium-ion batteries with Weibull distribution function—A consideration of random effect model in charge capacity decay estimation. *Batteries* **2017**, *3*, 32. [CrossRef]
22. Seaman, A.; Dao, T.-S.; McPhee, J. A survey of mathematics-based equivalent-circuit and electrochemical battery models for hybrid and electric vehicle simulation. *J. Power Sources* **2014**, *256*, 410–423. [CrossRef]
23. Fotouhi, A.; Auger, D.J.; Propp, K.; Longo, S.; Wild, M. A review on electric vehicle battery modelling: From Lithium-ion toward Lithium–Sulphur. *Renew. Sustain. Energy Rev.* **2016**, *56*, 1008–1021. [CrossRef]
24. Feng, X.; Weng, C.; He, X.; Han, X.; Lu, L.; Ren, D.; Ouyang, M. Online state-of-health estimation for Li-ion battery using partial charging segment based on support vector machine. *IEEE Trans. Veh. Technol.* **2019**, *68*, 8583–8592. [CrossRef]
25. Zhao, L.; Wang, Y.; Cheng, J. A hybrid method for remaining useful life estimation of lithium-ion battery with regeneration phenomena. *Appl. Sci.* **2019**, *9*, 1890. [CrossRef]
26. Richardson, R.R.; Birkl, C.R.; Osborne, M.A.; Howey, D.A. Gaussian process regression for in situ capacity estimation of lithium-ion batteries. *IEEE Trans. Ind. Inform.* **2018**, *15*, 127–138. [CrossRef]
27. Richardson, R.R.; Osborne, M.A.; Howey, D.A. Gaussian process regression for forecasting battery state of health. *J. Power Sources* **2017**, *357*, 209–219. [CrossRef]
28. Wang, J.; Deng, Z.; Li, J.; Peng, K.; Xu, L.; Guan, G.; Abudula, A. State of Health Trajectory Prediction Based on Multi-Output Gaussian Process Regression for Lithium-Ion Battery. *Batteries* **2022**, *8*, 134. [CrossRef]
29. Lin, H.; Kang, L.; Xie, D.; Linghu, J.; Li, J. Online State-of-Health Estimation of Lithium-Ion Battery Based on Incremental Capacity Curve and BP Neural Network. *Batteries* **2022**, *8*, 29. [CrossRef]
30. Lin, H.-T.; Liang, T.-J.; Chen, S.-M. Estimation of battery state of health using probabilistic neural network. *IEEE Trans. Ind. Inform.* **2012**, *9*, 679–685. [CrossRef]
31. Shahriari, M.; Farrokhi, M. Online state-of-health estimation of VRLA batteries using state of charge. *IEEE Trans. Ind. Electron.* **2012**, *60*, 191–202. [CrossRef]
32. Liu, H.; Wang, P.; Cheng, Z. A Novel Method Based on Encoder-Decoder Framework for Li-Ion Battery State of Health Estimation. *Proc. CSEE* **2021**, *5*, 1851–1859. [CrossRef]
33. Salehinejad, H.; Sankar, S.; Barfett, J.; Colak, E.; Valaee, S. Recent advances in recurrent neural networks. *arXiv* **2017**, arXiv:1801.01078. [CrossRef]
34. Pascanu, R.; Mikolov, T.; Bengio, Y. On the difficulty of training recurrent neural networks. In Proceedings of the International Conference on Machine Learning, Atlanta, GA, USA, 16–21 June 2013; pp. 1310–1318.
35. Staudemeyer, R.C.; Morris, E.R. Understanding LSTM—A tutorial into long short-term memory recurrent neural networks. *arXiv* **2019**, arXiv:1909.09586. [CrossRef]
36. Saha, B.; Goebel, K. *Battery Data Set, NASA Ames Prognostics Data Repository*; NASA Ames Research Center: Moffett Field, CA, USA, 2007. Available online: http://ti.arc.nasa.gov/project/prognostic-data-repository (accessed on 10 January 2021).
37. He, W.; Williard, N.; Osterman, M.; Pecht, M. Prognostics of lithium-ion batteries based on Dempster-Shafer theory and the Bayesian Monte Carlo method. *J. Power Sources* **2011**, *196*, 10314–10321. [CrossRef]

38. Xing, Y.; Ma, E.W.; Tsui, K.-L.; Pecht, M. An ensemble model for predicting the remaining useful performance of lithium-ion batteries. *Microelectron. Reliab.* **2013**, *53*, 811–820. [CrossRef]
39. Liu, D.T.; Zhou, J.B.; Liao, H.T.; Peng, Y.; Peng, X.Y. A Health Indicator Extraction and Optimization Framework for Lithium-Ion Battery Degradation Modeling and Prognostics. *IEEE Trans. Syst. Man Cybern. Syst.* **2015**, *45*, 915–928. [CrossRef]
40. Sekhar, R.; Shah, P.; Panchal, S.; Fowler, M.; Fraser, R. Distance to empty soft sensor for ford escape electric vehicle. *Results Control. Optim.* **2022**, *9*, 100168. [CrossRef]
41. Kingma, D.P.; Ba, J. Adam: A method for stochastic optimization. *arXiv* **2014**, arXiv:1412.6980. [CrossRef]

**Disclaimer/Publisher's Note:** The statements, opinions and data contained in all publications are solely those of the individual author(s) and contributor(s) and not of MDPI and/or the editor(s). MDPI and/or the editor(s) disclaim responsibility for any injury to people or property resulting from any ideas, methods, instructions or products referred to in the content.

*Article*

# Predicting the Cycle Life of Lithium-Ion Batteries Using Data-Driven Machine Learning Based on Discharge Voltage Curves

Yinfeng Jiang * and Wenxiang Song

School of Mechatronic Engineering and Automation, Shanghai University, Nanchen Road 133, Shanghai 200444, China; wxsong@shu.edu.cn
* Correspondence: jonyf@shu.edu.cn

**Abstract:** Battery degradation is a complex nonlinear problem, and it is crucial to accurately predict the cycle life of lithium-ion batteries to optimize the usage of battery systems. However, diverse chemistries, designs, and degradation mechanisms, as well as dynamic cycle conditions, have remained significant challenges. We created 53 features from discharge voltage curves, 18 of which were newly developed. The maximum relevance minimum redundancy (MRMR) algorithm was used for feature selection. Robust linear regression (RLR) and Gaussian process regression (GPR) algorithms were deployed on three different datasets to estimate battery cycle life. The RLR and GPR algorithms achieved high performance, with a root-mean-square error of 6.90% and 6.33% in the worst case, respectively. This work highlights the potential of combining feature engineering and machine learning modeling based only on discharge voltage curves to estimate battery degradation and could be applied to onboard applications that require efficient estimation of battery cycle life in real time.

**Keywords:** data driven; state of health; lithium-ion batteries; linear regression; Gaussian process regression; machine learning

## 1. Introduction

Lithium-ion batteries have been widely used in various applications, such as electric vehicles, battery energy storage systems (BESSs), and portable electronics, due to their high energy density, low cost, and low self-discharge rate [1]. However, similar to most complex mechanical, electrical, and chemical systems, the aging of lithium-ion batteries is inevitable due to side reactions occurring within their electrolyte and electrodes [2]. This aging process causes a decline in battery performance. Thus, it is essential to accurately predict the aging of lithium-ion batteries to ensure long-term stability and reliable operation.

Many approaches have been suggested to accurately predict the lifetime of lithium-ion batteries, including empirical models [3], equivalent circuit models [4–6], physical models [7], and data-driven models [2,8–12]. Empirical models assume that cells of the same chemistry age in the same manner [3], which may not always be the case. Equivalent circuit models are semiempirical and unable to represent various aging patterns [4], and the parameters are difficult to identify when considering different usage conditions, ambient temperatures, and load profiles [13–15]. Physical models consist of complex partial differential equations and require many parameters that are not easily obtainable [16–18]. While some studies have provided model parameters that accurately explain observed data, the accuracy of predictions may rapidly decline in the presence of uncertain mechanisms and aging rates under future usage conditions [8,18].

In contrast, data-driven models have many advantages, such as the ability to capture battery degradation mechanisms without complex chemical reaction knowledge. Recently, many studies [10,16,19,20] have used machine learning or deep learning tools for battery life

estimation. Feature extraction and selection are essential for machine learning approaches. Various studies have extracted features using charge voltage curves, raw data from battery cycle tests (i.e., voltage, current, temperature, and state of charge (SOC) data) [17,21,22], discharge voltage curves [23], and electrochemical impedance spectroscopy (EIS) [12,24]. Charge and discharge voltage curves can be obtained via the battery management system (BMS) in real time [23,25], while EIS data can only be measured with an electrochemical impedance analyzer. Extracting features based on the charge voltage curve is feasible because most charge protocols are typically constant current (CC) and constant voltage (CV) [10,11,21,23]. It is challenging to derive features through the discharge voltage curve because load behaviors vary among batteries. Feature selection typically relies on background knowledge or Pearson correlation analysis, with the aim of reducing the size of the input matrix and avoiding overfitting [10,21,26,27]. However, these approaches overlook the redundancy among features.

To achieve an accurate prediction of battery life, different fitting functions with optimizable parameters have been implemented. One such method is support vector regression (SVR) [28–30], which has been observed to have high accuracy; however, SVR is time-consuming for model training. In contrast, linear regression (LR) with an elastic net requires a much quicker training time [31,32], but its accuracy tends to decline for large datasets. Neural network (NN) models have also been used, with the performance improving as the number of hidden layers and neurons increases [33,34]; however, neural network models are hard to train, and it is difficult to choose a network structure. Gaussian process regression (GPR) has demonstrated promising accuracy and faster training speed than SVR [10,23,35,36]; however, its complexity remains problematic, hindering onboard deployment.

This paper proposed an innovative data-driven framework for accurately and promptly predicting battery cycle lives (as in Figure 1). Using pattern recognition and signal processing techniques, battery degradation features were extracted from discharge voltage curves. Next, using the maximum relevance minimum redundancy (MRMR) algorithm, 20 of 53 features were selected as the feature subset. Three different battery datasets were used to train and test the GPR and robust linear regression (RLR) algorithms. The test results suggested that GPR outperforms RLR in most cases, while RLR has a faster prediction speed than GPR. These results illustrate the power of combining feature extraction and selection with data-driven modeling based on discharge voltage curves to predict the degradation of lithium-ion batteries.

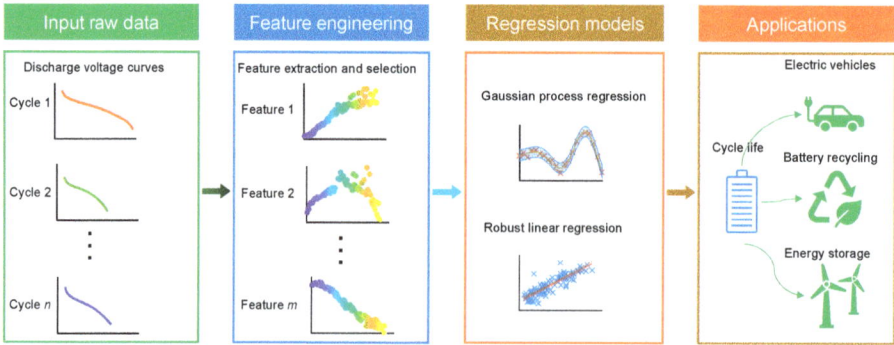

**Figure 1.** Schematic diagram of battery cycle life prediction based on discharge voltage curves. The colors of the discharge voltage curves indicate that they belong to different cycles, and the colors of the curves in the feature extraction and selection box suggest that their values change as the cycle number increases.

The main contributions of this article are listed as follows:
1. New features were developed using pattern recognition and signal processing techniques to capture degradation mechanisms using discharge voltage profiles.
2. The MRMR algorithm was proposed for feature selection, reducing the parameter size of the model and improving the prediction speed.
3. Two algorithms, GPR and RLR, were trained for battery cycle life prediction. GPR was found to have high accuracy but is time-consuming, making it best suited for battery pack manufacturing and battery recycling. Conversely, RLR requires less training time, and its accuracy is suitable for real-time battery management applications, making it ideal for onboard deployment.

The remainder of this article is organized as follows: Section 2 introduces the details of three lithium-ion battery datasets, Section 3 describes the machine learning framework, the results of feature extraction and battery cycle life prediction are presented in Section 4, and Section 5 discusses the test results. This article is concluded in Section 6.

## 2. Design of Battery Datasets

We deployed our methods on three different battery datasets due to the varying degradation mechanisms of lithium-ion batteries. Dataset I [33] incorporates 39 cells, cells 1 to 30 were used as the training set, and cells 31 to 39 served as the test set. The positive electrode material of the cells is a blend of lithium cobalt oxide (LCO) and ternary nickel cobalt lithium manganese (NCM), and the negative electrode material is graphite. The rated capacity is 2.4 Ah, with an upper voltage threshold and a lower voltage threshold of 4.2 V and 3.0 V, respectively, for all cells in Dataset I. All cells were cycled in two-stage degradation tests. The first stage included 20 preliminary cycles, with CCCV charging at a C-rate of 0.5 and CC discharging at a C-rate of 2. The second stage incorporated two different dynamic cycle profiles. The first profile consisted of a CC charge and discharge at a rate of 1 C, 2 C, or 3 C. The secondary profile included a CC charge with a random current of 1 C, 2 C, or 3 C and a CC discharge at a rate of 3 C. Cell 31, cells 33–34, cells 36–37, and cell 39 were cycled with the secondary profile, while cell 32, cell 35, and cell 38 were cycled with the first profile. All tests were conducted at 25 °C. The average total cycle number of the training cells and test cells was 120 cycles.

Dataset II [37] consists of eight commercial cells that were operated in identical dynamic cycle tests. The negative electrode material of the cells is graphite, and the positive electrode material is a blend of lithium cobalt oxide (LCO) and lithium nickel cobalt oxide (NCO). All cells were cycled using the Artemis urban drive cycle [38] and characterization cycles, repeated every 100 cycles. The Artemis urban drive cycle consists of dynamic charging and regenerative charging with a maximum rate of 6.75 C. The charge cycle was CC at a rate of 2 C. The characterization procedure consisted of low-rate discharge and charge cycles for OCV. The lower voltage threshold and the upper voltage threshold were 2.7 V and 4.2 V, respectively. All cell tests were conducted in thermal chambers at 40 °C. The average total cycle number of the training cells and test cells was 8100 cycles.

Dataset III [39] incorporates 14 cells under four different discharge profiles. The positive electrode material of the cells is a blend of lithium cobalt oxide (LCO) and ternary nickel cobalt lithium aluminate (NCA), and the negative electrode material is graphite. All cells were charged with the CCCV protocol with an identical rate of 0.75 C during the CC stage and an identical voltage of 4.2 V, with a cut-off current of 20 mA during the CV stage. B5, B6, and B7 were discharged at a CC level of 1 C until their cell voltages fell to 2.7 V, 2.5 V, and 2.2 V, respectively. B33 and B34 were discharged with the CC profile with a rate of 2 C until their cell voltages fell to 2.0 V and 2.2 V, respectively. B38 and B39 were discharged under multiple load current rates of 0.5 C, 1 C, and 2 C and stopped at 2.2 V and 2.5 V, respectively. B41 to B44 used two fixed load current rates of 2 C and 0.5 C, respectively, and the lower voltage thresholds were 2 V, 2.2 V, 2.5 V, and 2.7 V, respectively. B5-B7 and B33 and B34 were discharged at a room temperature of 24 °C. B38 and B39 were tested at ambient temperatures of 24 °C and 44 °C. B41–B44 were cycled at an ambient temperature of 4 °C. The average total cycle number of the training cells was 119 cycles, and the total cycle number of the test cells was 131 cycles.

## 3. Machine Learning Framework

### 3.1. Feature Development

Lithium-ion battery aging is a complex process that can result in capacity degradation and reduced power capability. There are many factors that can contribute to battery aging, such as the formation of a solid electrolyte interphase (SEI) film at the electrode/electrolyte surface, destruction of the electrode structure, lithium deposition, a phase change of the electrode material, dissolution of the active material, and electrolyte decomposition [40]. As the cycle number increases, charge/discharge voltage curves, incremental capacity curves, and electrochemical impedance spectroscopy can all be altered. Many machine learning algorithms extract features for battery health estimation based on these curves. In this section, we focus on using signal processing techniques to extract features from the discharge voltage curves.

For each discharge cycle, we defined the discharge voltage sample values as a signal $x = (v_1, v_2, \ldots, v_n)^T$. The main equations of the developed features were defined as follows.

#### 3.1.1. Root-Sum-of-Squares Level

The root-sum-of-squares (RSS) level of a vector $x$ is

$$RSS = \sqrt{\sum_{n=1}^{N} |x_n|^2} \tag{1}$$

where $x_n$ is the element of vector $x$ and the RSS level is also known as the $\ell_2$ norm. In this study, we used the discharge voltages as vector $x$.

#### 3.1.2. Distance between Signals Using Dynamic Time Warping

Two signals were considered:

$$x = (x_1, x_2, x_3, \ldots, x_m), y = (y_1, y_2, y_3, \ldots, y_n) \tag{2}$$

where $x$ has $m$ samples, $y$ has $n$ samples, and $d_{mn}(x, y)$ is defined as the distance between the $m$th sample of $x$ and the $n$th sample of $y$. The following equations are four types of distance definitions.

Here, we define a line as $y$, and $x$ is the discharge voltage vector per cycle.

The square root of the sum of squared differences is also known as the Euclidean or $\ell_2$ metric:

$$d_{mn}(x, y) = \sqrt{\sum_{k=1}^{K} (x_m - y_n) * (x_m - y_n)} \tag{3}$$

The sum of absolute differences is also known as the Manhattan, city block, taxicab, or $\ell_1$ metric:

$$d_{mn}(x, y) = \sum_{k=1}^{K} |x_m - y_n| = \sum_{k=1}^{K} \sqrt{(x_m - y_n) * (x_m - y_n)} \tag{4}$$

The square of the Euclidean metric is composed of the sum of squared differences:

$$d_{mn}(x, y) = \sum_{k=1}^{K} (x_m - y_n) * (x_m - y_n) \tag{5}$$

The symmetric Kullback–Leibler metric is only valid for real and positive values of $x$ and $y$.

$$d_{mn}(x,y) = \sum_{k=1}^{K}(x_m - y_n) * (\log x_m - \log y_n) \tag{6}$$

where $x_m$ is the element of $x$ and $y_n$ is the element of $y$, as defined in Equation (2).

### 3.1.3. Zero-Crossing Rate

The zero-crossing rate refers to the ratio of sign changes in a signal, for instance, a signal changing from positive to negative or vice versa. This feature has been widely used in the fields of speech recognition and music information retrieval and is a key feature for classifying percussion sounds. The ZCR is formally defined as:

$$zcr = \frac{1}{m-1}\sum_{t=1}^{m-1}\prod\{x_t x_{t-1} < 0\} \tag{7}$$

where $x$ is a signal with a length of $M$, and the function $\prod\{x\}$ is equal to 1 when the parameter $x$ is true, and 0 otherwise.

### 3.1.4. Mid-Reference Level

The mid-reference level in a bilevel waveform with a low state level of $S_1$ and a high state level of $S_2$ is

$$y_{50\%} = S_1 + \frac{1}{2}(S_2 - S_1) \tag{8}$$

Mid-reference level instant:
We let $y_{50\%}$ denote the mid-reference level.
We let $t_{50\%-}$ and $t_{50\%+}$ denote the two consecutive sampling instances corresponding to the waveform values nearest in value to $y_{50\%}$.
We let $y_{50\%-}$ and $y_{50\%+}$ denote the waveform values at $t_{50\%-}$ and $t_{50\%+}$, respectively. The mid-reference level instant is

$$t_{50\%} = t_{50\%} + \left(\frac{t_{50\%+} - t_{50\%-}}{y_{50\%+} - y_{50\%-}}\right)(y_{50\%+} - y_{50\%-}) \tag{9}$$

### 3.1.5. Standard Error

For a finite-length vector $x$ consisting of $N$ scalar observations, the standard deviation is defined as

$$S = \sqrt{\frac{1}{N-1}\sum_{i=1}^{N}|x_i - \mu|^2} \tag{10}$$

where $\mu$ is the mean of $x$:

$$\mu = \frac{1}{N}\sum_{i=1}^{N}x_i \tag{11}$$

The standard deviation is the square root of the variance.

### 3.1.6. Band Power

Band power is a measure of the amount of energy in a particular frequency band of a signal $x$ and is calculated as:

$$P_{band} = \int_{f_1}^{f_2}P(f)df \tag{12}$$

$$P(f) = 2\int[R(\tau)\cos(2\pi f \tau)]d\tau \tag{13}$$

where $P(f)$ is the estimated power spectral density estimate at frequency $f$; $f_1$ and $f_2$ are the lower bound and upper bound, respectively, of the frequency band of interest; and $R(\tau)$ is the autocorrelation function at the time lag $\tau$.

3.1.7. Mean Squared Error

The mean squared error is calculated using the following formula:

$$loss = \frac{1}{2N}\sum_{i=1}^{N}(x_i - t_i)^2 \tag{14}$$

where $x_i$ is the $i$th element of vector $x$, $t_i$ is the $i$th element of reference vector $t$, and $N$ is the total number of observations in $x$. In this case, $x$ is defined as the discharge voltage of each cycle and $t$ is defined as the discharge voltage of the first cycle.

3.1.8. Occupied Bandwidth

The occupied bandwidth is defined as:

$$B = \Delta f = f_H - f_L \tag{15}$$

where $f_H$ and $f_L$ are the upper frequency limit and lower frequency limit, respectively, of the band.

In this study, we calculated the 99% bandwidth:

$$\%B_F = 99\%\frac{\Delta f}{f_C} \tag{16}$$

where $f_C$ is defined as the arithmetic mean of the upper and lower frequencies:

$$f_C = \frac{f_H + f_L}{2} \tag{17}$$

3.1.9. Structural Similarity Index for a Vector (SSIM)

The SSIM was originally used to assess image quality, but here, we used it to assess the similarity of two vectors. The $SSIM$ is defined as:

$$SSIM(x,y) = \frac{(2\mu_x\mu_y + C_1)(2\sigma_{xy} + C_2)}{\left(\mu_x^2 + \mu_y^2 + C_1\right)\left(\sigma_x^2 + \sigma_y^2 + C_2\right)} \tag{18}$$

where $\mu_x$ and $\mu_y$, $\sigma_x$ and $\sigma_y$, and $\sigma_{xy}$ are the local means, standard deviations, and cross-covariance, respectively, for vectors $x$ and $y$. In this case, $x$ is defined as the discharge voltage of each cycle and $y$ is defined as the discharge voltage of the first cycle.

3.2. MRMR Feature Selection

To reduce the size of the model, eliminate redundant features, and reduce model complexity, we performed feature selection on all extracted features. We used the MRMR algorithm to search for a subset of features that minimized redundancy while maximizing relevancy with the response. This algorithm calculated pairwise mutual information between features and the response variable to quantify redundancy and relevancy [41,42].

Assuming there are $m$ features in total, the MRMR algorithm provides the importance of a given feature $X_i$ ($i \in \{1, 2, \ldots, m\}$).

$$f^{MRMR}(X_i) = I(Y, X_i) - \frac{1}{|S|}\sum_{X_S \in S} I(X_S, X_i) \tag{19}$$

where $Y$ represents the response variable, $S$ is the selected feature set, $|S|$ denotes the size of the feature set (i.e., number of features), $X_S \in S$ represents a feature in feature set $S$, $X_i$ represents a feature not in $S$: $X_i \notin S$, and $I(\cdot, \cdot)$ represents the mutual information.

$$I(Y, X) = \int_{\Omega_Y} \int_{\Omega_X} p(x, y) \log\left(\frac{p(x, y)}{p(x) p(y)}\right) \tag{20}$$

In the MRMR feature selection process, at each step, the feature with the highest importance score $\max f^{MRMR}(X_i)$, which is not already in the selected feature set $S$, is added to $S$. For discrete features, the mutual information difference (MID) is the original feature importance:

$$f^{MID}(X_i) = I(Y, X_i) - \frac{1}{|S|} \sum_{X_S \in S} I(X_S, X_i) \tag{21}$$

The mutual information quotient (MIQ) is defined as:

$$f^{MIQ}(X_i) = \frac{I(Y, X_i)}{\frac{1}{|S|} \sum_{X_S \in S} I(X_S, X_i)} \tag{22}$$

For continuous time features, the F-statistic is used to represent the correlation. The corresponding correlation difference is represented as:

$$f^{FCD}(X_i) = F(Y, X_i) - \frac{1}{|S|} \sum_{X_s \in S} \rho(X_S, X_i) \tag{23}$$

where $\rho(X_S, X_i)$ represents the Pearson correlation and $F(Y, X_i)$ represents the F-statistic. The Pearson correlation is represented as:

$$\rho(X_S, X_i) = \frac{cov(X_S, X_i)}{\sigma_{X_s} \sigma_{X_i}} \tag{24}$$

$$cov(X_S, X_i) = \mathbb{E}[(X_S - \mu_{X_S})(X_i - \mu_{X_i})] \tag{25}$$

$$\rho(X_S, X_i) = \frac{\mathbb{E}[(X_S - \mu_{X_S})(X_i - \mu_{X_i})]}{\sigma_{X_s} \sigma_{X_i}} \tag{26}$$

where $\rho(X, Y)$ is the Pearson correlation coefficient between $X$ and $Y$, $cov(X_s, X_i)$ represents the covariance of $X_s$ and $X_i$, $\sigma_{X_s}$ is the standard error of $X_s$, $\sigma_{X_i}$ is the standard error of $X_i$, $\mu_{X_S}$ is the mean of $X_s$, and $\mu_{X_i}$ is the mean of $X_i$.

Similarly, the correlation quotient is defined as:

$$f^{FCQ}(X_i) = \frac{F(Y, X_i)}{\frac{1}{|S|} \sum_{X_s \in S} \rho(X_S, X_i)} \tag{27}$$

### 3.3. Robust Linear Regression

Robust linear regression is designed to handle data that contain outliers, an issue commonly observed in raw data. This method uses iteratively reweighted least squares (IRLS) to assign a weight to each data point, allowing the algorithm to weigh the influence of data points based on their distance from the model's prediction. This iterative approach produces more accurate regression coefficients than the typical ordinary least squares (OLS) approach used in standard linear regression.

The IRLS algorithm includes multiple iterations. First, the algorithm assigns equal weights to all data points and calculates model coefficients using OLS. Second, in each iteration, the algorithm recalculates the weights for each data point, with those further from the model's prediction receiving lower weights. Using these new weights, the algorithm then calculates a new set of coefficients using weighted least squares. This process continues, with the algorithm iterating until the coefficient estimates converge within a specified tolerance. This iterative, simultaneous approach of fitting data using least squares methods, while minimizing the effect of outliers, makes IRLS a powerful algorithm.

A simple linear regression model of the form

$$y_i = x_i^T \beta + \varepsilon_i \qquad (28)$$

was proposed, where $y_I$ is the predicted cycle life for a battery $i$, $\varepsilon_i$ is the bias, $x_i$ is a p-dimensional feature vector for battery $i$, and $\beta$ is a p-dimensional model coefficient vector.

The ordinary least squares residual is

$$r_i = y_i - x_i^T \beta \qquad (29)$$

The weighted least squares method using the adjusted residuals is expressed as follows:

$$r_{adj} = \frac{r_i}{\sqrt{1 - h_i}} \qquad (30)$$

where $r_i$ is the ordinary least squares residual and $h_i$ is the least squares fit leverage value.

The leverage $h_i$ is the value of the $i$th diagonal term of the hat matrix $H$. The hat matrix $H$ is defined in terms of the data matrix $X$:

$$H = X \left( X^T X \right)^{-1} X^T \qquad (31)$$

The standardized adjusted residuals are defined as

$$u = \frac{r_{adj}}{K_s} = \frac{r_i}{K_s \sqrt{1 - h_i}} \qquad (32)$$

where $K$ is a tuning constant and $s$ is an estimate of the standard deviation of the error term given by $s = MAD/0.6745$. MAD is the median absolute deviation of the residuals from their median. The constant 0.6745 ensures that the estimates are unbiased from the normal distribution.

The robust weights $w_i$ are achieved using a bisquare weights function

$$w_i = \begin{cases} \left(1 - u_i^2\right)^2 & , |u_i| < 1 \\ 0 & , |u_i| \geq 1 \end{cases} \qquad (33)$$

Then, the weighted least squares estimate the coefficient $\beta$

$$\beta = \left( X^T W X \right)^{-1} X^T W y \qquad (34)$$

where $W = \text{diag}(w_1, \cdots, w_n)$, $X = (x_1, \cdots, x_n)^T$, and $y = (y_1, \cdots, y_n)'$.

The estimated weighted least squares error is

$$e = \sum_1^n w_i \left( y_i - x_i^T \beta \right)^2 = \sum_1^n w_i r_i^2 \qquad (35)$$

where $w_i$ are the weights, $y_i$ are the observed responses, and $r_i$ are the residuals.

### 3.4. Gaussian Process Regression

GPR is a nonparametric and Bayesian approach to regression that defines a probability distribution over functions rather than random variables. Using GPR, the regression problem is defined as

$$f(x) = k(x)^T(K + \lambda I_N)^{-1} t \qquad (36)$$

where $K$ is the Gram matrix with elements $K_{nm}$ and $k(x)$ is a vector with elements $k_n(x) = k(x_n, x)$. $K_{nm}$ is defined by

$$K_{nm} = k(x_n, x_m) \qquad (37)$$

and $k(x, x')$ is the kernel function.

Gaussian process regression methods use kernel functions to determine the covariance. In this case, we used the Matern covariance functions.

The Matern class of covariance functions is defined as follows:

$$k_{Matern}(r) = \frac{2^{1-v}}{\Gamma(v)} \left(\frac{\sqrt{2vr}}{\ell}\right)^v K_v\left(\frac{\sqrt{2vr}}{\ell}\right) \qquad (38)$$

where $v$ and $\ell$ are positive and $K_v$ is the modified Bessel function. The frequency density of the covariance function is

$$S(s) = \frac{2^D \pi^{\frac{D}{2}} \Gamma\left(v + \frac{D}{2}\right)(2v)^v}{\Gamma(v)\ell^{2v}} \left(\frac{2v}{\ell^2} + 4\pi^2 s^2\right)^{-(v+\frac{D}{2})} \qquad (39)$$

where $D$ is the dimension.

When $v$ is a half integer, the Matern covariance function is:

$$k_{v=p+\frac{1}{2}}(r) = exp\left(-\frac{\sqrt{2vr}}{\ell}\right) \frac{\Gamma(p+1)}{\Gamma(2p+1)} \sum_{i=0}^{p} \frac{(p+i)!}{i!(p-i)!} \left(\frac{\sqrt{8vr}}{\ell}\right)^{p-i} \qquad (40)$$

Most machine learning methods commonly use $v = 3/2$ and $v = 5/2$:

$$k_{v=\frac{3}{2}}(r) = \left(1 + \frac{\sqrt{3}r}{\ell}\right) exp\left(-\frac{\sqrt{3}r}{\ell}\right) \qquad (41)$$

$$k_{v=\frac{5}{2}}(r) = \left(1 + \frac{\sqrt{5}r}{\ell} + \frac{5r^2}{3\ell^2}\right) exp\left(-\frac{\sqrt{5}r}{\ell}\right) \qquad (42)$$

In this study, we used $v = 5/2$.

Figure 2 illustrates the main workflow of the proposed method. Figure 2a describes the feature extraction and selection, as explained in Sections 3.1 and 3.2. Figure 2b,c explain the main equations of Gaussian process regression (Equation (36)) and robust linear regression (Equation (34)) algorithms, respectively.

Considering the battery's early aging process before capacity degradation, we used the cycle life indicator to describe the battery's health state. The cycle life indicator is defined as

$$CI = \frac{C}{C_0} \qquad (43)$$

where $C$ is the current cycle number and $C_0$ is the total cycle number of the cycle test or the cycle number given by the battery manufacturers. The range of $C_0$ is from several hundred cycles to several thousand cycles due to various material and operation conditions.

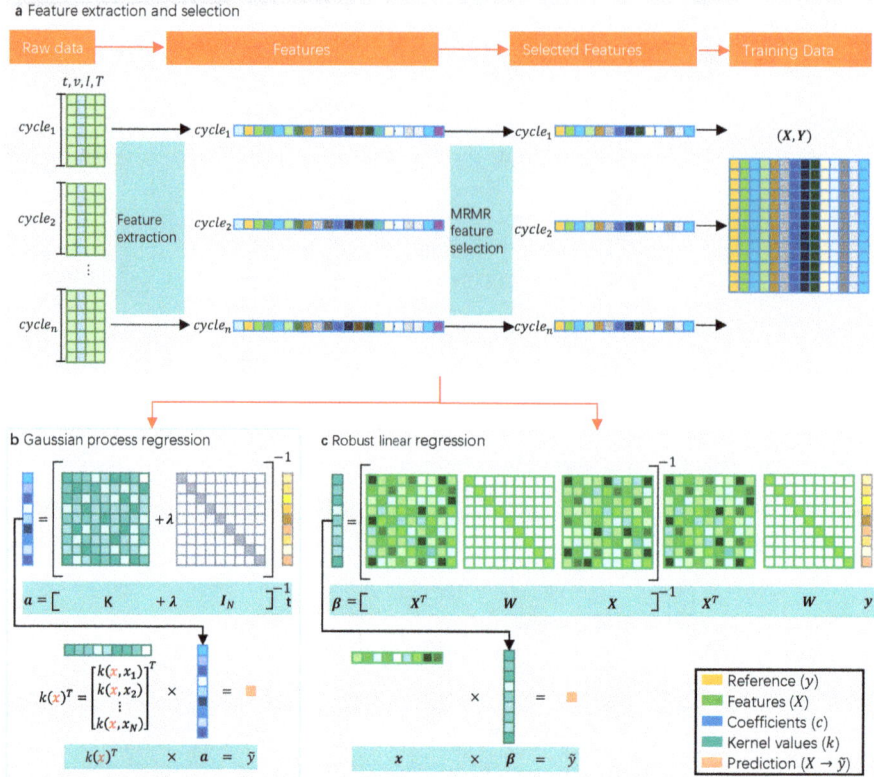

**Figure 2.** The main framework of the proposed method. (**a**) Schematic of feature extraction and selection from cycle data consisting of time ($t$), voltage ($v$), current ($I$), and temperature ($T$). First, each cycle data matrix is condensed into a vector through feature extraction. Next, a subset is selected out of the original features using the MRMR algorithm. Finally, the raw cycle data matrix is transformed into a feature matrix, which is used as the input of the machine learning models. (**b**) Linear expression of Gaussian process regression. (**c**) Visualization of robust linear regression.

As the cycle life of various cells is distinct, we defined the root-mean-square error (RMSE) and the mean absolute error (MAE) to metric the performance of the RLR and GPR models. The RMSE and MAE are defined as

$$\text{RMSE} = \frac{\sqrt{\frac{1}{n}\sum_{i=1}^{n}(y_i - \hat{y}_i)^2}}{C_0} \times 100\% \qquad (44)$$

$$\text{MAE} = \frac{\frac{1}{n}\sum_{i=1}^{n}|y_i - \hat{y}_i|}{C_0} \times 100\% \qquad (45)$$

where $y_i$ is the observed cycle number, $\hat{y}_i$ is the predicted cycle number, $n$ is the total number of samples, and $C_0$ is the total cycle number of the cycle test or the cycle number given by the battery manufacturers.

## 4. Results

In this study, we explored two algorithms, robust linear regression (RLR) and Gaussian process regression (GPR), with three different datasets of lithium-ion batteries. First, we extracted 53 features based on raw discharge voltage curves. Second, we used the MRMR algorithm to select the top 20 features with the highest median scores as the feature subset to compare with the full feature set (53 features). The GPR algorithm and the RLR algorithm were deployed on the subset of features and on the full set of features, respectively. The results showed that all algorithms could accurately predict the battery cycle life with a low error. Specifically, RLR achieved a maximum average RMSE of 6.90% and a maximum average MAE of 4.77% for the selected feature subset, whereas the GPR model achieved a maximum average RMSE of 6.33% and a maximum average MAE of 3.91% for the same feature subset. The GPR algorithm exhibited greater prediction accuracy than the RLR algorithm, while the RLR algorithm demonstrated faster prediction speed than the GPR algorithm for both the full features and the feature subset.

*4.1. Feature Extraction and Selection*

Features were created on Datasets I through III. Figure 3 illustrates the typical features that were created on Dataset II. To the best of our knowledge, all features in Figure 3, except for skewness and kurtosis coefficients, were developed by us for the first time to predict the battery cycle life using machine learning methods. Most features in Figure 3 show some correlation with the cycle number. For instance, certain features, such as the zero-crossing rate, standard error, and mean frequency, increased as the cycle life increased. Conversely, features such as the root-sum-of-squares (RSS) level, Euclidean metric, absolute metric, and peak signal-to-noise ratio (PSNR) decreased as the cycle number increased. Furthermore, specific features, including the coefficient of skewness, root-mean-square (RMS) level, and band power, fluctuated over cycles during the first 100 cycles. However, despite most of the proposed features exhibiting a correlation with the cycle number, their values can greatly differ, varying by orders of magnitude, as illustrated in Figure 3.

Feature selection simplifies machine learning models, reduces overfitting, and improves model interpretability. The MRMR algorithm was selected to search for the optimal feature subset among the 53 pre-extracted features. The ranking of the features, arranged in descending order based on their median scores computed with the MRMR algorithm, is shown in Figure 4. Some of the new features from Figure 3, such as the mean frequency of the discharge voltage curve (dsgMeanFreq), the squared metric, and the Euclidean metric between the discharge voltage curve and the reference line (dsgDistSqr and dsgDistEucl), were among the top 20 features in the correlation ranking (as shown in Figure 4), indicating that the proposed features in Section 3 can serve as optimal inputs for machine learning models. Traditional features, such as total discharge capacity (dsgTotalAh), discharge voltage at the start (dsgVbegin), total discharge energy (dsgTotalWh), and discharge time (dsgTime), also had high scores, which is not unexpected, given their physical meaning associated with battery degradation. Additionally, numerical partial derivatives of voltage concerning the SOC (dsgDeltaV_dSOC80, dsgDeltaV_dSOC50, and dsgDeltaV_dSOC90 in Figure 4) were also found to be significant, confirming prior studies.

The remaining features in Figure 4 are the kurtosis coefficient of the discharge voltage (dsgKurt), the discharge capacity (dsgQ), the occupied bandwidth of the discharge voltage curve (dsgOccupiedband), the symmetric Kullback–Leibler metric between the discharge voltage curve of cycle $i$ and the reference line (dsgDistSym), the structural similarity index for the discharge voltage (dsgSsim), the mean square error between the discharge voltage of cycle $i$ and the discharge voltage of the first cycle (dsgMse), the zero-crossing rate of the discharge voltage of cycle $i$ (dsgZerorate), the band power of the discharge voltage of cycle $i$ (dsgPowerband), the middle reference level for the discharge voltage of cycle $i$ (dsgMidcross), the standard error between the discharge voltage of cycle $i$ and the discharge voltage of the first cycle (dsgStd), and the Euclidean metric between the discharge voltage curve of cycle $i$ and the reference line (dsgDistEucl).

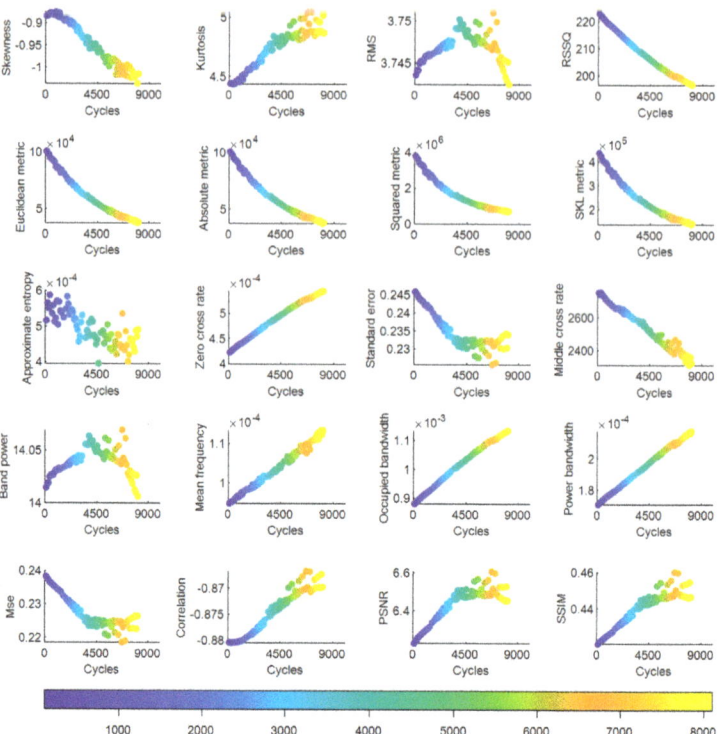

**Figure 3.** Typical features of Dataset II.

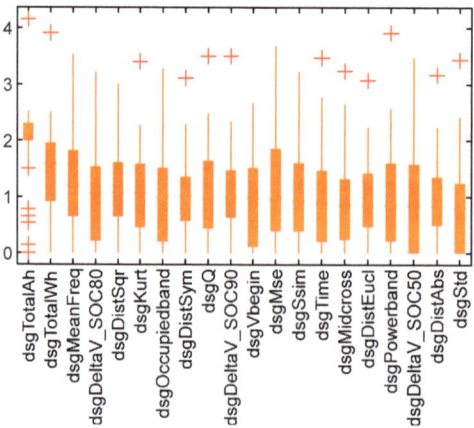

**Figure 4.** Top 20 features ranked by the median of their scores according to the MRMR algorithm. The mark of "+" indicates an outlier.

The MRMR algorithm computes relevance scores for all features, while attempting to reduce redundancy. This study presented the use of the first 20 features as an example. However, determining the optimal number of features to use in practice depends on the requirements of accuracy in prediction and efficiency in computation for a particular field. Notably, the features based on discharge voltage proposed in this study are statistical analyses of the variations in the battery discharge voltage curve and may not have any practical physical significance.

### 4.2. Performance of Models Based on Full Features

To further evaluate the performance of our proposed method, we conducted a 5-fold cross-validation using two algorithms: Gaussian process regression (GPR) and robust linear regression (RLR). To validate the models' performance on various load profiles and operating conditions, we assigned secondary test sets for all datasets. The training/testing partitions for Datasets I to III are summarized in Table 1. We tested the models using two feature sets: 53 features, which we named the full features, and a subset of the top 20 features selected using the maximum relevance minimum redundancy (MRMR) algorithm, which we referred to as the feature subset. The results demonstrated that both algorithms can accurately predict the battery cycle life with an error margin that is small compared to the actual cycle life, indicating that our proposed approach can yield reliable results and be used in applications that require accurate predictions of battery cycle life.

**Table 1.** Selection and allocation of training and test datasets, including the charge protocols and discharge profiles.

| Dataset | Number of Cells | Charge | Discharge | Positive Electrode | Negative Electrode | Training Set | Test Set |
|---|---|---|---|---|---|---|---|
| I | 72 | CC | CC with 3C rate or random | NCM | Graphite | Cells 1–30 | Cells 31–39 |
| II | 8 | CCCV | ARTEMIS dynamic driving profile or CC with 1 C rate | NCO | Graphite | Cells 1–2 | Cells 3–8 |
| III | 11 | CCCV | CC or random | NCA | Graphite | B5, B6, B33, B34, B38, B39, B41, B42, B43 | B7, B40, B36, B18, B44 |

The performance of the GPR and RLR algorithms on the full features of Datasets I-III is summarized in Tables 2–4. Both algorithms demonstrated promising performance across all datasets. The RLR algorithm achieved an average RMSE (ARMSE) of 6.90% and an average MAE (AMAE) of 4.77% on the test set of Dataset III, which was the model's worst-case scenario. The GPR model's worst performance was also observed on the test set of Dataset III, with an average RMSE and an average MAE of 6.33% and 3.91%, respectively. Figures 5–7 provide a comparison between the predicted cycle life and the actual cycle life for the test batteries from Datasets I-III on the GPR and RLR algorithms.

**Table 2.** Test results for the RLR and GPR models trained on the full feature set of Dataset I.

| Model | Metric | Battery ID | | | | | | | | | | Average RMSE/MAE |
| | | #31 | #32 | #33 | #33 | #34 | #35 | #36 | #37 | #38 | #39 | |
|---|---|---|---|---|---|---|---|---|---|---|---|---|
| RLR | RMSE | 2.06% | 6.65% | 1.62% | 1.47% | 6.56% | 1.97% | 2.92% | 6.70% | 2.42% | 1.48% | 3.60% |
| | MAE | 1.31% | 5.36% | 1.24% | 1.20% | 5.28% | 1.37% | 1.48% | 5.50% | 1.74% | 1.21% | 2.57% |
| GPR | RMSE | 0.82% | 5.78% | 0.89% | 1.03% | 5.94% | 0.78% | 3.24% | 6.91% | 1.14% | 0.85% | 2.95% |
| | MAE | 0.59% | 4.04% | 0.63% | 0.80% | 4.25% | 0.54% | 0.98% | 4.92% | 0.78% | 0.70% | 1.82% |

**Table 3.** Test results for the RLR and GPR models trained on the full feature set of Dataset II.

| Model | Metric | Battery ID | | | | | | Average RMSE/MAE |
|---|---|---|---|---|---|---|---|---|
| | | #3 | #4 | #5 | #6 | #7 | #8 | |
| RLR | RMSE | 1.31% | 0.71% | 0.47% | 0.79% | 1.26% | 1.44% | 1.00% |
| | MAE | 1.25% | 0.61% | 0.35% | 0.68% | 1.16% | 1.36% | 0.90% |
| GPR | RMSE | 0.50% | 2.22% | 0.47% | 1.19% | 0.96% | 0.86% | 1.03% |
| | MAE | 0.44% | 1.98% | 0.34% | 1.03% | 0.76% | 0.74% | 0.88% |

**Table 4.** Test results for the RLR and GPR models trained on the full feature set of Dataset III.

| Model | Metric | Battery ID | | | | | Average RMSE/MAE |
|---|---|---|---|---|---|---|---|
| | | #7 | #18 | #36 | #40 | #44 | |
| RLR | RMSE | 2.34% | 3.52% | 5.06% | 8.45% | 17.06% | 7.29% |
| | MAE | 1.76% | 3.13% | 4.51% | 7.75% | 10.63% | 4.76% |
| GPR | RMSE | 3.82% | 2.64% | 13.52% | 6.80% | 4.23% | 6.20% |
| | MAE | 3.01% | 2.51% | 11.90% | 5.79% | 2.78% | 4.45% |

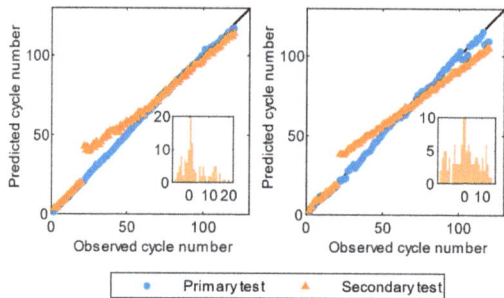

**Figure 5.** Test results of the full feature models of Dataset I. The left plot shows the predictions of the GPR algorithm, and the right plot shows the predictions of the RLR algorithm. Cell 31 is the primary test set, and cell 32 is the secondary test set.

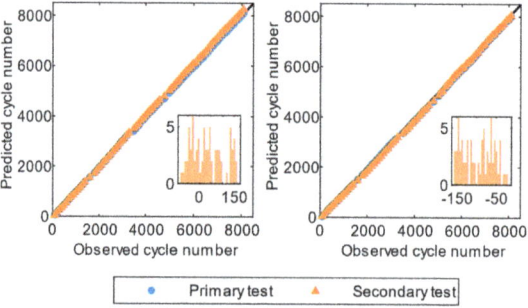

**Figure 6.** Test results of the full feature models of Dataset II. The left plot shows the predictions of the GPR algorithm, and the right plot shows the predictions of the RLR algorithm. Cell 3 is the primary test set, and cell 4 is the secondary test set.

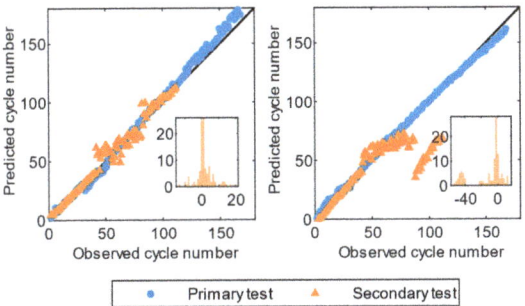

**Figure 7.** Test results of the full feature models of Dataset III. The left plot shows the predictions of the GPR algorithm, and the right plot shows the predictions of the RLR algorithm. Cell 7 is the primary test set, and cell 44 is the secondary test set.

An interesting observation in the test set of Dataset I, as depicted in Figure 5, is the sudden fluctuation of predictions at approximately cycle 20. This notable rise can be attributed to the finding that the initial 20 cycles were characterized by a constant current discharge, whereas subsequent cycles were characterized by a random current discharge, resulting in considerable fluctuations in the prediction. Nevertheless, the GPR algorithm showed a gradual decrease in the residuals, eventually confining them to a small range. In contrast, RLR's prediction diverged from the real cycle life after reaching a point of convergence, due to its limited ability to capture the nonlinearity of the degradation mechanisms. The predictions of cell 31 in Dataset I did not show any fluctuations near cycle 20, regardless of the analyzed GPR or RLR model, as cell 31 was cycled using the same constant current discharge profile.

It was evident that the RLR and GPR models achieved the best predictions in Dataset II, which contains cycle data from multiple batteries across all datasets. The average RMSE was 1.00% for the RLR algorithm and 1.03% for the GPR algorithm. Figure 6 illustrates that most predictions were near the diagonal, indicating a perfect match between the actual value and the predicted value. This result can largely be attributed to the finding that cells in Dataset II were cycled using the identical discharge profile. However, the distributions of residuals for GPR and RLR were distinct. As illustrated by the residual histograms in Figure 6, RLR exhibited a multimodal distribution, with all errors being negative, indicating that there may be several underlying sources of errors contributing to its overall performance. GPR had a moderately skewed distribution with a long rail to the right, and the largest peak was centered at zero, indicating that it was more prone to making large positive errors.

The predictions of the GPR and RLR models had a few outliers after cycle 50 during secondary testing in Dataset III, while the errors at primary testing were lower and did not present any outliers, as depicted in Figure 7.

The residual histograms of RLR in the secondary tests showed a few instances of large residuals at the tails of the distributions, suggesting that the model has difficulty handling certain extreme cases. The GPR model had a roughly bell-shaped distribution with a high peak at approximately zero, indicating that the model is better at capturing than RLR.

Overall, the GPR algorithm trained on Datasets I, II, and III is suggested to be more accurate in the tests, as it achieved lower relative MAE values and, in most cases, lower RMSE values compared to those of the RLR algorithm. However, there was an exception in the primary test of Dataset II, where RLR achieved an average RMSE of 1.00%, which was lower than GPR's RMSE of 1.03%. This result may be attributed to the discharge profile of cells in Dataset II being the same during the cycle test.

## 4.3. Performance of Models Based on Feature Subsets

We also explored GPR and RLR algorithms using 20 selected features (as shown in Figure 4). Tables 5–7 summarize the test results of the GPR and RLR algorithms. Figures 8–10 illustrate the battery cycle life predictions versus observations and the residual histograms based on 20 features from Datasets I-III. Both GPR and RLR exhibited lower prediction errors on all datasets. Specifically, RLR achieved an average RMSE of 0.75% and an average MAE of 0.52% on Dataset II. In contrast, GPR achieved an average RMSE and MAE of 0.67% and 0.54%, respectively, on the same dataset, indicating that GPR outperforms RLR on Dataset II. GPR also performed better than RLR on the other two datasets.

**Table 5.** Test results for the RLR and GPR models trained on the feature subset of Dataset I.

| Model | Metric | Battery ID | | | | | | | | | | Average RMSE/MAE |
|---|---|---|---|---|---|---|---|---|---|---|---|---|
| | | #31 | #32 | #33 | #33 | #34 | #35 | #36 | #37 | #38 | #39 | |
| RLR | RMSE | 2.01% | 6.25% | 1.69% | 1.44% | 6.69% | 1.55% | 2.25% | 6.26% | 1.69% | 0.80% | 3.31% |
| | MAE | 1.25% | 4.58% | 1.28% | 1.21% | 4.92% | 1.18% | 1.13% | 4.55% | 1.28% | 0.65% | 2.20% |
| GPR | RMSE | 0.57% | 2.96% | 0.86% | 1.56% | 1.93% | 0.97% | 0.70% | 2.66% | 1.51% | 0.82% | 1.52% |
| | MAE | 0.38% | 1.95% | 0.53% | 0.98% | 1.29% | 0.64% | 0.48% | 1.87% | 0.93% | 0.56% | 0.96% |

**Table 6.** Test results for the RLR and GPR models trained on the feature subset of Dataset II.

| Model | Metric | Battery ID | | | | | | Average RMSE/MAE |
|---|---|---|---|---|---|---|---|---|
| | | #3 | #4 | #5 | #6 | #7 | #8 | |
| RLR | RMSE | 0.38% | 0.65% | 0.89% | 0.31% | 0.56% | 0.64% | 0.75% |
| | MAE | 0.32% | 0.87% | 0.56% | 0.40% | 0.42% | 0.55% | 0.52% |
| GPR | RMSE | 0.45% | 0.67% | 0.20% | 0.35% | 0.76% | 0.89% | 0.67% |
| | MAE | 0.39% | 0.81% | 0.27% | 0.43% | 0.63% | 0.75% | 0.54% |

**Table 7.** Test results for the RLR and GPR models trained on the feature subset of Dataset III.

| Model | Metric | Battery ID | | | | | Average RMSE/MAE |
|---|---|---|---|---|---|---|---|
| | | #7 | #18 | #36 | #40 | #44 | |
| RLR | RMSE | 4.04% | 2.37% | 4.14% | 2.87% | 20.98% | 6.90% |
| | MAE | 3.45% | 1.85% | 4.08% | 2.03% | 16.45% | 4.77% |
| GPR | RMSE | 3.40% | 1.29% | 8.40% | 5.32% | 13.17% | 6.33% |
| | MAE | 2.69% | 1.08% | 6.03% | 4.80% | 8.21% | 3.91% |

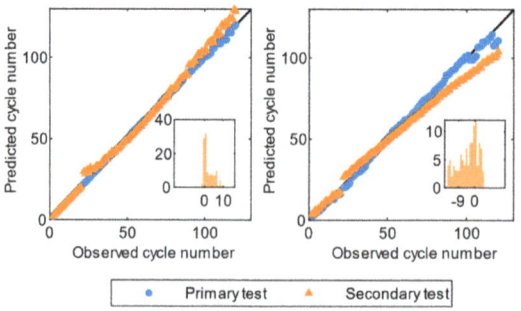

**Figure 8.** Test results of the feature subset models of Dataset I. The left plot shows the predictions of the GPR algorithm, and the right plot shows the predictions of the RLR algorithm. Cell 31 is the primary test set, and cell 32 is the secondary test set.

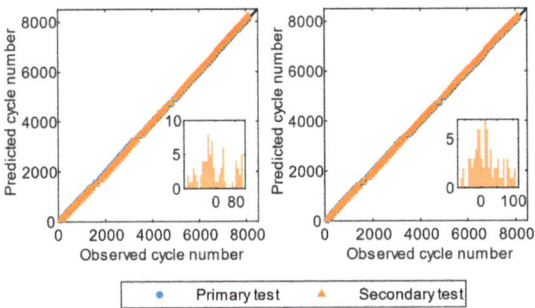

**Figure 9.** Test results of the feature subset models of Dataset II. The left plot shows the predictions of the GPR algorithm, and the right plot shows the predictions of the RLR algorithm. Cell 3 is the primary test set, and cell 7 is the secondary test set.

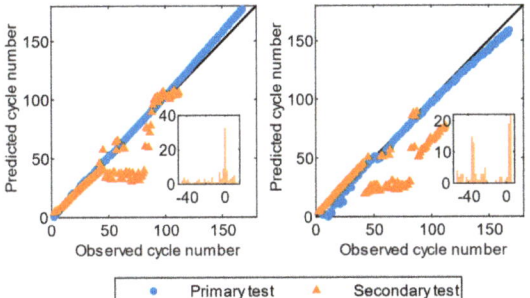

**Figure 10.** Test results of the full feature models of Dataset III. The left plot shows the predictions of the GPR algorithm, and the right plot shows the predictions of the RLR algorithm. Cell 7 is the primary test set, and cell 44 is the secondary test set.

Both GPR and RLR achieved an average RMSE and MAE of less than 3.4%. Comparing the residual histograms of the two algorithms on feature subsets of Dataset I, we discovered that GPR has a more negatively skewed distribution with a right tail, indicating that GPR is more likely to have positive errors. Conversely, the residual histogram of RLR showed a positively skewed distribution with a left tail, indicating that RLR is prone to having negative errors. Comparing the test results of the full features on the same dataset, we discovered that both GPR and RLR based on feature subsets output more accurate predictions than those based on full features (Table 2).

Cells in Dataset II were cycled with the ARTEMIS dynamic driving profile, followed by characterization cycles. It is evident from Figures 6 and 9 that the performance of tests in Dataset II was dominated by RLR, according to both RMSE and MAE. The largest RMSE achieved by both models was 0.89%, which is less than that of Dataset I. The cells in Dataset II had been cycled up to 8000 cycles, and both GPR and RLR achieved an average RMSE and MAE of less than 0.75% of the entire cycle life. Tables 3 and 6 show that both models based on feature subsets outperformed the models based on the full features of Dataset II, indicating that feature selection by MRMR could improve the prediction accuracy on the dataset. The high performance achieved by GPR and RLR in Dataset II may be attributed to the low variability in the charge and discharge conditions.

Both GPR and RLR based on feature subsets of Dataset III achieved the highest RMSE and MAE across all datasets. Both histograms of the residuals of GPR in Figures 7 and 10 show skewed distributions. Specifically, GPR on full features showed a negatively skewed distribution with a long tail to the right, and the peak center was approximately zero, indicating that it is prone to outputting positive errors. Conversely, GPR on feature subsets exhibited a positively skewed distribution with a long tail to the left, and the peak center was also approximately zero, indicating that it is prone to having negative errors. The residuals of RLR exhibited a multimodal distribution on the feature subset, indicating that there may be several underlying sources of errors contributing to its overall performance. The residual histogram of RLR on full features also showed two peaks, but the second peak was lower than that of RLR on the feature subset. GPR and RLR on the feature subset achieved a lower average RMSE and MAE than those on full features, suggesting that the feature selection could avoid overfitting.

The prediction speed of the two algorithms on both full features and feature subsets of Datasets I to III are summarized in Table 8. All models were trained and tested on a computer with two Intel Xeon 2666 V3 CPUs and an Nvidia 2080Ti GPU.

Table 8. Training time and prediction speed of the full feature models and feature subset models.

| Algorithms | Full Feature Models | | | | | | Feature Subset Models | | | | | |
|---|---|---|---|---|---|---|---|---|---|---|---|---|
| | Training Time (s) | | | Prediction Speed (obs/s) | | | Training Time (s) | | | Prediction Speed (obs/s) | | |
| | I | II | III | I | II | III | I | II | III | I | II | III |
| Robust linear | 1.905 | 1.640 | 1.365 | 58,075 | 3782 | 27,870 | 1.400 | 1.161 | 1.738 | 78,441 | 6939 | 44,937 |
| Matern 5/2 GPR | 372.840 | 1.313 | 20.478 | 22,750 | 5575 | 26,299 | 160.290 | 2.390 | 20.834 | 31,785 | 4022 | 31,797 |

As expected, the feature subset models showed a significantly higher prediction speed than the full feature models, primarily due to a reduction of more than half of the variables. RLR particularly emphasized this point, demonstrating a minimum of twice the prediction speed of the full feature set models, except for Dataset I, which showed an almost 50% faster prediction speed. For GPR, all three datasets showed an increase in the prediction speed of less than 50%, except for Dataset II. This discrepancy is attributed to the complex random process of the GPR algorithm, which impacts the overall prediction speed.

The results of using feature subsets, instead of full features, in GPR for Dataset I yielded considerable reductions in training time. Conversely, for Datasets II and III, the difference in training time between the models using feature subsets and full features was limited to 2 s. As Table 1 describes, Dataset I consists of most cells of the three datasets, so the training time of GPR was the largest, with a maximum of 372.840 s. The training time of GPR for Datasets II and III was smaller than that of GPR for Dataset I, and the training speed of the two algorithms did was not significantly improve.

## 5. Discussion

The proposed battery cycle life prediction approach promises to enhance battery management systems, allowing for highly accurate estimation of battery degradation. This proposed method is distinct in that it can estimate cycle life using only discharge voltage curves and can accommodate various operational conditions, such as random or high discharge rates. Future work could be extended to random partial discharge/charge scenarios and batteries with different designs and chemistries.

The algorithms based on full features had strong performance, as they achieved a low RMSE and MAE, but the large feature set was too complicated for onboard application and likely contained some redundancies. To address this issue, we used the MRMR algorithm for feature selection. The score distribution of each feature indicated that the importance of features is not consistent across the different datasets. This lack of consistency could be attributed to the various aging mechanisms and modes present in the different battery datasets, which were caused by the varying cycle conditions and charge/discharge protocols. Therefore, it is essential to select features using the MRMR algorithm for each respective battery dataset prior to model training to achieve a satisfactory trade-off between accuracy and computational efficiency.

To meet real-time requirements, a subset of 20 features was selected from 53 features as a paradigm of feature selection; these features could be extracted from every cycle discharge profile. The aim of the proposed method was to optimize a process suitable for on-board applications that emphasize computation efficiency and real-time accuracy over precision. Therefore, multicycle features were excluded, as they require the extraction of multiple cycle data, and we used only features that can be calculated for each cycle.

Our investigation of two algorithms, GPR and RLR, for three datasets revealed that feature selection has a positive effect on the performance of both algorithms for Datasets I and III, except for Dataset II. Specifically, both algorithms achieved relatively low average RMSEs and MAEs for all datasets, and GPR outperformed RLR in terms of RMSE and MAE for both feature subsets and full features of Datasets I and III, indicating that GPR is the optimal algorithm for large battery datasets with complex discharge profiles. Conversely, RLR output accurate predictions with a lower RMSE and MAE for Dataset II compared to GPR, owing to identical discharge profiles. As discussed in Section 3, lithium-ion battery aging is a nonlinear process with a multitude of potential factors. It can be seen from Figure 4 that almost all features demonstrate nonlinear correlations with the cycle number. The GPR model incorporates a nonlinear kernel function, which is used to fit the correlation between input and target. This kernel function makes GPR perform better than RLR for battery cycle life prediction, especially under dynamic load profiles.

Table 9 compared the proposed method and 10 different data-driven methods for battery degradation estimation. Compared to previous methods, we developed some new features, such as the warp distance of discharge voltages, which makes it possible to extract useful information from dynamic discharge profiles. The main reason for the discrepancy in results between our methods and those of other literature can be attributed to the difference in targets of machine learning models. As seen in Table 9, our model uses the cycle life index (CI) as the target, the denominator of which is the total cycle number of the cycle tests. In contrast, the equation of the remaining useful life (RUL) reported by other literature has a different denominator, namely the cycle life given by the manufacturer. For instance, in Dataset III, the total cycle number of tests averages 131 cycles, while the cycle life given by the manufacturer ranges between 300 and 500 cycles. The difference in denominators of the targets thus affects the RMSE of the two methods. Another reason for the discrepancy in the results between our methods and other methods is the use of a linear regression model, which is less accurate than other machine learning algorithms in dynamic load profiles. Training linear regression models requires less computational resources than most machine learning models, and it is simple to implement linear regression models, which makes it possible to apply machine learning algorithms to onboard battery management systems in electric vehicles. Many studies [1,8,31,32] have demonstrated that linear regression is good at fitting simple battery degradation with minimal variance in charge and discharge conditions. After considering both the prediction speed and the training cost, we determined that the RLR algorithm is optimal for battery life estimation in onboard applications with inadequate computing resources and high real-time requirements, whereas the GPR algorithm is better suited for battery pack manufacturing and recycling, due to the high prediction accuracy requirements and sufficient computational power.

Table 9. Comparison of various data-driven methods for battery degradation estimation.

| Method | Positive Electrode | Target | Main Features | Precision |
|---|---|---|---|---|
| RNN [33] | NMC, LFP | RUL | Capacity–voltage matrix | RMSE $\leq$ 2.4% |
| BRR, GPR, RF, dNNe [1] | LCO, NCA | SOH | Energy ratio, entropy, skewness, kurtosis, Hausdorff distance of the CCCV curve | dNNe: RMSPE $\leq$ 4.26%<br>RF: RMSPE $\leq$ 2.70%<br>GPR: RMSPE $\leq$ 3.70%<br>BRR: RMSPE $\leq$ 5.54% |

Table 9. Cont.

| Method | Positive Electrode | Target | Main Features | Precision |
|---|---|---|---|---|
| Linear regression with lasso and elastic net regularization [8] | LFP | Cycle life | dV/dQ, dQ/dV, $\Delta Q(V)$ | 'Full' model: mean percentage error: 9.1% |
| RBF [43] | NCA | SOH | KL distance based on the hidden Markov model, KL distance based on kernel density estimation | RMSE $\leq$ 1.13% MAE $\leq$ 1.05% |
| Deep transfer learning [34] | LFP | Capacity, RUL | Difference in the charge voltage curve between each cycle and the 10th cycle, difference in the charge capacity curve between each cycle and the 10th cycle | Capacity: RMSE $\leq$ 0.328% RUL: RMSE $\leq$ 9.90% |
| Elastic net, SVR, transfer learning model [11] | NCM, NCA | Capacity | Variance, skewness, excess kurtosis of relaxation voltage | RMSE $\leq$ 1.7% |
| SVM, GPR [23] | LFP | SOH | Discharge capacity differences of two cycles | MAE $\leq$ 1% RMSE $\leq$ 1.3% |
| AdaBoost–PSO–SVM [30] | NCA | SOH | SOC, time, voltage | RMSE $\leq$ 2.316% |
| Multivariate regularized linear regression [44] | NMC | Lifetime | Low-SOC resistance, capacity variance between each cycle and the 10th cycle | Mean standard deviations: $\leq$15.2 cycles |
| Extratrees, NuSVR [26] | NMC | Cycle life | OCV, dQ/dV, dV/dQ, resistance | MAE $\leq$ 102 cycles |
| Proposed method | NMC, NCA, NCO | CI | Distance between discharge voltage curve and reference curve using time warping, entropy, SSIM | RMSE $\leq$ 6.33% MAE $\leq$ 3.91% |

RNN, recurrent neural network; BRR, Bayesian ridge regression; RF, random forest; dNNe, deep neural network; RBF, radiant-based function; PSO, particle swarm optimization; NuSVR, Nu support vector regression.

## 6. Conclusions

Data-driven models are widely adopted for diagnosing and prognosticating the behavior of lithium-ion batteries. In this study, we proposed a data-driven framework to accurately predict battery cycle life using various discharge profiles. This method offers several advantages over conventional methods, including adaptability to random and high discharge rates, robustness to changes in discharge mode, and prediction based solely on discharge profiles.

We extracted 53 features from battery discharge profiles, 18 of which were newly proposed for battery cycle life prediction models. The MRMR algorithm was used for feature selection. We explored two machine learning models: GPR and RLR. All models were evaluated using the error metrics RMSE and MAE. GPR achieved a maximum RMSE of 6.33% and a maximum MAE of 3.91%, while RLR attained a maximum RMSE of 6.90% and a maximum MAE of 4.77%. GPR was preferred for battery pack manufacturing and recycling, while RLR was preferred for on-board battery cycle life prediction.

Overall, our work highlights the value of combining machine learning techniques with discharge profiles for battery cycle life estimation. Moreover, although the estimation accuracy is not always improved, the algorithm should be subjected to feature selection before being deployed in the field. We demonstrate that feature selection can improve the prediction accuracy and reduce the computational cost. We infer that this framework should also be effective with charge profiles. In future work, it would be beneficial to combine features extracted from both charge profiles and discharge profiles and to use this method to prognosticate batteries with different materials.

**Author Contributions:** Conceptualization, Y.J. and W.S.; methodology, Y.J.; software, Y.J.; validation, Y.J.; formal analysis, Y.J.; investigation, Y.J.; resources, Y.J.; data curation, Y.J.; writing—original draft preparation, Y.J.; writing—review and editing, W.S.; visualization, Y.J.; supervision, W.S.; project administration, Y.J.; funding acquisition, W.S. All authors have read and agreed to the published version of the manuscript.

**Funding:** This research received no external funding.

**Data Availability Statement:** The battery datasets used in this study are available at https://data.mendeley.com/datasets/kw34hhw7xg (accessed on 4 August 2023) for Dataset I, https://ora.ox.ac.uk/objects/uuid:03ba4b01-cfed-46d3-9b1a-7d4a7bdf6fac (accessed on 4 August 2023) for Dataset II, and https://phmdatasets.s3.amazonaws.com/NASA/5.+Battery+Data+Set.zip for Dataset III (accessed on 4 August 2023).

**Conflicts of Interest:** The authors declare no conflict of interest.

# References

1. Roman, D.; Saxena, S.; Robu, V.; Pecht, M.; Flynn, D. Machine Learning Pipeline for Battery State-of-Health Estimation. *Nat. Mach. Intell.* **2021**, *3*, 447–456. [CrossRef]
2. Ng, M.-F.; Zhao, J.; Yan, Q.; Conduit, G.J.; Seh, Z.W. Predicting the State of Charge and Health of Batteries Using Data-Driven Machine Learning. *Nat. Mach. Intell.* **2020**, *2*, 161–170. [CrossRef]
3. Tran, M.-K.; Mathew, M.; Janhunen, S.; Panchal, S.; Raahemifar, K.; Fraser, R.; Fowler, M. A Comprehensive Equivalent Circuit Model for Lithium-Ion Batteries, Incorporating the Effects of State of Health, State of Charge, and Temperature on Model Parameters. *J. Energy Storage* **2021**, *43*, 103252. [CrossRef]
4. Weng, C.; Sun, J.; Peng, H. A Unified Open-Circuit-Voltage Model of Lithium-Ion Batteries for State-of-Charge Estimation and State-of-Health Monitoring. *J. Power Sources* **2014**, *258*, 228–237. [CrossRef]
5. Bian, X.; Liu, L.; Yan, J.; Zou, Z.; Zhao, R. An Open Circuit Voltage-Based Model for State-of-Health Estimation of Lithium-Ion Batteries: Model Development and Validation. *J. Power Sources* **2020**, *448*, 227401. [CrossRef]
6. Bian, X.; Wei, Z.; Li, W.; Pou, J.; Sauer, D.U.; Liu, L. State-of-Health Estimation of Lithium-Ion Batteries by Fusing an Open Circuit Voltage Model and Incremental Capacity Analysis. *IEEE Trans. Power Electron.* **2022**, *37*, 2226–2236. [CrossRef]
7. Li, J.; Adewuyi, K.; Lotfi, N.; Landers, R.G.; Park, J. A Single Particle Model with Chemical/Mechanical Degradation Physics for Lithium Ion Battery State of Health (SOH) Estimation. *Appl. Energy* **2018**, *212*, 1178–1190. [CrossRef]
8. Severson, K.A.; Attia, P.M.; Jin, N.; Perkins, N.; Jiang, B.; Yang, Z.; Chen, M.H.; Aykol, M.; Herring, P.K.; Fraggedakis, D.; et al. Data-Driven Prediction of Battery Cycle Life before Capacity Degradation. *Nat. Energy* **2019**, *4*, 383–391. [CrossRef]
9. Ma, Z.; Yang, R.; Wang, Z. A Novel Data-Model Fusion State-of-Health Estimation Approach for Lithium-Ion Batteries. *Appl. Energy* **2019**, *237*, 836–847. [CrossRef]
10. Deng, Z.; Hu, X.; Li, P.; Lin, X.; Bian, X. Data-Driven Battery State of Health Estimation Based on Random Partial Charging Data. *IEEE Trans. Power Electron.* **2022**, *37*, 5021–5031. [CrossRef]
11. Zhu, J.; Wang, Y.; Huang, Y.; Bhushan Gopaluni, R.; Cao, Y.; Heere, M.; Mühlbauer, M.J.; Mereacre, L.; Dai, H.; Liu, X.; et al. Data-Driven Capacity Estimation of Commercial Lithium-Ion Batteries from Voltage Relaxation. *Nat. Commun.* **2022**, *13*, 2261. [CrossRef] [PubMed]
12. Zhang, Y.; Tang, Q.; Zhang, Y.; Wang, J.; Stimming, U.; Lee, A.A. Identifying Degradation Patterns of Lithium Ion Batteries from Impedance Spectroscopy Using Machine Learning. *Nat. Commun.* **2020**, *11*, 1706. [CrossRef]
13. Vermeer, W.; Chandra Mouli, G.R.; Bauer, P. A Comprehensive Review on the Characteristics and Modeling of Lithium-Ion Battery Aging. *IEEE Trans. Transp. Electrif.* **2022**, *8*, 2205–2232. [CrossRef]
14. Tian, H.; Qin, P.; Li, K.; Zhao, Z. A Review of the State of Health for Lithium-Ion Batteries: Research Status and Suggestions. *J. Clean. Prod.* **2020**, *261*, 120813. [CrossRef]
15. Shahjalal, M.; Roy, P.K.; Shams, T.; Fly, A.; Chowdhury, J.I.; Ahmed, M.R.; Liu, K. A Review on Second-Life of Li-Ion Batteries: Prospects, Challenges, and Issues. *Energy* **2022**, *241*, 122881. [CrossRef]
16. Chen, M.; Ma, G.; Liu, W.; Zeng, N.; Luo, X. An Overview of Data-Driven Battery Health Estimation Technology for Battery Management System. *Neurocomputing* **2023**, *532*, 152–169. [CrossRef]
17. Vanem, E.; Salucci, C.B.; Bakdi, A.; Sheim Alnes, Ø.Å. Data-Driven State of Health Modelling—A Review of State of the Art and Reflections on Applications for Maritime Battery Systems. *J. Energy Storage* **2021**, *43*, 103158. [CrossRef]
18. Che, Y.; Hu, X.; Lin, X.; Guo, J.; Teodorescu, R. Health Prognostics for Lithium-Ion Batteries: Mechanisms, Methods, and Prospects. *Energy Environ. Sci.* **2023**, *16*, 338–371. [CrossRef]
19. Sui, X.; He, S.; Vilsen, S.B.; Meng, J.; Teodorescu, R.; Stroe, D.-I. A Review of Non-Probabilistic Machine Learning-Based State of Health Estimation Techniques for Lithium-Ion Battery. *Appl. Energy* **2021**, *300*, 117346. [CrossRef]
20. Jiang, S.; Song, Z. A Review on the State of Health Estimation Methods of Lead-Acid Batteries. *J. Power Sources* **2022**, *517*, 230710. [CrossRef]

21. Li, Y.; Stroe, D.-I.; Cheng, Y.; Sheng, H.; Sui, X.; Teodorescu, R. On the Feature Selection for Battery State of Health Estimation Based on Charging–Discharging Profiles. *J. Energy Storage* **2021**, *33*, 102122. [CrossRef]
22. Luo, K.; Chen, X.; Zheng, H.; Shi, Z. A Review of Deep Learning Approach to Predicting the State of Health and State of Charge of Lithium-Ion Batteries. *J. Energy Chem.* **2022**, *74*, 159–173. [CrossRef]
23. Deng, Z.; Hu, X.; Lin, X.; Xu, L.; Che, Y.; Hu, L. General Discharge Voltage Information Enabled Health Evaluation for Lithium-Ion Batteries. *IEEE/ASME Trans. Mechatron.* **2021**, *26*, 1295–1306. [CrossRef]
24. Messing, M.; Shoa, T.; Habibi, S. Estimating Battery State of Health Using Electrochemical Impedance Spectroscopy and the Relaxation Effect. *J. Energy Storage* **2021**, *43*, 103210. [CrossRef]
25. Pradhan, S.K.; Chakraborty, B. Battery Management Strategies: An Essential Review for Battery State of Health Monitoring Techniques. *J. Energy Storage* **2022**, *51*, 104427. [CrossRef]
26. Paulson, N.H.; Kubal, J.; Ward, L.; Saxena, S.; Lu, W.; Babinec, S.J. Feature Engineering for Machine Learning Enabled Early Prediction of Battery Lifetime. *J. Power Sources* **2022**, *527*, 231127. [CrossRef]
27. Gou, B.; Xu, Y.; Feng, X. State-of-Health Estimation and Remaining-Useful-Life Prediction for Lithium-Ion Battery Using a Hybrid Data-Driven Method. *IEEE Trans. Veh. Technol.* **2020**, *69*, 10854–10867. [CrossRef]
28. Zhou, Z.; Duan, B.; Kang, Y.; Shang, Y.; Cui, N.; Chang, L.; Zhang, C. An Efficient Screening Method for Retired Lithium-Ion Batteries Based on Support Vector Machine. *J. Clean. Prod.* **2020**, *267*, 121882. [CrossRef]
29. Zhang, J.; Wang, P.; Gong, Q.; Cheng, Z. SOH Estimation of Lithium-Ion Batteries Based on Least Squares Support Vector Machine Error Compensation Model. *J. Power Electron.* **2021**, *21*, 1712–1723. [CrossRef]
30. Li, R.; Li, W.; Zhang, H. State of Health and Charge Estimation Based on Adaptive Boosting Integrated with Particle Swarm Optimization/Support Vector Machine (AdaBoost-PSO-SVM) Model for Lithium-Ion Batteries. *Int. J. Electrochem. Sci.* **2022**, *17*, 220212. [CrossRef]
31. Shi, M.; Xu, J.; Lin, C.; Mei, X. A Fast State-of-Health Estimation Method Using Single Linear Feature for Lithium-Ion Batteries. *Energy* **2022**, *256*, 124652. [CrossRef]
32. Vilsen, S.B.; Stroe, D.-I. Battery State-of-Health Modelling by Multiple Linear Regression. *J. Clean. Prod.* **2021**, *290*, 125700. [CrossRef]
33. Lu, J.; Xiong, R.; Tian, J.; Wang, C.; Hsu, C.-W.; Tsou, N.-T.; Sun, F.; Li, J. Battery Degradation Prediction against Uncertain Future Conditions with Recurrent Neural Network Enabled Deep Learning. *Energy Storage Mater.* **2022**, *50*, 139–151. [CrossRef]
34. Ma, G.; Xu, S.; Jiang, B.; Cheng, C.; Yang, X.; Shen, Y.; Yang, T.; Huang, Y.; Ding, H.; Yuan, Y. Real-Time Personalized Health Status Prediction of Lithium-Ion Batteries Using Deep Transfer Learning. *Energy Environ. Sci.* **2022**, *15*, 4083–4094. [CrossRef]
35. Wang, Z.; Yuan, C.; Li, X. Lithium Battery State-of-Health Estimation via Differential Thermal Voltammetry With Gaussian Process Regression. *IEEE Trans. Transp. Electrific.* **2021**, *7*, 16–25. [CrossRef]
36. Guo, W.; Sun, Z.; Vilsen, S.B.; Meng, J.; Stroe, D.I. Review of "Grey Box" Lifetime Modeling for Lithium-Ion Battery: Combining Physics and Data-Driven Methods. *J. Energy Storage* **2022**, *56*, 105992. [CrossRef]
37. Birkl, C.R.; Roberts, M.R.; McTurk, E.; Bruce, P.G.; Howey, D.A. Degradation Diagnostics for Lithium Ion Cells. *J. Power Sources* **2017**, *341*, 373–386. [CrossRef]
38. André, M. The ARTEMIS European Driving Cycles for Measuring Car Pollutant Emissions. *Sci. Total Environ.* **2004**, *334–335*, 73–84. [CrossRef]
39. Goebel, K.; Saha, B.; Saxena, A.; Celaya, J.; Christophersen, J. Prognostics in Battery Health Management. *IEEE Instrum. Meas. Mag.* **2008**, *11*, 33–40. [CrossRef]
40. Xiong, R.; Pan, Y.; Shen, W.; Li, H.; Sun, F. Lithium-Ion Battery Aging Mechanisms and Diagnosis Method for Automotive Applications: Recent Advances and Perspectives. *Renew. Sustain. Energy Rev.* **2020**, *131*, 110048. [CrossRef]
41. Peng, H.; Long, F.; Ding, C. Feature Selection Based on Mutual Information Criteria of Max-Dependency, Max-Relevance, and Min-Redundancy. *IEEE Trans. Pattern Anal. Mach. Intell.* **2005**, *27*, 1226–1238. [CrossRef] [PubMed]
42. Zhao, Z.; Anand, R.; Wang, M. Maximum Relevance and Minimum Redundancy Feature Selection Methods for a Marketing Machine Learning Platform. In Proceedings of the 2019 IEEE International Conference on Data Science and Advanced Analytics (DSAA), Washington, DC, USA, 5–8 October 2019.
43. Lin, M.; Zeng, X.; Wu, J. State of Health Estimation of Lithium-Ion Battery Based on an Adaptive Tunable Hybrid Radial Basis Function Network. *J. Power Sources* **2021**, *504*, 230063. [CrossRef]
44. Weng, A.; Mohtat, P.; Attia, P.M.; Sulzer, V.; Lee, S.; Less, G.; Stefanopoulou, A. Predicting the Impact of Formation Protocols on Battery Lifetime Immediately after Manufacturing. *Joule* **2021**, *5*, 2971–2992. [CrossRef]

**Disclaimer/Publisher's Note:** The statements, opinions and data contained in all publications are solely those of the individual author(s) and contributor(s) and not of MDPI and/or the editor(s). MDPI and/or the editor(s) disclaim responsibility for any injury to people or property resulting from any ideas, methods, instructions or products referred to in the content.

MDPI
St. Alban-Anlage 66
4052 Basel
Switzerland
www.mdpi.com

*Batteries* Editorial Office
E-mail: batteries@mdpi.com
www.mdpi.com/journal/batteries

Disclaimer/Publisher's Note: The statements, opinions and data contained in all publications are solely those of the individual author(s) and contributor(s) and not of MDPI and/or the editor(s). MDPI and/or the editor(s) disclaim responsibility for any injury to people or property resulting from any ideas, methods, instructions or products referred to in the content.

www.ingramcontent.com/pod-product-compliance
Lightning Source LLC
LaVergne TN
LVHW070442100526
838202LV00014B/1646